LEGAL GEOGRAPHY

This book is the first legal geography book to explicitly engage with method. It complements this by also bringing together different perspectives on the emerging school of legal geography. It explores human–environment interactions and showcases distinct environmental legal geography scholarship.

Legal Geography: Perspectives and Methods is an innovative book concerned with a new relational and material way of examining our legal-spatial world. With chapters examining natural resource management, Indigenous knowledge and political ecology scholarship, the text introduces legal geography's modes of analysis and critique. The book explores topics such as Indigenous environmental rights, the impacts of extractive industries, mediation of climate change, food, animal and plant patents, fossil fuels, mining and coastal environments based on empirical, jurisdictional and methodological insights from Australia, New Zealand and the Asia-Pacific to demonstrate how space and place are invoked in legal processes and contestations, and the methods that may be employed to explore these processes and contestations.

This book examines the role of legal geographies in the 21st century beyond the simple "law in action", and it will thus appeal to students of socio-legal studies, human geography, environmental studies, environmental policy, as well as politics and international relations.

Tayanah O'Donnell has over ten years' experience focused on the law and the legal geographies of climate change adaptation. Her papers and research cover themes such as property rights, land use planning, climate law, coastal policy and management, and the legal, political and cultural impacts of climate change regulation.

Daniel F. Robinson has more than 15 years' experience focused on the regulation of nature and knowledge. His papers and books cover themes including "biopiracy", access and benefit-sharing relating to biological resources, appropriation and regulation of Indigenous knowledge, Indigenous/customary laws and biocultural protocols, ethical biotrade, political ecology, environmental policy and management.

Josephine Gillespie is an academic, and former lawyer, based at the University of Sydney, Australia. She is an environmental legal geographer interested in the complex intersection of geography and law. Her research investigates environmental protection and human-environment geographies throughout Australia and the Asia-Pacific.

LEGAL GEOGRAPHY

Perspectives and Methods

Edited by Tayanah O'Donnell,
Daniel F. Robinson
and Josephine Gillespie

Routledge
Taylor & Francis Group

LONDON AND NEW YORK

First published 2020
by Routledge
2 Park Square, Milton Park, Abingdon, Oxon OX14 4RN

and by Routledge
52 Vanderbilt Avenue, New York, NY 10017

Routledge is an imprint of the Taylor & Francis Group, an informa business

© 2020 selection and editorial matter, Tayanah O'Donnell, Daniel F. Robinson, and Josephine Gillespie; individual chapters, the contributors

The rights of Tayanah O'Donnell, Daniel F. Robinson, and Josephine Gillespie to be identified as the authors of the editorial material, and of the authors for their individual chapters, has been asserted in accordance with sections 77 and 78 of the Copyright, Designs and Patents Act 1988.

All rights reserved. No part of this book may be reprinted or reproduced or utilised in any form or by any electronic, mechanical, or other means, now known or hereafter invented, including photocopying and recording, or in any information storage or retrieval system, without permission in writing from the publishers.

Trademark notice: Product or corporate names may be trademarks or registered trademarks, and are used only for identification and explanation without intent to infringe.

British Library Cataloguing-in-Publication Data
A catalogue record for this book is available from the British Library

Library of Congress Cataloging-in-Publication Data
Names: O'Donnell, Tayanah, editor. | Robinson, Daniel F., editor. | Gillespie, Josephine, editor.
Title: Legal geography: perspectives and methods / edited by Tayanah O'Donnell, Daniel F. Robinson, and Josephine Gillespie.
Description: Abingdon, Oxon; New York, NY: Routledge, 2020. | Includes bibliographical references and index.
Identifiers: LCCN 2019050123 (print) | LCCN 2019050124 (ebook) | ISBN 9781138387379 (hardback) | ISBN 9781138387386 (paperback) | ISBN 9780429426308 (ebook)
Subjects: LCSH: Law and geography. | Law–Indo-Australian Region. | Law–Australasia.
Classification: LCC K487.G45 .L44 2020 (print) | LCC K487.G45 (ebook) | DDC 340–dc23
LC record available at https://lccn.loc.gov/2019050123
LC ebook record available at https://lccn.loc.gov/2019050124

ISBN: 978-1-138-38737-9 (hbk)
ISBN: 978-1-138-38738-6 (pbk)
ISBN: 978-0-429-42630-8 (ebk)

Typeset in Bembo
by Deanta Global Publishing Services, Chennai, India

We dedicate this book to our colleague, mentor and dear friend, Dr Stewart Williams.

CONTENTS

CONTRIBUTORS

Maria Bargh is Associate Professor in Te Kawa, a Māui/School of Māori Studies, Victoria University of Wellington Maria completed her PhD in Political Science and International Relations at the Australian National University in 2002. Her research interests focus on Māori politics, including constitutional change and Māori representation, voting in local and general elections and the Māori economy including hidden and diverse economies such as Mā ori in the private military industry. She also researches on matters related to Māori resource management, such as freshwater, mining and renewable energy.

Robyn Bartel is an Associate Professor in Geography and Planning at the University of New England. Robyn has qualifications in law, science and education. Her teaching and research strengths are in the areas of sustainability, environmental law and geography. Her research encompasses the interrelationships between formal and informal law and regulation, within the cultural, social, institutional and natural landscapes in which all are situated. Her work in legal geography has been influential in shaping the direction of this rich new field of scholarly exploration, particularly in terms of recognising beyond-human agency, and appreciating place itself as a law-maker and shaper.

Dr Cobi Calyx is a Postdoctoral Fellow with the Centre for Social Impact (CSI) at UNSW Sydney. She also works with the Climate and Sustainability Policy Research (CASPR) group at Flinders University. Cobi is a Climate Reality Leader who has lived on five continents. Cobi has worked for organisations ranging from the United Nations in Geneva to the Country Fire Service in South Australia. Cobi completed her PhD in science communication, concerning deliberative engagement, at the Australian National University and has earlier qualifications in health promotion, international studies and journalism. Cobi has worked in international development in Asia and the Pacific and as an environmental law scholar at Melbourne Law School.

Dr Josephine Gillespie is an academic at the University of Sydney. Jo is a legal geographer interested in the complex intersection of geography and law. This research investigates environmental protection and human-environment geographies throughout Australia and the Asia-Pacific. As a former lawyer, Jo draws on her industry experience in the law to inform both her research and teaching. She is a Commissioner on the IUCN Commission on Environment, Economic and Social Policy, a Committee member of the New South Wales Geographical Society and a Co-Convenor of the Legal Geography Study Group of the Institute of Australian Geographers.

Professor Lee Godden is the Director of the Centre for Resources, Energy and Environmental Law at Melbourne University. She researches in environmental resources law, natural resources law, water law, and indigenous people's land and resources rights. Recent publications include *Environmental Law: Scientific Policy and Regulatory Dimensions* 2010 (with J. Peel), *Comparative Perspectives on Communal Lands and Individual Ownership: Sustainable Futures* 2010 (with M. Tehan) and *Australian Climate Law in Global Context* 2013 (with A. Zahar and J. Peel). The impact of her work extends beyond Australia with comparative research on environmental law and sustainability, property law and resource trading regimes, water law resources and Indigenous land rights issues, in countries as diverse as Canada, New Zealand, UK, South Africa, and the Pacific.

Dr Nicole Graham is an Associate Professor at the University of Sydney Law School. She teaches and researches in the fields of property law and theory, and legal geography. Nicole has written on the relationship between law, environment and culture with a particular focus on property rights, natural resource regulation and the concept of place. Among numerous other publications, Nicole is the author of *Lawscape: Property, Environment, Law* which was published in 2010. Nicole is a member of various professional associations, including the Sydney Institute of Agriculture, the Australian Centre for Environmental Law, the Sydney Environment Institute, the National Environmental Law Association, the Australasian Property Law Teachers Association, the Institute of Australian Geographers and the Australian and New Zealand Legal History Society.

Francis Hickey has worked as Coordinator, Traditional Resource Management Program at Vanuatu Cultural Centre for many years, and has published several papers on traditional resource management in Vanuatu.

Dr Brad Jessup is a Lecturer at Melbourne Law School. Brad is a geographer and an environmental law specialist who offers global, national, comparative and local perspectives in his research. Brad is a member of the University of Melbourne's

Centre for Resources, Energy and Environmental Law and has a particular interest in interdisciplinary scholarship that traverses areas of law, environment, society and policy. Brad joined Melbourne Law School in 2012 from the Australian National University, where he had been teaching and researching within the ANU College of Law since 2007. Previously, Brad worked as a lawyer within Herbert Smith Freehills' planning and environmental law practice. Brad has a Masters in geography from the University of Cambridge and completed a PhD on the topic of environmental justice and law at the Australian National University.

Connor Jolley is currently undertaking a PhD at RMIT University, researching workers at the intersection of climate mitigation and adaptation pressures in the field of environmental labour geography.

Donna Kalfatak is the Director of Vanuatu's Department of Environmental Protection and Conservation. Donna studies climate impacts to fisheries, small scale fisheries, and ecosystem-based fisheries management.

Dr Paul McFarland is a Lecturer in the School of Humanities, Arts, and Social Sciences at the University of New England. Paul specialises in a range of topics, including land use planning regulatory systems, peri-urban development, local government planning, land use policies and growth in the context of non-urban land. Paul has over 20 years' experience in local government town planning. He has worked at a variety of Councils in the Sydney metropolitan area, the Sydney urban / rural fringe and in rural New South Wales.

Dr Tayanah O'Donnell is currently a Director with Future Earth, a global sustainability research and innovation network, with whom she leads the Australian hub based at the Australian Academy of Science. Tayanah is an inter- and transdisciplinary researcher with an academic background in law and in human geography, and a career spanning private and public practice. Her research examines sociocultural and legal dimensions of climate change and sustainability, with a particular contextual focus on urban spaces, human settlements, coastal governance, and adaptive capacity. She is also highly regarded for her ability to break down siloes, and in bringing teams together towards common goals. She is an Honorary Senior Lecturer with the Climate Change Institute at the Australian National University and an Adjunct Principal Research Fellow at RMIT University. She is an Honorary Senior Lecturer with the Fenner School at the Australian National University, an Adjunct Principal Research Fellow at RMIT University, and chair of the Australian Legal Geography Study Group of the Institute of Australian Geographers.

Dr Margaret Raven is Yamatji-Nyoongar and has a Bachelor of Science (Geography) from the University of Western Australia. Margaret is currently a Research Fellow at the Indigenous Policy and Dialogue Research unity with the Social Policy

Research Centre at the University of New South Wales. Throughout the course of her PhD, Margaret was a PhD Fellow with the United Nations University Institute of Advanced Studies (UNU-IAS) in Yokohama, Japan. She has also engaged in international negotiations through the United Nations Convention on Biodiversity (CBD), and the World Intellectual Property Organisation (WIPO). She was also Indigenous Fellow with the Office of the High Commissioner for Human Rights (OHCHR) in Geneva. In 2009 she moved to Sydney to become the inaugural Coordinator of the Indigenous Human Rights Network Australia (IHRNA) currently hosted by the Australian Human Rights Commission.

Lauren Rickards is a co-leader of the Climate Change Transformations research programme of the Centre for Urban Research, and Associate Professor in the School of Global, Urban and Social Studies at RMIT University, Melbourne, Australia. Lauren is an interdisciplinary researcher with a background in human geography. Her research examines the sociocultural dimensions of climate change and broader environmental challenges. A Rhodes Scholar with extensive industry experience and networks, Lauren is a lead author in the Intergovernmental Panel on Climate Change's Sixth Assessment Report, and a Senior Research Fellow with the Anthropocene and Resilience networks of the Earth System Governance programme.

Dr Daniel F. Robinson is an Associate Professor in the School of Humanities at the University of New South Wales. Daniel's research focuses on the regulation of nature and knowledge. His papers and books cover themes including 'biopiracy', access and benefit-sharing relating to biological resources, appropriation and regulation of Indigenous knowledge, Indigenous/customary laws and biocultural protocols, ethical biotrade, political ecology, environmental policy and management. Daniel has acted as a researcher and policy advisor for the International Centre for Trade and Sustainable Development (ICTSD) and with their joint project with UNCTAD, The United Nations Development Programme (UNDP) and Global Environment Facility (GEF), the German development implementation agency (GIZ), AusAid and Department of Environment, the Union for Ethical BioTrade (UEBT), The Pacific Islands Forum Secretariat, the National Human Rights Commission of Thailand, amongst others.

Dr Christine Schenk is a lecturer with the Department of Religious Studies and Department of Geography at the University of Zurich. Christine Schenk completed her PhD in Geography at the University of Geneva, Switzerland in 2016, co-funded by the Swiss National Science Foundation. Her PhD investigated the role of Muslim leaders and civil rights activists in legal reforms to women's rights and Sharia law in Aceh, Indonesia. Christine had been an academic visitor to the London Schools of Economics and Political Science (LSE, 2013), University of Oxford (2016 – 17, including teaching assignments) and University of Colombo (2017).

Dr Meg Sherval is a Senior Lecturer in the School of Environmental and Life Sciences (Geography and Environmental Studies) at the University of Newcastle. Meg specialises in a range of topics, including: community engagement, energy development, environmental law, justice and ethics, governance, land-use transformation and the social and economic impacts of climate change. Meg's geographic research interests are place-based and revolve around issues of land-use change and development of new and emerging energy sources both locally and internationally. In researching these issues, Meg seeks to understand the complicated dynamics associated with energy development – how it is framed materially and discursively, the strategic decision-making around it and the contestation that exists over access to and uses of land and water sources.

Mona Sihombing is the Communication Programme Coordinator at the Asia Indigenous Peoples Pact. Previously, Mona was the Media Relations Officer between 2013 and 2014 and Director of Information and Communication between 2016 and 2017 for the Aliansi Masyarakat Adat Nusantara (AMAN), which is a group representing more than 2,000 Indigenous community members in Indonesia. Mona has qualifications in media and communications from the University of Indonesia and the University of Melbourne.

Liesel Spencer is a Senior Lecturer and Director of Research in the School of Law at the Western Sydney University. Liesel's research interests are in public health law and food systems governance. Her PhD took a comparative law and legal geographical approach to analysing the food security and public health impacts of Australia's place-based income management trial, and the US suite of food welfare law and policy. Liesel has also published research on the public and environmental health implications of urban agriculture, and on aspects of local government law as public health regulation.

Trinison Tari is Principal Officer – Provincial Outreach, Information and Communication for the Vanuatu Department of Environmental Protection and Conservation.

Hai-Yuean Tualima is a UNSW Scientia PhD candidate. She was previously the World Intellectual Property Organization (WIPO) Indigenous Fellow from 2015–2017.

Dr David J. Turton is an Honorary Lecturer for the Fenner School of Environment and Society at the Australian National University. Prior to commencing his PhD in 2013, David worked for the Commonwealth Department of Veterans' Affairs, in procurement, research and front-line service delivery roles. As part of the Department's graduate year in Canberra in 2010, David completed a Diploma of Government through the Australian Public Service Commission, before taking 'the path less travelled' to Townsville, where he served the veteran and Defence Force communities directly for two years. David has continued to explore the theme of public perceptions of policy in his ongoing research, which is focused on various aspects of Australia's coal seam gas debate.

Estair van Wagner researches and teaches in the areas of land use planning, natural resource and property law and is co-director of Osgoode's Environmental Justice and Sustainability Clinic. Estair joined Osgoode from the Victoria University of Wellington Faculty of Law where she taught Property, Natural Resource and Resource Management Law. She is currently the primary investigator on an inter-disciplinary Social Science and Humanities-funded research project examining the relationship between private property and Indigenous title to forest lands in British Columbia, Canada. Estair is also a collaborator on a SSHRC-funded project examining consultation and Indigenous consent in the context of mining activities in the Far North of Ontario, Canada.

Stewart Williams was a human geographer with expertise in qualitative research methods. He held a deep interest in critical social and political theory which informs his work in higher education, research and community engagement. As a geographer, loved to be involved in diverse projects that address real-world problems around health, hazards and housing, and with an emphasis on creating more socially just as well as sustainable and resilient communities. Stewart died suddenly and tragically in February 2019.

ACKNOWLEDGEMENTS

This book project has been a labour of love and collegiality since its inception in 2016 over a few drinks during which we moaned, yet again, the dearth of a concerted effort to explicitly engage with method in legal geography scholarship. We didn't get serious about this book until March 2018, following the regular annual convening of the Institute of Australian Geographer's legal geography study group workshop, this time at the University of Canberra. By then, the "Australian" cohort had been busy not only publishing in various journals, but also expanding our reach across the region and further afield in and across the global south. We decided then that it was time for an edited collection, and this book is the end result.

There are many people to thank for their contributions to this book, not the least of whom are the authors within, who have each made significant contributions. Importantly, this volume would not have taken shape without the unwavering efforts of Tayanah, as lead editor, in herding us together. Without her enthusiasm and ability to draw our collective work into sharp focus this project would have fallen at the first hurdle. We are also grateful for the financial support from the Institute of Australian Geographers and from the Australian Climate Change Adaptation Research Network for Settlements and Infrastructure (ACCARNSI), and from the director of ACCARNSI, Professor Ron Cox, which enabled the convening of the Canberra workshop that initially brought this collection together.

The Australian legal geography study group workshops have become a permanent fixture on the Australian scholarly circuit. They have become the foundation not only of critical thought, but also of collegiality and mentoring of researchers we hope to entice to our exciting field. One of our earlier workshops was hosted by our dear colleague and friend Stewart Williams. In these early days of (organised) legal geography scholarship in Australia, Stu took on a mentoring role for many of us who followed, and became a dear friend. Stu's tragic and sudden passing in February 2019 has left a hole in our hearts, and we were all unanimous in our commitment to not only dedicate this book to Stu, but to also include his scholarship *in memoriam*.

We must also extend deepest appreciation to research assistants Elizabeth Makin (UNSW) and Georgie Juszczyk (ANU), whose professionalism and expertise in getting the manuscript finalised for publication are greatly appreciated. We have also enjoyed working with our editors at Routledge, who have been accommodating and helpful throughout the manuscript preparation process, and who believed in and supported our vision for this book. The research for Chapters 8 and 12 was funded by the Australian Government through the Australian Research Council's Discovery Projects funding scheme (DP190101373). The views expressed therein are those of the author and are not necessarily those of the Australian Government or the Australian Research Council.

Finally, the most important thanks of all: to our families, who understand better than most an academic's life. We love you.

PART 1

Introduction

1

AN AUSTRALASIAN AND ASIA-PACIFIC APPROACH TO LEGAL GEOGRAPHY

Tayanah O'Donnell, Daniel F. Robinson and Josephine Gillespie

Introduction

Legal geography has developed as a field of study over a few decades, gaining increasing recognition for its role in providing a critical forum for analysis of law-space-society relations. Although there are antecedents in kin disciplines of critical legal studies, law and society and legal anthropology (e.g. Davies 2017; von Benda-Beckmann et al. 2009), legal geography has drawn its own lines on the scholarly map for some time now. Nicholas Blomley's (1994) seminal work *Law, Space, and the Geographies of Power* was perhaps the first comprehensive review of scholarship that brought together the concept of legal geography/geographies. Blomley (1994, p.51) explains that these critical geographies "seek to reconstruct the law-space nexus so as to accord proper recognition to both and to affirm the complex interplay of the two, evaluating the manner in which legal practice serves to produce space yet, in turn, is shaped by a sociospatial context". A recent succession of review papers (Bartel et al. 2013; Bennett & Layard 2015; Delaney 2015a; Delaney 2015b; Delaney 2017), journal special issues (Bartel et al. 2013; Graham & Bartel 2016; Robinson & Graham 2018) and books including Braverman (2014), have highlighted the development of this field since its clear culmination in the mid-1990s. Recent entanglements with scholarship in, for example, feminist geography (Cuomo & Brickell 2019, Brickell & Cuomo 2019; Perry & Gillespie 2019), environmental law research methods (Brooks & Philippopoulos-Mihalopoulos 2017), political ecology (Andrews & McCarthy 2014; Salgo & Gillespie 2018; O'Donnell 2019) and feminist political ecology (Gillespie & Perry 2019), point to an increasingly conceptually mature approach. As a scholarly field, legal geography has now moved beyond its adolescence, into a richer level of maturity. Our aim with this book is to add to this enrichment in legal geography research.

Legal geographers compel us to consider that in the "world of lived social relations and experience, aspects of the social that are analytically identified as either legal or spatial are conjoined and co-constituted" (Braverman 2014, p.1). Connecting the diverse works that underpin the legal geographical endeavour, and indeed distinguishing this field, is its "fine-grained, detailed attention to the complex processes of legal constitutivity and a desire to understand the reciprocal or mutual constitutivity of the legal and the spatial" (Delaney 2015a, p.98). The legal geography field, then, is used to disrupt ideas about the "closure" of law as a discrete, formalistic or even archaic set of institutions (statutes, courtrooms, case-law and contracts etc.); it highlights the political nature, social relations and power relations of law-making and law enforcement (Blomley 1994). One manifestation of this process has been revealed in the publication of papers about the reality of legally plural landscapes (Robinson & Graham 2018).

Of the spatial, Doreen Massey (1992, p.66), has written that the socio-spatial "is by its very nature, full of power and symbolism, a complex web of relations of domination and subordination, of solidarity and cooperation". If we apply this kind of thinking to the law, Hogg (2002, p.39) explains that this "is one path to subverting its imperial claims of objectivity, generality and sovereignty and to recognise the subsistence of other legal orders and other legal possibilities". Thus, legal geography helps ground the law in the "world" and in assemblages of socio-spatial-material relations. It is not only about understanding "law-making", but it is also about understanding "world-making", and the interaction between the two – which Delaney describes as *nomosphericity* (Delaney 2010).

Legal geography also requires analysis of the relationships between economies and nature or the environment (often commodified as "natural resources") and the role of humans in generating a range of often perverse and "unnatural" outcomes. Humans attempt, on the one hand, to respond to climate change, biodiversity loss, land and water degradation and other challenges of the Anthropocene while, on the other hand, the politics underlying the many legal and market failures inherent in these challenges are clearly evident in legal geography scholarship (Bartel et al. 2013; Bartel & Graham 2016; Gillespie 2018; Graham 2011; O'Donnell 2016; O'Donnell 2019; Robinson & Forsyth 2016). Groundswells of public and grassroots activism at a localised level push for responses to these global environmental issues, often with mixed results on international law and the global stage (e.g. see Bavikatte & Robinson 2011). This leads us to another aspect of the legal geographic study: the way we have come to think about *scale*. Authors such as Valverde (2015), with her concept of "chronotopes" encourage us to think about the assemblages of relations involved in law-making and that give us concepts such as jurisdiction and scale, which are used as methods of layering in the process of legal ordering. Legal geography scholarship thereby offers scholars, and as well as activists, paradigmatic strategies for approaching the spatial and material dimensions of social and environmental justice (Delaney 2015b; Jessup 2013), the governmentalities of decision-making, social order and the regulation of publics (e.g. Barkan 2011; Layard 2010; O'Donnell 2016; Robinson & Graham 2018).

Australian researchers have innovated the legal geography field with a specific, and deliberate, focus on human-environment relations. Prominent in the "Australian field", Bartel et al. (2013) argued for a specificity towards understanding and exploring the place-basedness of law. Innovative scholars such as Nicole Graham, as early as the mid and late 2000s, were working through concepts such as "lawscape" (2011, though the term appeared in her dissertation some years earlier) to better explicate human-environment relations and the attendant necessity for Anglo people in particular to reconsider our relationships with land and nature. Place specificity as argued by Bartel et al. (2013) built on that and other contributions, with a resulting explosion of scholarship that we claim has its origins in a distinctly "Australian" school of legal geography scholarship. Some Australian geographers have, in fact, been at the forefront of early incarnations of the legal geography project. Professor Gordon Clark, a distinguished Australian geographer, was an early up-taker with his 1985 book, *Judges and the Cities. Interpreting Local Autonomy*, Along with Blomley (1994), Blomley, Delaney and Ford (2001) and Delaney (1998) from the United States of America (USA)/Israel and Holder and Harrison (2003) from the United Kingdom (UK), Clark was also influential in these early legal geography works, thus proving that the legal geography remit was not immune from a critical Australian perspective.

We suggest that Australian legal geography scholarship might usefully be categorised/thought of as "environmental legal geography". This is not a spatially or locationally restrictive description. Australian legal geography scholarship has been at the forefront of various research programs throughout the Asia-Pacific region in recent years. Beyond our own Australian shores, our iconic legal geography scholarship embraces our regional Asia-Pacific perspectives, and thus contributions within this volume purposefully represent the variety of research efforts from our region. The distinction is evidenced by way of comparison to the Anglo-American legal geography approach, especially to prominent American legal geography scholarship, which is itself often concerned with critical analyses of power relations as between human-institution, or other times human-human interactions. This is akin to UK legal geography scholarship, which is also similarly though not exclusively human focussed. We note, in particular, that there has overwhelmingly been a tendency in legal geography reported scholarship to emphasise the urban as the foremost spatial site. We see in this book a deliberate turn away from this trend, in that we want to open spaces for a consideration of methodology in "other" settings and to keenly appreciate and consider Indigenous knowledge; reflectivity and positionality; environmental rights; mediation of human-nature relations including the "more-than-human" that has dominated other subdisciplines of human geography; and a focus on extractive industries and the effects of climate change. And we see this focus as being complementary to dominant urban focussed perspectives associated with much legal geography scholarship. Our primary aim here, then, is to broaden the legal geographer-as-scholar's toolkit. Exposing the complexity of the spatial and temporal implications of the law-society-environment matrix is our target.

What unites us is a coherency of purpose in reading land- and waterscapes for regulatory and social impacts. A distinctive Australian legal geography methodology is born in this endeavour.

The genesis of this book, and its focus on methods in legal geography, has been stimulated through the numerous discussions over the years within the Legal Geography Study Group of the Institute of Australian Geographers (IAG), buoyed by provocations about methodology by authors in the field such as Braverman's (2014) "Who's afraid of methodology" in *The Expanding Spaces of Law: A Timely Legal Geography* and Bennett and Layard's (2015) "Becoming spatial detectives" in *Geography Compass*. Recent incursions by Bennett include an attentiveness to our mindfulness in legal psycho-geography worldly interactions (Bennett 2018).

For the past decade, we have convened a series of legal geography conference sessions and workshops throughout Australia and New Zealand. Ranging from Cairns to Christchurch, Armidale, Sydney, Melbourne, Canberra and then Adelaide, like-minded graduate students, early career researchers and more established academics and practitioners have gathered to contemplate the appeal in unpacking the geo-legal entanglement. Many of these events have, importantly, revealed a methodological consistency in applied research built on grounded empirical accounts which underscore the importance of synergising law and geography approaches in braided lines of enquiry (following the observations in Braverman 2014). Our colleagues and peers from outside our region have encouraged us to advance the field beyond its adolescence, through a critical reflection on the methodologies and approaches used in the field (Delaney 2017). In doing so, we have also been reminded of the distinctiveness of the origins of Australian legal geography scholarship and our preoccupation with human-nature relations. This book is as much a collection of this clustering of these relations, as it is a book that reflects specifically on method and methodology within the remit of legal geography scholarship. We have therefore sought to engage attentively with both the perspectives and common methods in legal geography.

Reflecting on method

The broad diversity of methods utilised in legal geography might be thought of as challenging to the identity of the field but in this book, we argue that such diversity is a strength. Many of our chapters draw from research in Australasia, as well as parts of the Asia-Pacific and further abroad. Discussion throughout the chapters applies different methods best able to interrogate the "world-making" (following Delaney 2010) in those places underpinned by legal-socio-spatial processes. Cumulatively, the collection of chapters in this collection reveals the critical relevance of an explicit legal geography approach to scholarly research grounded in a common concern with human-nature-law-relations.

The rubric of legal geography highlights the law-place-people nexus across space and time. Self-identified legal geography methodologies reinforce three core components. First, one tool for legal geographers is to map the spatiality of law, but that often exposes a simplistic rendering of the law/place dynamic that deeper analysis reveals to be more nuanced and complex than foreseeable at first glance. This deeper analysis often requires in-depth understanding and analysis of doctrinal law. Second, and necessarily following this first point, the depth in research is brought about through engaging with methods and methodologies that unpack this complexity. Additional key insights are revealed in clarifying researcher positionality and in becoming reflexive about legal geography research processes. Third, legal geographers are well-placed to engage with material dimensions of real-world problems by incorporating multiple methods, and a variety of disciplinary perspectives. This usually occurs with the use of different methods applied to case studies, of which we see several examples in this book. Because of this, case studies are often used to underpin, or frame, the multiple methods utilised as well as to engage them empirically. Moreover, we argue that multiple methods are utilised precisely because of the appeal of inter- and multidisciplinary research enabled through a legal geography lens.

Many of the chapters in this collection use a case study approach to unpack how law, in its various guises, shapes and is shaped by the world around us – our geography. Governance arrangements and land-use practices collide in a series of real-world examples where the law-place dynamic no longer takes for granted rigid categorisations. Exploration of the ways in which people think about places and the regulatory practices that become embedded in or, sometimes, emplaced upon place is a strength of the legal geography approach, which is revealed throughout this collection.

While case studies provide an important "tie-in" to broader perspectives and methodological debates, this collection is not merely a collection of case studies. As is shown, the legal geography case-study approach is complemented by a range of empirical, normative, discourse and doctrinal analyses. Moreover, within the empirical "social research" frame, which tends to dominate this literature, we see in legal geography research a range of methods including interviews, ethnography and mixed-method surveys. As is demonstrated in this collection, these are often supplemented with critical readings of text. Recent innovations, such as that of Spencer (Chapter 9) show a blending of comparative legal analysis with empirical social research methods.

Turning attention specifically to methodology enables progress in this field by overtly recognising the scholarly innovation that is possible through a legal geography lens. This is not to overstate matters; nor has this occurred in a scholarly vacuum. Rather, the legal geography field has benefitted enormously from interdisciplinary scholarship. This has evolved to critical points in the scholarly debate that examine the cornerstones of methodological concerns including positionality, reflexivity and materiality.

Positionality, reflexivity and materiality

Consistently throughout this collection we see chapters concerned with defining and describing the role and influence of the researcher in research design and practice. To date, legal geographers use the tools of human, cultural and urban geographers, anthropologists and lawyers, and yet there has been limited critical reflection on the suitability of these approaches in specific areas of legal geography work, or how they might be better adapted to the particular interests of the work of legal geographers. We argue that there is a growing awareness of the need for critical reflection in this area and we hope that this book goes some way towards enabling that reflection. Indeed, one aspect of the stimulus for the book itself is to encourage these discussions, following Braverman's (2014) recognition of this growing need. In doing so, we are particularly concerned to make space for Indigenous ways of knowing and being, and to centre Indigenous voices in forward-thinking legal geography scholarship.

This additional focus on the researcher as a participant in the research process uncovers how we all influence and are, in turn, influenced by practice, knowledge and relations between humans and "the other" (whether this other refers to the more-than-human, nature, institutions, power, or many other ontologies). In a long-established human geography tradition, best-practice qualitative research embraces methodology that puts both the researcher and researched front and centre. The forefront of this evolution was led, at least partially by feminist geography scholars (e.g. Rose 1997, or the early work of McDowell 1992). Few, if any, writing within this tradition would reject the inherent critical perspective embedded within these scholars' research practices as they strive to reveal power relationships in research processes, to understand how multi-faceted identities shape the way we approach data collection, to comprehend the influence of our upbringings, educations, world-views, gender and experiences. Paying attention to the way we do our legal geography research, to the people we engage with and to the landscapes we inhabit, make our scholarship more meaningful. The issues we investigate have both a biophysical and social construction that demands we pay attention to them. Political ecological scholarship has recognised this, beginning with its very inception in Blaikie's (1985) work (1985) and continuing to this day (see for example papers in the *Journal of Political Ecology*). The complexity and dynamism of our socio-ecological systems requires innovation and a need to become more reflexive in our research practices. Within this collection, Australian legal geographers have risen to this challenge.

Legal geography scholarship tends to take a holistic rather than a doctrinal approach to law, though this is not a universal representation and, indeed, recent scholarship is showing how blended doctrinal and empirical research can result in rich and fruitful theoretical endeavours. Part of what makes legal geography distinctive and yet accessible is its ambitious remit within the broader constellation of social sciences as well as its overt engagement with legal scholarly practices.

Because legal geography draws from a multitude of disciplinary backgrounds, it requires more than merely applying methods of social science from geography or legal analysis; the field's methodological breadth is more demanding and thus requires a particular kind of reflection in order for it to reach a state of maturity. This collection aims to respond to Braverman's (2014) call for more specificity on method and methodology, and thus the breadth and depth of this collection is a timely contribution not only on method and methodology, but also serves to shine a bright light on the work in legal geography across the Global South and in Australia.

Finally, a focus on sensitivity to place demonstrated throughout this volume is based in part on how the material world (air, water, soil, animals, insects and other "stuff", however defined) is as important in shaping and re-shaping the work as culture (people). Overt recognition of this is a mainstay of our region's legal geography scholarship over the past decade, and many of our contributors have a growing body of work to evidence this point. Our hope is to draw attention to this in a more systematic way; that is, all legal geographers ought be sensitive to place *because* we recognise that places are different given their soil, water, topography (etc.) – all create physical environments that change across space and time – which law (per se) does not always recognise or appreciate. Materiality of place should be as important to a legal geographer as the social dimension of place. It is unquestionably the case that our physical scapes influence how law is developed through, for example, physical land features marking legal boundaries (Blomley 2008; Graham 2011). At the same time, law has an enormous power to transform the material (including landscapes) through fixing, labelling, restricting and ordering. Overlaying all of this are powerful and unknown (in terms of specific manifestations) impacts of environmental and climatic change on our material environments. There is, therefore, a lot to think about through a legal geography lens. If you've found this book, then you will likely be well aware of this complexity. We therefore encourage scholars to embrace this distinctive perspective, and we hope that this book is of use to you, as you do so.

Outline of this book

This book is organised into three substantive sections: (1) investigating the legal geographies of Indigenous peoples and their environments; (2) investigating the legal geographies of regulation; and (3) investigating the legal geographies of extractive industries. Each is situated in a particular case-study context that is representative of Australia and the Asia-Pacific region, though a small number of chapters traverse to the UK and the USA to undertake comparative analyses. In addition, authors were asked to dedicate one third of the discussion to specific method(s) that we argue comprise "core" legal geography methods. Some chapters engage specifically with positionality and reflexivity. On this, see Chapters 6 and 7 in particular.

During the compilation of this book, our colleague Stewart Williams passed away tragically and suddenly. This prompted us to add the Part 5, *in memoriam*, in recognition of his recent scholarship and contribution to the field of legal geography. We deliberately chose his 2016 paper on the legal geographies of supervised injecting rooms as a serious nod to legal geography that is concerned with regulating both the urban and personhood.

In inviting close and deliberate reflection on the legal geographies of Indigenous and local peoples and their environments, the first section begins with Gillespie's thorough reflection upon what it means to undertake fieldwork which explores legalities in remote locations, and where cross-cultural and language barriers may exist and persist (see Chapter 2). Gillespie artfully explores these intersections and is attentive to the importance of multiple methods in such contexts. Specifically, she draws attention to defects in exploring law without exploring the subjects of such laws; intertwined with this is a necessary focus on the normative regulatory practises of the communities she has engaged with – in this instance focused on Cambodia, Southeast Asia.

This first section then moves to an innovative endeavour in Chapter 3, in which Calyx, Jessup and Sihombing explore how Indigenous communities use technology and their local knowledge to advance legal rights to land, by digitally mapping their lands and documenting their rights to those lands. Their methods comprise activism, interviews and legal document analysis - all intertwined to tell the stories of Indonesian communities struggling against the state in asserting their rights. Chapter 3 also offers reflection on how the research team was formed through their common connection to an Indonesian court case.

Chapter 4 (Schenk) is also contextualised in Indonesia, exploring the role of Islam in shaping both customary and judicial law. Here, the theoretical and empirical contribution is to offer an examination of the relationship between politics, religion and law; the contribution to legal geography methodology is one that teases out not only what it means to look behind public documents, as well as the importance of taking positionality seriously. Schenk reminds us that it is not just the positionality of ourselves as researchers that is relevant, but also that of research participants.

Chapter 5 is a multi-authored contribution led by Robinson. The empirical context of the chapter is an analysis of Indigenous knowledge and how western systems of law, in this instance involving patents and intellectual property laws, have sometimes aided the appropriation of Indigenous knowledge. The chapter focuses specifically on Vanuatu, a country which has a rich system of *Kastom* – traditions and rules which might otherwise be understood as "customary laws". Robinson et al. highlight that some of the species being patented are covered by existing customary law and also focus on where the patenting or monopolisation of uses of a plant species might be bio-culturally offensive and unjust. Legal responses are also discussed along with the possibilities of greater recognition of customary law and customary legal ideas, including totemic plants/species, within other legal frameworks, such as through biodiversity laws under the Nagoya Protocol.

In Chapter 6, Bargh and van Wagner examine the consultation of Māori in New Zealand's minerals and mining regime. Theirs is a deeply reflective and personal account of their own research collaboration. The authors' genuine collaborative efforts to meaningfully engage with Indigenous communities, without creating burdens, is revealed in their comprehensive account of the ways in which settler state legal processes can act to constrain Māori people-place dynamics. This chapter gives us the unscripted and honest description of their identities, which depend on "friendship and mutual trust" to bring their research to life. Such accounts are to be commended to all legal geography scholars - for this forthright approach regarding reflexivity and positionality lends enormous credibility to their research work.

Chapter 7 begins the next section in our collection, being concerned broadly with the legal geographies of regulation. In Chapter 7, O'Donnell writes about her insider/outsider experiences in her coastal climate change adaptation research. This chapter is set in O'Donnell's ongoing and highly influential original research into the legal geographies of coastal processes, sea level rise and climate change. The focus for her chapter concerns her positionality in being a lawyer/researcher in exploring coastal climate change adaptation governance. Her story begins by situating herself as an insider employed in the courtroom as a Tipstaff and acknowledges her background prior to the present, working in the legal profession. O'Donnell's tale takes us on a journey, stimulated by her exposure to land use planning law litigation (the *Vaughan* litigation) that explores the reflexive dimensions of legal geography scholarship.

Chapter 8 by Godden focuses on Indigenous Australian connections to Country, which derive from deep ancestral connections to their lands, often through Dreaming stories. The chapter focuses on the *Native Title Act* and the focus on "connection to Country" as a fundamental element of the claiming of native title to land and waters by Indigenous Australians, following the violent and oppressive history since Australia's colonisation by the British. Due to dispossession from Country, native title processes require a range of methods for proving this connection to Country including archival, historical and geographical (mapping) processes. Godden explains the challenges inherent in these processes, and some of the developments including having court hearings on Country, which provide at least some form of procedural connection to place and the materialities of Country.

Chapter 9 by Spencer focuses on "legal transplants", or the borrowing of legal ideas and solutions from other jurisdictions, often as a law reform technique. Legal transplants are, however, controversial and problematic where the borrowed laws of the "transplantor" jurisdiction are insufficiently adapted to the human and environmental contexts of the "transplantee" jurisdiction. This chapter builds on Kedar's proposal for a new hybrid methodology and considers how comparative legal geography can contribute critical depth to investigations of human and environmental contexts – the nuances of "place" – as extra-legal factors in legal transplants.

Chapter 10 by Bartel focuses on plant classifications made in the name of biodiversity conservation. This chapter argues for greater recognition of "place law" as revealing the inherent bias of the dominant legal system, imposing a hierarchy of plants (e.g. weeds and non-weeds), which is predicated on a human-nature binary and enforces its own class of primacy and privilege. Bartel argues that it is this "othering" perpetrated by "us" – the settler state – that must be problematised, rather than the plants.

Chapter 11 by McFarland focusses on the use of key informant interviews as a research method in legal geography. It takes a phenomenological approach to understanding underlying human perceptions of the world, and analyses the expert interviews relating to differences in approaches to peri-urban planning. McFarland argues that phenomenological research using key informant interviews is eminently suitable for case studies in legal geography. His use of this method is argued to be highly appropriate to the examination of policy development and implementation for peri-urban land use in Oregon, as a basis of comparison with Australian examples. In so doing, the method reveals the strengths and weaknesses of the regulatory system based on multiple perspectives. It exposes the human factor in the development and operation of law in relation to place and space.

Chapter 12 by Graham provides a legal geographical analysis of Sydney's drinking water catchment. Her approach allows us to understand this place as neither a "natural" nor "cultural" place, but rather as a dynamic "lawscape" of interrelated rivers, swamp ecologies, groundwater systems, violent colonial dispossession and conflict, dams, underground coal mines, Special Areas, National Parks and the source of drinking water for five million people. The chapter explains how, over two centuries, Anglo-Australian laws have transformed this place into a resource frontier by conceptually and physically fragmenting it into different categories of lands and separable component parts, principally water and coal. This chapter critiques the atomism of these laws through the lens of legal geography, which emphasises the need for a more relational, holistic and congruous legal approach.

Chapter 13 by Turton considers some of the challenges and opportunities facing legal geography researchers who wish to examine the insights of lawyers in their work. Drawing on coal seam gas (CSG) literature, especially including lawyer perspectives, this chapter surveys obstacles to securing access to lawyer voices in legal geography research – while also addressing how this challenge might be negotiated to ensure fruitful outcomes. Building on this, the chapter then addresses the possibility of investigating lawyer perspectives through the prism of submissions authored by legal professionals for a CSG-related government inquiry – raising socio-spatial questions about the geographical scope of compensation for landholders in the process. This case study reveals lawyers as not only active participants in Australia's CSG debate, but also the purveyors of legal advice for a variety of interest groups, who may then seek to utilise lawyers' opinions to support their own advocacy efforts on the subject.

Chapter 14 by Sherval explores the mobilisation of women's political subjectivities in response to government rhetoric that frames the development of the hydrocarbon industry as a "bridge" towards a low carbon future. Reflecting on the recent movement by governments to construct a narrative which suggests that hydrocarbons such as shale and coal seam gas are the 'greenest' of fossil fuels, this chapter explores how recent technological advances in energy extraction have transformed what were once considered 'unconventional' energy sources into accessible, but also highly politicised materialities. By considering the links between the geopolitics of energy placement, government rationalisation of space and community perceptions of government decision making, this chapter considers how power can be mobilised amongst local subjects and potentially renegotiated in decision making about the future of rural spaces.

Chapter 15 by Rickards and Jolley draws on Delaney's "nomosphere" and also Jacques Rancière's "politics of aesthetics", which describe how, through law, certain worlds are made visible and sensible while others are not. The authors argue that understanding and addressing the complexities of the fossil fuel regime and its connections to climate change requires a legal geography attuned to the non-representational dimensions of law. In doing so, they frame research as a political tool that contributes to determining what is made visible, what problems are recognised and what arguments and responses are legitimated. They further explain that this requires reflexivity on the part of researchers, with respect to the political and aesthetic acts, orders and regimes to which their work contributes.

Finally, Chapter 16 is a republication of Williams' 2016 journal paper in *Space and Polity*. There, Williams explores how law created nested hierarchies of order and control, and how debate and law reform regarding the establishment of safe injecting rooms in Sydney, Australia disrupted this dominant narrative and the implications of this for both legal geography scholarship and for public health outcomes.

References

Andrews, E. and McCarthy, J. 2014, 'Scale, shale, and the state: Political ecologies and legal geographies of Shale gas development in Pennsylvania', *Journal of Environmental Studies and Sciences*, vol. 4, no. 1, pp.7–16.

Barkan, J. 2011, 'Law and the geographic analysis of economic globalization', *Progress in Human Geography*, vol. 35, pp.589–607.

Bartel, R., & Graham, N. 2016, 'Property and place attachment: A legal geographical analysis of biodiversity law reform in New South Wales', *Geographical Research*, vol. 54, pp.267–284.

Bartel, R., Graham, N., Jackson, S., Prior, J. H., Robinson, D. F., Sherval, M., & Williams, S. 2013, 'Legal geography: An Australian perspective', *Geographical Research*, vol. 51, pp.339–353.

Bavikatte, K. S., & Robinson, D. F. 2011, 'Towards a people's history of the law: Biocultural Jurisprudence and the Nagoya protocol on access and benefit sharing', *Law, Environment & Development Journal*, vol. 7, pp.35–55, <http://www.lead-journal.org/content/11035.pdf>.

Bennett, L., & Layard, A. 2015, 'Legal geography: Becoming spatial detectives', *Geography Compass*, vol. 9, no.7, pp.406–422.

Bennett, L. 2018, 'Towards a legal psychogeography: Pragmatism, affective-materialism and the spatio-legal', *Revue Geographique de l'Est*, vol. 58, no. 1–2, pp.1–46, <https://journals.openedition.org/rge/7534>.

Blaikie, P. 1985, *The Political Economy of Soil Erosion in Developing Countries*, Longman, New York and London.

Blomley, N. 1994, *Law, Space, and the Geographies of Power*, The Guilford Press, New York.

Blomley, N., Delaney, D., & Ford, R. (eds), 2001, *The Legal Geographies Reader*, Blackwell, Oxford.

Blomley, N. 2008, 'Simplification is complicated: Property, nature and the rivers of law', *Environment and Planning A*, vol. 40, pp.1825–1842.

Brickell, K., & Cuomo, D. 2019, 'Feminist geolegality', *Progress in Human Geography*, vol. 43, no. 1, pp.104–122, <https://doi.org/10.1177/0309132517735706>.

Braverman, I. 2014, 'Who's afraid of methodology? Advocating a methodological turn in legal geography', in *The Expanding Spaces of Law: A Timely Legal Geography*, eds I. Braverman, N. Blomley, D. Delaney, & A. Kedar, Stanford University Press, Redwood City, California, pp.120–141.

Brooks, V., & Philippopoulos-Mihalopoulos, A. (eds), 2017, *Research Methods in Environmental Law: A Handbook*, Edward Elgar, Cheltenham, UK.

Clark, G. 1985, *Judges and the Cities. Interpreting Local Autonomy*, University of Chicago Press, Chicago and London.

Cuomo, D., & Brickell, K. 2019, 'Feminist legal geographies', *Environment and Planning A*, vol. 51, no. 5, pp.1043–1049.

Davies, M. 2017, *Law Unlimited: Materialism, Pluralism, and Legal Theory*, Routledge, Oxon.

Delaney, D. 1998, *Race, Place and the Law*, University of Texas Press, Austin, Texas.

Delaney, D. 2010, *The Spatial, the Legal and the Pragmatics of World-Making: Nomospheric Investigations*, Routledge, Oxon.

Delaney, D. 2015a, 'Legal geography I: Constitutivities, complexities, and contingencies', *Progress in Human Geography*, vol. 39, pp.96–102.

Delaney, D. 2015b, 'Legal geography II: Discerning injustice', *Progress in Human Geography*, vol. 40, pp.267–274, <https://doi.org/10.1177/0309132515571725>.

Delaney, D. 2017, 'Legal geography III: New worlds, new convergences', *Progress in Human Geography*, vol. 41, no. 5, pp.667–675.

Gillespie, J. 2018, 'Wetland conservation and legal layering: Managing Cambodia's great lake', *Geographical Journal*, vol. 184, no. 1, pp.31–40.

Gillespie, J., & Perry, N., 2019, 'Feminist political ecology and legal geography: A case study of the Tonle Sap protected wetlands of Cambodia', *Environment and Planning A: Economy and Space*, vol. 51, no. 5, pp.1089–1105, <https://doi.org/10.1177/0308518X18809094>.

Graham, N., 2011, *Lawscape*, Routledge, New York and London.

Graham, N., & Bartel, R. 2016, 'Legal Geography/ies', *Geographical Research*, vol. 54, no. 3, pp.231–232.

Hogg, R. 2002, 'Law's other spaces', *Law Text Culture*, vol. 6, pp.29–38.

Holder, J. and Harrison, C, (eds), 2003, *Law and Geography*, Oxford University Press, Oxford.

Jessup, B. 2013, 'Environmental justice as spatial and scalar justice: A regional waste facility or a local rubbish dump out of place?', *McGill International Journal of Sustainable Development Law and Policy*, vol. 9, pp.71–107.

Layard, A. 2010, 'Shopping in the public realm: A law of place', *Journal of Law and Society*, vol. 37, pp.412–441.

Massey, D. 1992, 'Politics and space/time', *New Left Review*, no. 196, pp.65–84.

McDowell, L. 1992, 'Doing gender: Feminism, feminists and research methods in human geography', *Transactions of the Institute of British Geographers*, vol. 17, no. 4, pp.399–416.

O'Donnell, T. 2016, 'Legal geography and coastal climate change adaptation: The Vaughan litigation', *Geographical Research*, vol. 54, no. 3, pp.301–312.

O'Donnell, T. 2019, 'Coastal management and the political-legal geographies of climate change adaptation in Australia', *Ocean and Coastal Management*, vol. 175, pp.127–135.

Perry, N., & Gillespie, J. 2019, 'Restricting spatial lives? The gendered implications of conservation in Cambodia's protected wetlands', *Environment and Planning E: Nature and Space*, vol. 2, no. 1, pp.73–88.

Robinson, D. F., & Forsyth, M. 2016, 'People, plants, place, and rules: The Nagoya Protocol in pacific island countries', *Geographical Research*, vol. 54, no. 3, pp.324–335.

Robinson D. F., & Graham, N. 2018, 'Legal pluralisms, justice and spatial conflicts: New directions in legal geography', *Geographical Journal*, vol. 184, no. 1, pp.3–7.

Rose, G. 1997, 'Situating knowledges: Positionality, reflexivities and other tactics', *Progress in Human Geography*, vol. 21, no. 3, pp.305–320.

Salgo, M., & Gillespie, J. 2018, 'Cracking the code: A legal geography and political ecological perspective on vegetation clearing regulations', *Australian Geographer*, vol. 49, no. 4, pp.483–496.

Valverde, M. 2015, *Chronotopes of Law: Jurisdiction, Scale and Governance*, Routledge, London.

von Benda-Beckmann, F., von Benda-Beckmann, K. & Griffiths, A. 2009, *Spatialising Law: An Anthropological Geography of Law in Society*, Ashgate, England & USA.

PART 2

Investigating the legal geographies of Indigenous peoples and local communities and their environments

2

CHALLENGES IN LEGAL GEOGRAPHY RESEARCH METHODOLOGIES IN CROSS-CULTURAL SETTINGS

Josephine Gillespie

Introduction

Legal geography methods range from an analysis of legal regulation in the form of codified laws and informal norms through to field-based primary data collection using qualitative research techniques. Research in locations outside a researcher's cultural context requires particularly careful attention to research design and execution. In my work I have found that written material pertaining to environmental law or regulation, particularly in English, is limited. Moreover, in many remote places in rural Cambodia, where much of my research has been based, any written codified material articulating laws in Khmer is also limited. Another key challenge here is a relatively high level of illiteracy in rural populations in Cambodia, with UNESCO suggesting 36% illiteracy rates in rural Cambodia (UNESCO n.d.). In order to understand the ways in which law and geography intertwine in this context there is a need to ask people – on the ground – about how they understand the law's impacts. Thus, doing in-country research as an outsider using a legal geography lens requires significant pre-fieldwork preparation. Finding English-source material for local laws and understanding the implications of custom, or lore, is challenging as an outsider. The necessity in undertaking these pursuits links legal geography work to cognate work under the rubric of legal pluralism (Robinson & Graham 2018). Engagement with legal pluralism by legal geographers is important in unpacking the complexities of law as an entity in the law–place–people dynamic, and there has been a recent surge of engagement, see particularly a collection of papers published in *The Geographical Journal* (Schenk 2018; Gillespie 2018; Robinson & McDuie-Ra 2018).

Pre-field investigative work aids in-field qualitative work. In these circumstances, it is essential to draw on a breadth of literature to enhance the data collection approach. This includes many secondary data sources and incorporates legislation, reports, internal working documents, theses, books, leaflets, workshop events and print, digital and archival materials. For legal geographers this should also include maps, aerial images and conventional photographic material. Following Crang (2005), the need to gather textual material moves beyond written "texts" alone to include these additional sources. Much of this material is considered "grey literature" and includes technical and policy documents. The online sourcing of materials is critical to preliminary research efforts. There is, however, an obvious shortcoming in relying too heavily on English-translation documentation. Therefore, again, it is essential to supplement this material with grounded empirical data collection (Braverman 2014, p.131; Bennett & Layard 2015). For social scientists the data collection experience may draw on both quantitative and qualitative methods/techniques. However, in this chapter the focus is on using qualitative methods in legal geography research for projects in non-Western country settings. The chapter draws on my field experiences from Southeast Asia, working within villages in-situ to explore law's various impacts through the localisation of regulatory obligations.

Qualitative methods

Literature on the use of qualitative methods or techniques in human geography helps to inform the data collection component of in-country legal geography research. Qualitative data collection methods, and the variety of techniques that accompany this category of research, are a standard component of social science research and form part of the repertoire of human geography's catalogue of commodities (Hay 2016). Moreover, in-country field-based social research techniques permeate much of the report-based literature produced by many NGOs working in less-developed country settings. In this chapter my legal geography perspective arises within a broader human geography approach using qualitative methods; yet this is but one way to label such work. As this book demonstrates, legal geographers use a wide variety of methods and draw on a variety of traditions to flesh out their work. Similarly, scholars working within a cognate socio-legal tradition use the very same methods to report their research. Moreover, such scholarship is often imbued with a critical lens as it aims to uncover the ways in which law (in all its guises) enables/disables social relations or exposes power dynamics, a perspective that is often associated with the critical legal theory movement (Unger 1983). In this chapter the legal geography commentary is set within a human geography qualitative methods prism, yet the broader conceptual questions surrounding the need for a thorough, and thus critical, account of the law–place–people dynamic is acknowledged, even if this is not always explicitly expressed.

The decision to conduct fieldwork, and to acquire primary data, must be based on meeting the demands of research questions. The sentiments expressed by Rundstom and Kenzer (1989, p.294), when they suggest that primary data collection is critical in geographical studies to generate first-hand information *in situ*, are reflected in much legal geography research. However, another significant motivator for in-country data collection is often the absence of any secondary data sources on the issues raised by the research questions. Although interviewing remains a "time consuming, labor-intensive task" (Rundstrom & Kenzer 1989, p.301), it is also among the best ways of accumulating data in a less-developed country setting, which may lack the administrative infrastructure to produce reliable secondary data sources, such as census data. Qualitative methods extrapolate how people understand their world and provide the opportunity for people to give extensive, wide-ranging accounts of their situation. In my fieldwork based at Angkor, Cambodia, data collected during fieldwork seasons based in two villages within the highly regulated core of the World Heritage Park took the form of semi-structured in-depth interviews (Gillespie 2009; Gillespie 2012; Gillespie 2013). Other field-based work in and around the floating villages of the Tonle Sap in central Cambodia also used the interview format (see Gillespie & Perry 2018; Perry & Gillespie 2019). Drawing on the interview responses, the remarks of residents provide verbally rich and nuanced accounts of individual experiences which help to shape the law–place–people dynamic (Bennett & Layard 2015). Part of the process of framing legal geography research through one-on-one interviewing also requires compliance with meeting precise standards for research based on views, beliefs and opinions. This is an issue to which we now turn.

Research design and rigour

Framing legal geography research within a qualitative structure should be informed by a consideration of the appropriate method for conducting field-based research activities, in addition to a deliberation about the theoretical setting in which the research takes place (adopting Bradshaw & Stratford 2005, p.69). It is important to recognise that the methods used throughout any study ought to fit into a conceptual setting (or methodology), where such methodology (*sensu* Shurmer-Smith 2002) is the epistemological context which guides the methods. Accordingly, the difference between "methods" and "methodology" merits consideration. In my research I adopt Hoggart et al. (2002, pp.48, 310) definition wherein

> method refers to the 'process' or technical means of collecting data …

and

> methodology embraces issues of methods of data collection and analysis when these are grounded in the bedrock of a specific view on the nature of 'reality' (ontology) and the basis of which knowledge claims are made (epistemology).

Qualitative research methods are not immune from criticism. For semi-structured interviews, a central issue of contention surrounds the problem that narratives derived from conversations are selective and anecdotal. While the fruitfulness of this technique lies in the possibility for expansive answers and a broadening of the researcher's knowledge (Valentine 2005), there are consistent calls for social science researchers to be wary about the rigour (or trustworthiness) of their approach and/or the validity of their results (Baxter & Eyles 1997). Crang (2002) suggests,

> (i)n geography, there has also been debate about ways of ensuring the rigour and evidential quality of qualitative work, set in motion by Baxter's and Eyles's (1997) critique of the lack of methodological transparency in published papers based on interviews.

This sentence is critical to the position adopted in my legal geography in-country research – the methods used should be subject to a degree of attention which will enable them to pass a test of rigour. This call for transparency is not restricted to the (human) geography discipline, and examples of social science research across the board embrace this call (for example, de Wet & Erasmus 2005, pp.27–40, write under the auspices of sociology). Baxter and Eyles (1999) apply a rigour yardstick to their own study in which face-to-face, semi-structured interviews were used to accumulate knowledge about how residents felt about the locating of landfill in their neighbourhood. They argue that to meet the standards of rigour in qualitative research an evaluation criterion must be incorporated into the research design, and that often the failure to maintain rigour is brought about by an unwillingness by researchers to articulate, in full, the methods adopted for their study (Baxter & Eyles 1999, pp.310–11, see also Bradshaw & Stratford 2005). For this reason, the details about methods should be described in as much detail as is practically possible. As Baxter and Eyles (1999, pp.311–15) dictate, consideration should be given to, inter alia, expressing the history of the research, data collection and analysis techniques, sampling strategies, the significance of the way in which results are presented and the transferability of findings. Others, such as Bradshaw and Stratford (2005), have also reiterated the point that a researcher has a responsibility to "share, interpret, and represent others' experiences" and that this obligation implies that the results need to pass a test of evaluation,

> it is vital that we document all stages of our research process. Such documentation allows members of our interpretative and participant communities to check all of these stages, so our work might be considered dependable.
>
> *(Bradshaw & Stratford 2005, p.75)*

Compliance with this prescription is the intention of in-country legal geography research in order that the findings may be tested, expanded, nuanced and so forth by further research.

Developing nuanced data sets through qualitative methods in social science research in a Southeast Asian setting remains challenging. In relation to their research experiences in Vietnam, Scott et al. (2006) observe that there has been a pre-existing preference for studies that use positivist practices – and they identify surveys, questionnaires and mapping – in preference to techniques such as semi-structured interviews. Understanding the differences between methods within the suite of qualitative tools available to research in this tradition is important. Paying close attention to the tools we choose to execute our research is essential to expanding the legal geography remit, and Braverman makes a similar point in calling for reflection in crafting the research agenda of legal geographers (Braverman 2014). To this end, I have found that using in-depth semi-structured interviews as a method for building my information about law's various impacts (in cross-cultural settings) is a useful way to build personal interactions, often requiring exchanges to be friendly and informal with much extraneous talking to help establish rapport, and to a lesser extent, a trust between interviewee and interviewer. Given my approach, the next part describes the integral, and complex, role the researcher plays in the fieldwork process.

Reflection: positionality and reflexivity

Before detailing fieldwork methods for cross-cultural settings, it is important to reiterate the well-established notion that fieldwork does not take place in a vacuum, free from the social and cultural filters brought to the study by the researcher (Howitt & Stevens 2005). Conducting research in a country such as Cambodia, which is often regarded as a post-conflict society, presents a plethora of issues. For a non-Khmer researcher who brings an outsider's perspective, field-based research is even more complex. This is neither a localised, or otherwise unique phenomenon, as field-based researchers around the world simultaneously are enchanted with the newness and charm of novel places, but become exhausted with frustrations associated with, inter alia, poor living/working conditions and labyrinth-like administrative hurdles (see Steinberg 2006, p.14 for a compelling account of this contradiction).

Understanding the role of the researcher leads us to consider the twin concepts of "positionality" and "situatedness" (Haraway 1991; Rose 1997; Chacko 2004). These labels enable us to recognise that research is a product of the time, space and social setting/s in which it takes place and that the researcher is an inherent part of the research process. Some salient points about positionality were made by Chacko (2004) in her graduate research in rural India. She observed that while self-reflexivity is important as a process through which the research produced is informed, "active measures" taken during fieldwork which attempt to equalise any power imbalance between the researcher and the researched may do more to validate the research in the eyes of the researched than would otherwise be the case. In other words, the way in which fieldwork is conducted is as important as an analysis of the researcher's individual traits (Chacko 2004, p.55).

While endorsing Chacko's observations, it is timely to consider the position of this researcher vis-à-vis the research and the researched.

The researcher

There is no doubt that the representations of the people who are the subject of the study are filtered through the researcher – her knowledge, experiences and the way in which she communicates all play a significant role in the how the participants are depicted. As a tertiary educated, white, female researcher, neither native to Cambodia nor fluent in Khmer, there are a multitude of variables which may act to influence the way in which accounts from local Khmer are interpreted. In Cambodia I am a "barang" – a foreigner. From this perspective, the research is framed, organised and executed by an "outsider". It attempts to embrace Crang's (2003, p.496) wariness "of work that divides positionality formulaically into being insiders (good but impossible) and outsiders (bad but inevitable)", and Rose's rejection of clear "'inward' and 'outward' reflexivity demanded by transparent reflexion" (1997, p.316). There are shortcomings associated with not being Indigenous, but there are advantages for researchers too.

Mullings (1999, p.341) describes her insider/outsider dilemma as "a black woman of British/Jamaican heritage, from a North American University" in the setting of interviews conducted in Jamaica, and suggests that she is neither insider nor outsider, challenging the assumption of binary insider/outsider positions in cross-cultural settings. It is Mulling's (1999) use of "outsider" status – to the advantage of her research – which is particularly interesting. She concedes that she consciously and actively promoted "outsider" status by aligning herself with her British past and North American University position, rather than as someone with a Jamaican heritage, to "represent [herself] in the least threatening way" (Mullings 1999, p.344). In the course of conducting interviews in Cambodia, I have taken advantage of the "outsider" label for the very same reason – to avoid affiliation with Cambodian authority in an attempt to bring a perception of neutrality to the interview (in practice this meant that the affiliation with an Australian university and the pure "research" elements of this work were often reiterated during the course of an interview with local residents). On the other hand, when interviewing managers (both local and international), an element of empathy for the challenges facing management required the researcher to understand, and articulate, the questions from the perspective of "insider" – in as much as questions were prefaced by outlining a sympathetic position towards management. A warning bell is sounded, however, by Hubbell (2003) who writes about an inherent deceptiveness involved when a researcher adopts a flexible approach about presenting him- or herself to participants. Rather than being deceptive, I argue that this flexibility is a product of the semi-structured in-depth interview, where the researcher seeks to build some rapport with the respondent in a personal, face-to-face interaction. As Mandel (2003) observes, it seems impossible to become an insider, but this does not allow one to forego the necessity of

complying with local expectations regarding dress standards or appropriate culturally sensitive conduct whilst in the field in order to build rapport or empathy.
Crang (2003, p.497) observes,

> (w)hile deception can and does occur, from both parties, it is also quite
> important to recognize that our projects are often unstable entities which
> are not only presented, but actually exist, in multiple versions given to
> funders, colleagues, friends, family, peers and (different) respondents, one
> of which need be necessarily the 'true one'.

Is it not impossible in the course of an interview, which may last any length
of time and in my own research range between five and ten minutes and two
hours or more, for the researcher to give a thorough account of their positionality. However, given that the point of the exercise is to understand the way the
respondent – not the researcher – thinks, understands and believes in relation to
the research questions, an account of the interviewer's experiences is not necessarily required.

In legal geography, Braverman (2014) explicitly writes about the conundrum
of insider and outsider positioning. Braverman has positioned herself as a "halfie"
(Braverman 2014, p.129) in some of her research. Her status as Jewish-Israeli
woman enabled her to gain access to people/places that others would not in
the context of the very conflicted Israel/Palestine territories. However, in her
work on zoos and bureaucracies in North America she encountered resistance as
an outsider. Braverman explains this came as a result of zoo workers becoming
"wary of strangers asking provocative questions" (Braverman 2014, p.129). These
vastly different research contexts throw up different challenges, and one person's
different identities both enable and disable research encounters.

Elements of method

The following part details primary data collection methods for social science
research in a legal geography tradition. Considerations relating to questions about
where, who and how the primary fieldwork data are collected are addressed.

Where: site selection and the spatial setting

At the individual scale I often conduct face-to-face interviews on-site, *in situ*.
Interviews have taken place beneath houses, on open-platform bed frames outside houses, sitting or standing outside dwellings, on boats, in floating homes, in
floating businesses and on *motos* (motorbikes) and tuk-tuks as well as in the more
usual setting of offices in government buildings.

The setting of the interview can be of some significance for research results
(Hoggart et al. 2002; Valentine 2005). Elwood and Martin (2000) provide some
useful insights into the importance of the location of the interview, and suggest

that the "microgeographies of interview locations" enable researchers to "'read'... important insights into social geographies of places being studied" (Elwood & Martin 2000, p.652). Other commentators (for example, Sin 2003) suggest the interview setting can add extra insights and make the obvious, but crucial, point that home spaces make an interviewee feel at ease during the interview (see also Valentine 2005). Moreover, adverse interviewing conditions can be both negative (due to background noise, lack of privacy in communal settings) and positive (for shedding light on people's social/work settings and their status in the wider community) (Sin 2003). In conducting research which aims to reveal how resident communities understand the regulations imposed upon them, their physical spaces (their yards, homes or fishing lots) can and do reflect the extent to which they comply with the restrictions. In my research on protected areas in Cambodia, the need to understand the impact of rules often requires face-to-face interactions. For instance, when residents explain that boundary markers are traditionally in the form of cactus rows the interview location moves to the boundary and discussion ensues on the use of cactus or banana trees for this purpose. In another example, when discussing how residents interpret the rules about how big their new house or extension may be, this is explained by walking around the existing house and pointing out (literally through hand-waving explanation) that the house can only "be as big as this" (while explaining that it also must accommodate ten people). Both examples illustrate how qualitative methods and *in situ* site/location fieldwork becomes essential in creating more discerning, astute and thoughtful understandings about the effect of the World Heritage classification (Gillespie 2013). It is also the situation that, in many interview encounters at the village-level, interview locations in and around homes, on the street, in front gardens and outside shops/pagodas, amongst other locations, often (in fact, mostly) encourage multiple participants to listen, add, correct or debate both the questions and the answers during the interview process. Countless encounters with an uncle or aunt, cousin, friend or neighbour added to the interview experience; debates about boundaries and/or access or "who owned what" lead to (literally) hours of discussion amongst family and neighbours. My presence, being a non-tourist visitor from Australia, seemed to make the encounter, at times, an interesting village event. Additionally, many such interviews involved more than one family member, so intergenerational responses are frequently taken into account through the interview process. The resultant depth, nuance and richness of account in interviewee responses in these circumstances provide incredibly useful insights into the law–place–people dynamic. These *in situ* dynamics render the law–place–people dynamic complex, and recall Delaney's idea of "nomosphere" wherein the socio-spatial and legal combine to produce place (Delaney 2010).

Who: locals and gatekeepers

Following Scott et al. (2006), one of the major concerns for researchers in conducting fieldwork in Southeast Asia is the political context in which the research takes place. In the case of research in Vietnam described by Scott et al. (2006), the concern is to be aware of the way in which a society in transition to a market economy may impact the research. Cambodia, however, had significant disruption from the Khmer Rouge era, and thus the transition is less from a socialist structure than it is from genocide and chaos to a market economy. Nonetheless, the observations (Scott et al. 2006) that access and permissions to conduct interviews are restricted by a bureaucracy that is fundamentally hierarchical mean that researchers must have prior approval to conduct interviews from the "authorities". Whilst comments by Scott et al. (2006) about the political setting of research are valid, there may be additional factors to consider. It is arguable that societal issues, in addition to the political context, are at play in both site access and gaining permissions for researchers working in a cross-cultural setting. The significance of hierarchy in Cambodian society cannot be underestimated:

> In Cambodian society social stratification and differences in status are extremely important. Everyone knows, and needs to know, their place relative to that of others. This is exemplified through the everyday language people use to address each other which acknowledges their respective age and status. (O'Leary & Nee 2001, p.48)

A Cambodian proverb suggests:

> Kom bos san touch ro long phnom. Don't throw the fishing line over the mountain. You cannot do anything that is not following the 'proper' way, according to hierarchical structures. (O'Leary & Nee 2001, p.51)

Many ethnographic, anthropological and social/political commentators of Cambodian society write about its inherently hierarchical character (for example, O'Leary & Nee 2001; Chandler 1991; Ebihara 1968). Understanding that patrimonialism exists in Cambodia (Pak et al. 2007) becomes important when interviewees explain their actions. In my work around fisheries and wetland protection on the Tonle Sap, these complex power interactions inherent in pre-existing political and administrative structures are apparent (Gillespie 2018).

It is also important that the link between access, permissions and power should be acknowledged for:

> (o)ne goes to the field as a kind of "stranger" and draws on that status to see difference and ask questions that under other circumstances might seem (even more) intrusive, ignorant, or inane to those who answer them. The answers, and what one makes of them, have currency in other sites of enunciation – journals. (Katz 1994, p.68)

Yet, in the process of engaging with people in the field, in the actualities of being present in another place, establishing contacts, gaining rapport and taking advantage of affiliations, these connections may influence the researcher's position and the power relations inherent within fieldwork (Chacko 2004). In the case of research conducted at Angkor, affiliations with the management authority are a pre-requisite for access to the protected Park.

The role of gatekeepers in this type of research activity needs attention. Mandel (2003) has written about the role of (male) gatekeepers in providing access to, and giving authority for, her research during her fieldwork in Porto Novo, Benin, Africa, and admits to underestimating the role of the local gatekeepers. In my own Cambodian-based research, fieldwork conducted within the villages is limited by the same phenomenon – access to villages can be direct, but, even with passes from "higher" government or administrative authorities, it remains essential protocol to negotiate access to villages through the Village Chief as recognition of the role he (or she) plays in the structure of village life. Acknowledging and working within the constraints of this cultural norm is vital both to good fieldwork practice but also to the quality of the research. Similar experiences have been documented by others, including Vandergeest and Peluso (1995), in their work in rural Thailand.

Understanding the complexities of normative behaviours in complying with regulations is essential in the research work of legal geographers. My work in Cambodia has demonstrated that regulations imposed for the protection of critical habitat and species can, sometimes, work to exacerbate existing socioeconomic disadvantage, including gendered disadvantage. Unpacking the often-invisible gendered dimensions of law's impacts is also important in different cultural contexts for, in my work, protected area laws are not immune from gendered implications (Gillespie & Perry 2018; Perry & Gillespie 2019). The complexities of access to a range of varied-identity participants in qualitative research is, therefore, necessary to disentangle in order to provide for richer research outcomes. Legal geographers need to pay attention to the ways in which laws variously impact people from a wide range of backgrounds.

The approach adopted throughout my research uses a combination of sampling methods. These methods include identifying and liaising with key informants who are often playing the role of "gatekeeper", and using the "snowball" technique to recruit interviewees from the recommendations of those already interviewed and providing the opportunity to introduce another potential interviewee (Valentine 2005). Recruitment varied according to the time and place in which it took place. However, this is not to imply that recruitment was merely opportunistic. The selection of participants was based on an ever-evolving understanding of the cultural sensitivities of working in a cross-cultural setting, and the peculiarities of working in a Cambodian context (following, for example, O'Leary & Nee 2001; Miura 2004; Luco 2002). Accordingly, recruitment of potential participants tends to be made in deference to patterns of hierarchy, notions of patronage and an awareness of power relationships both between the respondent, gatekeepers and potential recruits.

A primary recruitment technique is based on a key informant approach. This is part of a purposive sampling technique which "aims to uncover information-rich phenomena/participants that can shed light on issues of central importance to the study" (Hay 2005, p.292). In my work, using a key informant approach, the views of the Village Chief are often sought first to identify people within the village who may be able to comment on issues relating to the impact of externally crafted laws. Therefore, the Village Chief acts in identifying other residents who could shed light on law's varied impacts. In the absence of advice from the Village Chief or Commune Representative, individuals are sought within households who may have been experiencing some difficulties in complying with the rules and regulations for the site. In my research based at the World Heritage site at Angkor, potential participants were identified by observing the appearance of homes/dwellings. The rules for building construction within the Park differ for those outside the Park, being more restrictive as to the type of material which should be used in construction of dwellings, their size, location and appearance. Observation, therefore, plays a part in identifying compliance and non-compliance issues with the regulations (Gillespie 2013) and represents the basis upon which recruitment for interviews was made. Observation, however, was not used in isolation. Rather, other villagers were often consulted as to whether (or indeed which) neighbours may be available for interviewing. This "snowball" recruitment (as per Monk & Bedford 2005) technique was a very useful method for identifying potential interviewees.

How: interviews

An interview is a face-to-face exchange where the interviewer asks, and the interviewee answers, questions. In a social science setting the interview is usually this and more, as recognition is given to the phenomena of the "co-construction" of data, meaning that the interplay between interviewer and interviewee results in an interview product. Researchers in the geographical tradition have commented extensively on fieldwork, which employs in-depth, semi-structured interview techniques. Lindsay (1997, p.56) says that interviewing "is not an easy option" for organising people, places and times and is very time-consuming. Dunn (2005) suggests that

> (i)nterviewing in geography is so much more than 'having a chat' and reiterates the organisational demands from making contact through to the dedicated hours of transcribing, all before even beginning any analysis. (p.79)

Nonetheless, human geographers and those working in a socio-legal tradition more generally continue to use interviews to collect primary data as it provides an opportunity to delve in-depth into issues. Duncan and Duncan (2001, p.401) observe that "(o)nly interviewing ... can extract the conflicts, interconnections, anxieties, and specificities (historical or emerging) that roil beneath the calm surface".

Bias, shocks and falsehoods

The inherently unpredictable nature of interviewing is undeniable; it is fundamental to the nature of the experience. Those familiar with the interview process understand that events often depart from any pre-prepared script – even if this script is designed to be as flexible as possible in order to take account of the daily exigencies of life. Semi-structured, in-depth interviews are considered amongst the more flexible approaches to data collection (Dunn 2005). But what happens when the interviewee and/or respondent react to questioning in a capricious way?

Sands and Krumer-Nevo (2006) address issues of shock in interviews. They define shock as "cognitive emotional reactions of the interviewer to the unexpected" (Sands & Krumer-Nevo 2006, p.950). In this article, the authors (ibid) use a "postmodern lens" to describe the influence of dominant ("master") narratives on the interview interaction and illustrate how interviewees may react against (consciously or unconsciously) this dominant narrative and adopt shock tactics to, effectively, sabotage it. In interviews conducted in a cross-cultural setting, the complications of shocks may be compounded. Three types of shock are identified by Sands and Krumer-Nevo (2006): (1) the violation of a social taboo, (2) shock over role reversal (for professionals) and (3) shock based on reversing stereotypes. However, shock over role reversal and stereotypes are, perhaps, unlikely to occur in interviews conducted outside the researcher's dominant master narrative because issues of stigma are not always obvious to a researcher working outside his or her own cultural settings. It may appear harsh to add "falsehoods" to the heading of this sub-chapter; yet, it is excessively naïve to assume that the information gathered from interviews is always a clear, comprehensive, balanced and a neutral account of events or a disinterested narrative of opinion. This point is well-made by Perramond (2001) in his account of fieldwork conducted in rural Mexico. He points out that all interactions camouflage a variety of nuances, many of which go unnoticed in "not-so-blissful ignorance for a fieldworker who is trying to grasp community social dynamics" (Perramond 2001, p.153). Perramond also identifies how the researcher, making premature assumptions, can easily fall victim to falsehoods – where it is nigh impossible to fathom or appreciate the motivations of respondents. These observations are not restricted to fieldwork conducted as an outsider, but these concerns do reinforce the need to be aware of accuracy in the data.

Recording and transcribing

In an interview situation, an interviewer's wish to keep the respondent's thoughts flowing means that assiduous note taking is not always possible – it could break the flow of the conversation and take excessive time in a free-flowing communication. The usefulness of recording interviews cannot be overstated and has been long recognised as having enormous benefits to the research and researcher alike. Despite the advantages of recording interviews there are negative elements associated with this process. One of these is the wariness of a respondent to the recorder, for people often seem far more comfortable chatting in the absence of a recording device. Of her experiences of interviewing in a cross-cultural setting using a tape recorder, Farrow (1995) found that taping interviews exacerbated power imbalances. However, digitally recording interviews was essential in this research to maintain rigour for the purposes of ensuring a measure of quality-control in a cross-cultural setting. Moreover, in recording translations, from English to Khmer then from Khmer to English, the translation could be later subject to further checking and verification using interpreter-translating services.

Translators

Much of my research throughout Southeast Asia, and Cambodia in particular, has relied on a translator in the field. The reality of not being fluent in the language of your field location presents obvious problems. Without a doubt, fluency in the language in which fieldwork is conducted makes for better research, and the calls for fieldworkers to familiarise themselves with the language of locals is well-intentioned (see, for example calls by Veeck 2001). Yet,

> (t)he ideal has to cede to the realistic. If the total field period is shorter than seven months, learning to speak a new tongue may not be a wise investment of time. Engaging an interpreter is the best overall solution for short-term projects, assuming one keeps in mind that the informant is always more in tune with the interpreter than with the researcher formulating the questions. (Gade 2001, p.376)

During my research, several translators have been employed – ranging from male tertiary students to experienced social science female researchers. I tend to use the nomenclature of "translator" rather than "interpreter" because, it is contended, there is a strong supposition that to "interpret" is to infer or read a meaning, even to deduce meaning, rather than to "translate", which ought to imply that the information is explained without alteration. In the context of research conducted by a foreign researcher, unfamiliar with the social context in which the interview is situated, it is important that the translator (as far as practicable) convert words literally from English to the local language, and vice versa. Many writers advocate that fieldwork is best conducted in the tongue of those being interviewed and argue that a linguistic understanding helps the researcher understand cultural and social nuances (for example, Hoggart et al. 2002, p.212). Undoubtedly, the use of translators has its limitations, but how much is lost in translation? Arguably, errors in translations occur even when those involved speak the same language. Watson (2004) writes that her rudimentary grasp of language did not enable her to be free from the shackles of "help":

> I continued to work with help. Otherwise the technical details and subtleties of the language often escaped me, and the regional variations were confusing. I knew enough to understand that sometimes people used my lack of language expertise to deny contradictions in what they had told me … The 'translator' … was also more than someone who just translated from one language to another. He was not trained in social science methods, but his longer-term engagement with the research gave him an overall perspective on it, and added a great deal to the research, as someone with whom I could discuss research progress, problems and ideas. (Watson 2004, p.66)

These remarks do not obviate the lack of language skill in the individual researcher, but they do justify the need for assistance through a translator. Accordingly, the role of the much-needed translators and their impact on the research process need to be examined and ought to not be "conspicuously absent" (Twyman et al. 1999, p.315). Arguably, this becomes a case of "interpreter/translator reflexivity". The interpretative nature of a study steeped in a human geographical tradition requires that the researcher reflect on his or her role in the research process (discussed above), but when the research is set in a cultural setting which is foreign to the researcher and which necessitates that the researcher rely on translators and interpreting services, the role of the translator in the research process ought also be examined. In other words, to show respect for those participating in the research (Hay 2005), the role of a local translator should be scrutinised. Scott et al. (2006) also comment that the positionality of the translator is often poorly reported. Muller (2007) also argues that the implications of conducting fieldwork through translators must be addressed in a more comprehensive manner than it appears to have been done to date. The benefit in using local people for translation lies in the value of having someone conversant in the local dialect and familiar with local customs and traditions. Conversely, however, the local translator may be a threat to potential participants in an interview for fear that divulging information to a local may be viewed as a breach of privacy. Although a researcher is often subject to the strictures of university ethics obligations and the need to conduct his or her research in an ethical manner more generally, a local translator hired to do a specific job may not be as aware of these obligations. In this context, time spent building professional working arrangements with translators, which stresses the ethical dimensions of social science research, is time exceedingly well spent.

Conclusion

Supported by robust social science methods, legal geography research can unpack complexity in the law/space nexus. The complexities and subtleties of land and water use regulation can be viewed more comprehensively in non-Western contexts when legal geographers move beyond textual analysis to embrace the difficult, but rewarding, task of engaged enquiry through using techniques such as interviewing. Moreover, research of this sort can provide an empirical basis for more effective and applied regulatory policy. Understanding the real implications of regulation in my work surrounding protected area laws in a Southeast Asian setting, through grounded field-based research, potentially results in better management that aims to achieve more just and sustainable outcomes.

In this chapter I embrace qualitative methods to draw out and pull apart the law–place–people web of entanglements through an explicitly legal geography approach. This approach highlights both the ways in which law crafts place but also the ways in which law creates place-based incongruities and tensions at a localised level. Deficiencies in regulatory objectives, when these collide with contrasting social norms and expectations, are brought to the fore with the use of in-depth semi-structured interviews. The challenges of working in a setting outside my usual milieu are also considered in this chapter with a view to providing reflection on the effort needed as we, as researchers, strive towards producing accountable research outcomes. To this end, documenting and understanding any researchers' limitations and strengths make for best-practice research design and execution. This chapter makes a call for all researchers to embrace cross-cultural settings, to document challenges and to recognise limitations in order to produce a thoughtful, insightful scholarship for all of us to build upon.

References

Baxter, J., & Eyles, J. 1997, 'Evaluating qualitative research in social geography: Establishing "rigour" in interview analysis', *Transactions of the Institute of British Geographers*, vol. 22, no. 4, pp.505–525.

Baxter, J., & Eyles, J. 1999, 'The utility of in-depth interviews for studying the meaning of environmental risk', *Professional Geographer*, vol. 51, no. 2, pp.307–320.

Bennett, L., & Layard, A. 2015, 'Legal geography: Becoming spatial detectives', *Geography Compass*, vol. 9, no. 7, pp.406–422.

Bradshaw, M., & Stratford, E. 2005, 'Qualitative research design and rigour', in *Qualitative Research Methods in Human Geography*, ed. I. Hay, Oxford University Press, Melbourne, pp.67–76.

Braverman, I. 2014, 'Who's afraid of methodology?', in *The Expanding Spaces of Law. A Timely Legal Geography*, eds I. Braverman, N. Blomley, D. Delaney, & A. Kedar, Stanford, California, pp.120–141.

Chacko, E. 2004, 'Positionality and praxis: Fieldwork experiences in rural India', *Singapore Journal of Tropical Geography*, vol. 25, no. 1, pp.51–63.

Chandler, D. 1991, *The Tragedy of Cambodian History: Politics, War and Revolution Since 1945*, Yale University Press, New Haven, CT.

Crang, M. 2002, 'Qualitative methods: The new orthodoxy?', *Progress in Human Geography*, vol. 26, no. 5, pp.647–655.

Crang, M. 2003, 'Qualitative methods: Touchy, feely, look-see?', *Progress in Human Geography*, vol. 27, no. 4, pp.494–504.

Crang, M. 2005, 'Qualitative methods: There is nothing outside the text?', *Progress in Human Geography*, vol. 29, no. 2, pp.225–233.

De Wet, J., & Erasmus, Z. 2005, 'Towards rigour in qualitative analysis', *Qualitative Research Journal*, vol. 5, no. 1, pp.27–40.

Delaney, D. 2010, *The Spatial, the Legal and the Pragmatics of World-making: Nomospheric Investigations*, Routledge-Cavendish, Abingdon, Oxfordshire.

Duncan, J. S., & Duncan, N.G. 2001, 'Theory in the field', *Geographical Review*, vol. 91, no. 1–2, pp.399–406.

Dunn, K. 2005, 'Interviewing', in *Qualitative Research Methods in Human Geography*, ed. I. Hay, Oxford University Press, Melbourne, pp.79–105.

Ebihara, M. 1968, *A Khmer Village in Cambodia*, unpublished doctoral dissertation, Department of Anthropology, Columbia University, New York.

Elwood, S., & Martin, D. G. 2000, '"Placing" Interviews: Location and scales of power in qualitative research', *Professional Geographer*, vol. 52, no. 4, pp.649–657.

Farrow, H. 1995, 'Researching popular theatre in Southern Africa: Comments on a methodological implementation', *Antipode*, vol. 27, no. 1, pp.75–81.

Gade, D. W. 2001, 'The languages of foreign fieldwork', *Geographical Review*, vol. 91, no. 1–2, pp.370–379.

Gillespie, J. 2009, 'Protecting world heritage: Regulating ownership and land use at Angkor Archaeological Park, Cambodia', *International Journal of Heritage Studies*, vol. 15, no. 4, pp.338–354.

Gillespie, J. 2012, 'Buffering for conservation at Angkor: Questioning the spatial regulation of a World Heritage property', *International Journal of Heritage Studies*, vol. 18, no. 2, pp.194–208.

Gillespie, J. 2013, 'World heritage management: Boundary-making at Angkor Archaeological Park, Cambodia', *Journal of Environmental Planning and Management*, vol. 56, no. 2, pp.286–304.

Gillespie, J. 2018, 'Wetland conservation and legal layering: Managing Cambodia's Great Lake', *Geographical Journal*, vol. 184, no. 1, pp.31–40.

Gillespie, J., & Perry, N. 2018, 'Feminist political ecology and legal geography: A case study of the Tonle Sap protected wetlands of Cambodia', *Environment and Planning A*, vol. 50, no. 7, pp.1–17.

Haraway, D. 1991, *Simians, Cyborgs and Women: The Reinvention of Nature*, Routledge, London.

Hay, I. 2005, *Qualitative Research Methods in Human Geography*, 2nd edition, Oxford University Press, Melbourne.

Hay, I. 2016, *Qualitative Research Methods in Human Geography*, 4th edition, Oxford University Press, South Melbourne.

Hoggart, K., Lees, L., & Davies, A. 2002, *Researching Human Geography*, Arnold, London.

Howitt, R., & Stevens, S. 2005, 'Cross-cultural research: Ethics, methods, and relationships', in *Qualitative Research Methods in Human Geography*, ed. I. Hay, Oxford University Press, South Melbourne, pp.30–49.

Hubbell, L. D. 2003, 'False starts, suspicious interviewees and nearly impossible tasks: Some Reflections on the difficulty of conducting field research abroad', *The Qualitative Reporter*, vol. 8, no. 2, pp.195–209.

Katz, C. 1994, 'Playing the field: Questions of Fieldwork in Geography', *Professional Geographer*, vol. 46, no. 1, pp.67–72.

Lindsay, J. 1997, *Techniques in Human Geography*, Routledge, London.

Luco, F. 2002, *Between a Tiger and a Crocodile. Management of Local Conflicts in Cambodia*, Phnom Penh, United Nations Educational Scientific Cultural Organization (UNESCO), Paris.

Mandel, J. L. 2003, 'Negotiating expectations in the field: Gatekeepers, research fatigue and cultural biases', *Singapore Journal of Tropical Geography*, vol. 24, no. 2, pp.198–210.

Miura, K. 2004, *Contested Heritage: People of Angkor*, Unpublished PhD thesis, School of Oriental and African Studies, University of London.

Monk, J., & Bedford, R. 2005, Writing a compelling research proposal. *Qualitative Research Methods in Human Geography*, ed. I. Hay, Oxford University Press, Melbourne, pp. 51–61.

Muller, M. 2007, 'What's in a word? Problematizing translation between languages', *Area*, vol. 39, no. 2, pp.206–213.

Mullings, B. 1999, 'Insider or outsider, both or neither: Some dilemmas of interviewing in a cross-cultural setting', *Geoforum*, vol. 30, pp.337–350.

O'Leary, M., & Nee, M. 2001, *Learning for Transformation: A Study of the Relationship between Culture, Values and Experience and Development Practice in Cambodia*, Krom Akphiwat Phum, Phnom Penh.

Pak, K., Horng, V., Eng, N., Sovatha, A., Sedara., K., Knowles, J., & Craig., D. 2007, *Accountability and Neopatrimonialism in Cambodia: A Critical Literature Review*, Working Paper 34, CDRI (Cambodian Development Resource Institute), Phnom Penh.

Perramond, E. P. 2001, 'Oral histories and Partial Truths in Mexico', *Geographical Review*, vol. 91, no. 1–2, pp.151–157.

Perry, N., & Gillespie, J. 2019, 'Restricting spatial lives? The gendered implications of conservation in Cambodia's protected wetlands', *Environment and Planning E*, vol. 2, no. 1, pp.73–88.

Robinson, D. F., & Graham, N. 2018, 'Legal pluralisms, justice and spatial conflicts: New directions in legal geography', *The Geographical Journal*, vol. 184, no. 1, pp.3–7.

Robinson, D. F., & McDuie-Ra, D. 2018, '(En) countering counterfeits in Bangkok: The urban spatial interlegalities of intellectual property law, enforcement and tolerance', *The Geographical Journal*, vol. 184, no. 1, pp.41–52.

Rose, G. 1997, 'Situating knowledges. Positionality, reflexivity and other tactics', *Progress in Human Geography*, vol. 21, no. 3, pp.305–320.

Rundstrom, R. A., & Kenzer, M. 1989, 'The decline of fieldwork in human geography', *Professional Geographer*, vol. 41, no. 3, pp.294–303.

Sands, R. G., & Krumer-Nevo, M. 2006, 'Interview shocks and shockwaves', *Qualitative Inquiry*, vol. 12, no. 5, pp.950–971.

Scott, S., Miller, F., & Lloyd, K. 2006, 'Doing fieldwork in development geography: Research culture and research spaces in vietnam', *Geographical Research*, vol. 44, no. 1, pp.28–40.

Schenk, C. G. 2018, 'Islamic leaders and the legal geography of family law in Aceh, Indonesia', *The Geographical Journal*, vol. 184, no. 1, pp.8–18.

Shurmer-Smith, P. 2002, 'Methods and methodology' in Do*ing Cultural Geography*, ed. P. Shurmer-Smith, Sage Publications, London, pp. 95–100.

Sin, C. H. 2003, 'Interviewing in 'place': The socio-spatial construction of interview data', *Area*, vol. 35, no. 3, pp.305–312.

Steinberg, M. K. 2006, 'Editorial introduction: Geographers in guatemala, fieldwork in a conflicted landscape', *Geoforum*, vol. 37, pp.13–14.

Twyman, C., Morrison, J., & Sporton, D. 1999, 'The final fifth: Autobiography, reflexivity and interpretation in cross-cultural research', *Area*, vol. 31, no. 4, pp.313–325.

Unger, R. 1983, 'The critical legal studies movement', *Harvard Law Review*, vol. 96, pp.561–675.

United Nations Educational Scientific and Culturual Organization, n.d., *Literacy*, viewed 10 January 2019, <http://www.unesco.org/new/en/phnompenh/education/learning-throughout-life/literacy/>.

Valentine, G. 2005, 'Tell me about..: Using interviews as a research methodology', in *Methods in Human Geography; A Guide for Students Doing a Research Project*, eds R. Flowerdew, & D. Martin, Pearson, Harlow, pp.110–127.

Vandergeest, P., & Peluso, N. L. 1995, 'Territorialization and state power in Thailand', *Theory and Society*, vol. 24, pp.385–426.

Veeck, G. 2001, 'Talk is cheap: Cultural and linguistic fluency during field research', *Geographical Review*, vol. 91, no. 1–2, pp.34–40.

Watson, E. W. 2004, 'What a dolt one is: Language learning and fieldwork in geography', *Area*, vol. 36, no. 1, pp.59–68.

3

ASSERTING LAND RIGHTS THROUGH TECHNOLOGY AND DEMOCRATIC EXPRESSION

The effect of the Indigenous Peoples Alliance of the Archipelago v Indonesia case

Cobi Calyx, Brad Jessup and Mona Sihombing

Introduction

In May 2013, the Indonesian Constitutional Court found parts of Indonesia's Law No 41 Year 1999 on Forestry (the "Forestry Law") unconstitutional due to Indigenous peoples' rights. Aspects of the law asserting that customary forests were state land were deemed inconsistent with the rule of law, human rights and Indigenous peoples' interests. An earlier government investigation in Kalimantan found that more than 90% of all plantation and mining companies operating in forested lands had committed a permit violation, while two-thirds were operating without legally required environmental and social impact assessments (White and Martin 2002: 4). The experience of state asserted ownership of these forests during the previous decade was a retreat from community and sustainable management of landscapes (Kleden et al 2009). In making its finding about the constitutionality of the Forestry Law, the court explained that customary ownership of forested land could lead to sustainable management of those forests and, in so doing, acknowledged that forest governance practices of Indigenous communities would likely improve environmental outcomes.

The Indonesian Constitutional Court struck out the word "*state*" from the law previously worded: "customary forests are *state* forests located in the areas of custom-based communities" (AMAN 2013). This change has opened up opportunities for greater recognition and documentation of Indigenous forest governance (Agence France-Presse 2013) within Indonesia. Yet there remains significant work to be done in translating this legal change into practical change on the ground (Parker 2014; Palangka Raya Declaration 2014). Mapping and communications technologies are being used by Indonesians, and both technologies are supporting these changes. Their limits – both technological and as a means to alter political ecologies – highlight where progress is lacking. Further transformation in support of human and environmental rights is needed for the court decision to fulfil international goals to reduce deforestation and empower communities to make decisions about their forests. In this context, this chapter offers an analysis of the court case and its implications in the context of the collective rights of Indigenous peoples, environmental citizenship and forest governance. This chapter adopts the rights perspective introduced by the court as the basis for its analytical framework, specifically exploring intersections of rights: those about land and indigeneity and civil and political rights, especially those of relevance to emergent democracy. The analysis draws on mixed methods, multiple disciplines and participatory action research methodologies (Kemmis et al 2013), which the authors have brought together through locational and temporal serendipity (Foster and Ford 2003; Fine and Deegan 1996).

Indigenous Peoples Alliance of the Archipelago v Indonesia case[1]

The change of law resulted from a case brought by the Aliansi Masyarakat Adat Nusantara (AMAN) and some of its members. AMAN is a group representing more than 2,000 Indigenous community members in Indonesia (CCMIN 2010). The organisation is known in English as the Indigenous Peoples Alliance of the Archipelago. As well as advocating for the thousands of Indigenous communities within Indonesia, AMAN is a member of regional and global organisations including the Thai-based Asia Indigenous Peoples Pact (AIPP) and UK-based Forest Peoples Programme. AMAN's role in international affairs was described in the case as being directed to uphold the rights of communities whose lands were being adversely affected by the expansion of palm oil plantations in Borneo and Malaysia (Tebtebba 2013: 6). Its stated mission is to realise just and prosperous lives for the Indigenous peoples of Indonesia.

The court framed the issue as a matter of rights and redress for past and ongoing exploitation. Noting that the Indonesian Constitution mandates the state to use natural resources for the greatest benefit of Indonesian people, the court highlighted Article 3 of the 1999 Forestry Law and its aim to provide "maximum prosperity for the people based on justice and sustainability", and suggested that the constitutional benefit is not to be viewed in purely economic or national terms (Tebtebba 2013: 157). Rather, benefit may be understood at other scalar levels as the aggregated fulfilment of rights, interests and the advancement of the welfare and wellbeing of communities. The court accepted that:

> for more than 10 years of enactment, the Forestry Act has been used as a tool by the state to take over the rights of indigenous peoples over their customary forests areas to become state forest, which then on behalf of the state were given/or handed over to capital owners, through various licensing schemes to be exploited without consideration to the rights and local wisdom of indigenous peoples in the region. (Tebtebba 2013: 4)

Having addressed questions of land and participation rights, a connection previously explored in the case of *Maya Indigenous Community of the Toledo District v Belize* (Interamerican Court of Human Rights 2004), the court then highlighted challenges for communities exercising civil and political rights. There was, the court said, a widespread conflict between Indigenous peoples and "entrepreneurs exploiting their customary forest", leading to Indigenous communities being excluded from, victimised by and occupying a position of illegality within the law (Tebtebba 2013: 5, 62).

The court responded within this rights framework and socio-political setting as follows. First, it affirmed the existence of Indigenous peoples' hereditary rights to forests as predating 1999, supported by data of persistence and dependence. It accepted that 70% of Indigenous peoples interacting with forests "heavily depend on forest resources" (Tebtebba 2013: 4), and that their form of forest governance was consistent with healthy ecosystems. The court observed a plurality of self-governance regimes over the Indonesian archipelago, acknowledged by the law, Dutch as it was then, since at least 1945. The court noted that:

> The existence of various forest management practices by indigenous peoples are known by various terms such as Mamar in East Nusa Tenggara, Lembo in the Dayak community in East Kalimantan, Tembawang in the Dayak community West Kalimantan, Repong in Coastal Community in Lampung, Tombak in Batak community in North Tapanuli, and in relation to Petitioner III [the Indigenous peoples of Kasepuhan Cisitu]; it is known as Titipan Forest (entrusted forest), i.e. a forest area which should not be disturbed or destroyed. This area is usually sacred. (ibid: 6)

Second, within its analysis of the rule of law in Indonesia and constitutional protection, the court asserted that the basic and primary function of the state was to protect, respect and fulfil human rights. The Indonesian Constitution created an obligation for the state to respect Indigenous peoples' rights, first expressed in the 1945 Constitution and affirmed as being a component of the "benefit" guaranteed by the Constitution as recently as 2010 (ibid: 34–5). Values of fairness for all people were discussed, providing context for ways in which Indigenous peoples were not treated fairly due to inconsistent rules and a lack of clarity and certainty regarding legality, predictability and transparency. While not expressly influential, the court noted the relevance of the United Nations Declaration on the Rights of Indigenous Peoples (2007) – an instrument that Indonesia is a signatory to, but over which it holds reservations (AMAN and AIPP 2016). The court also acknowledged the role of the United Nations (1992) Declaration on Environment and Development, particularly Principle 22 regarding Indigenous peoples, saying the "state should recognise and support entity, culture, and their interests as well as providing the opportunity to actively participate in achieving sustainable development" (Tebtebba 2013: 160).

Third, the court accepted that there was a scalar contradiction in the law, especially around who benefits, or whose interests are prioritised in the law (Jessup 2014). The conflict was between "custom and culture and societal values" on one hand and "national norms" on the other hand (Tebtebba 2013: 31). Experts noted that forested land was not "used for the maximum benefit of the people, but given to large private companies engaged in agriculture, plantation, or mining, which clearly aim to seek maximum profit" (ibid: 52). This led to the finding which was instrumental to the invalidation of the Forestry Law. The court said:

> Article 5 of the Forestry Law clearly has given a power beyond limit to the Government over something that is not within their authority ... the existence (or extinction) of an ethnic group should not be handed over to state officials or the Government because it is part of the human rights of a group of people who should have been guaranteed and protected by the constitution, which requires the government to actualize it. (ibid: 33)

The effect of the operation of the law inconsistent with the Constitution and ignorant of the human rights of Indigenous peoples was to render the lives of Indigenous communities illegal, uncertain and increasingly invisible (Mihalopoulos-Philippopoulos 2015; Mihalopoulos-Philippopoulos and FitzGerald 2008) to a government whose priorities were to engage with corporations at the expense of communities. Communities became subjects rather than participants in forest management (Tebtebba 2013: 71). Community representatives described for themselves their long-held connections with land, manifold forms of governance and the more recent conflicts with license holders over access, environmental condition and livelihoods of the communities (e.g. ibid: 81). They also retold histories of resistance, arrests and detainments since the 1960s for continuing to use and occupy land over which the state asserted control.

Past resistance and ongoing presence were implicitly accepted as being of contemporary importance. While acknowledging the existence and function of Indigenous customary "living laws" in forests, and the right to management in accordance with those laws, the court formed the view that "if in the development the indigenous peoples are no longer there, the right to manage the customary forest goes to the hand of the Government" (ibid: 160). The court discussed two contrasting possible scenarios where Indigenous groups will not be recognised as having rights over land: first, where Indigenous communities remain in existence but are not acknowledged as existent; second, where communities are no longer in existence but are acknowledged for their past connection. The effect of both of these scenarios would be to render invisible Indigenous groups' customary connection with land and their forestry management practices on land. The court elaborated on how the process of inexistence may arise, noting particularly the role of technology as a tool of displacement:

> The cultural identities and rights of traditional communities shall be respected in accordance with the development of times and civilizations … due to the influence of the development of science and technology, indigenous peoples sooner or later will also undergo a change. (ibid: 166)

Following the earlier Indonesian Constitutional Court decision concerning the establishment of Indonesia's Forest Zone, *Kawasan Hutan*, the court affirmed the role of those communities considered to be existent in directing their own land claims. It stated that: "determination of boundaries of state forest and customary forest cannot be determined only by the state but … should involve stakeholders in the area concerned" (Tebtebba 2013: 173).

The Secretary-General of AMAN has interpreted the verdict as offering certainty and legality for the communities who had previously seen access to, and control over, their ancestral lands granted by their government to resource industries, whose activities degraded landscapes and livelihoods, noting that:

> Good things come to those who wait. About 40 million Indigenous Peoples now are rightful over our customary forests because the state has become unable to expel us out of our customary forests that have become our source of livelihood from generation to generation. (Nababan and Sihombing 2013)

Human rights and a visible legal geography

The decision of the court has consequences for the law and legal geography of Indonesia's forests and beyond, as social and technological trends for recording community interests in forests crosses legal borders. The decision informs understanding of Indonesia's laws about the environment and Indigenous communities dependent on it. Due to the human rights–based approach adopted by the court, laws will only satisfy the constitutional goal of realising public benefit arising from decisions made under environmental or resources laws, like the Forestry Law, where those communities and their livelihoods are recognised and protected. Rights will not be realised, however, unless and until a community's connection to place is understood and acknowledged (Gillespie 2017).

Through rights, communities can influence (Bartel et al 2013) governance of ancestral spaces, while concepts of place, justice and environment become both local and communal (Carmalt 2018). As Gillespie (2016) explains, property rights in space and over time render a particular and contextualised legal geography. It is through property rights and land regulations that a lived landscape is formed. An understanding of who is within that landscape, and the customs and laws they hold, will inform the legal geography of that place.

Writing prior to the legal change and in the context of REDD+, the mechanism under the United Nations Framework Convention on Climate Change to achieve emissions reductions by reducing deforestation (Danielsen et al 2011),

Wright said that land tenure was the most crucial reform for Indonesia (2011: 130). Boyd (2010: 533) asked: "is it possible to imagine and build a set of enabling environments such that REDD can be made meaningful and valuable, in whatever form, for the Adat communities in Indonesia?" Part of this imagining of an enabling environment, of a new legal geography, has been led by AMAN, in bringing about this legal change in tenure over forested lands in Indonesia. Other parts of this imagining, however, depend on communities themselves recording and visualising their ownership. Mobilisation and networking by individual communities will construct a particular and unique legal landscape (Carmalt 2018) and, in so doing, reimagine (Scott 2013) and revisualise communities hidden (Braverman 2011) through the operation of the Forestry Law. The accumulation of data and information made public will materialise the environment over which Indigenous communities have power (Barry 2013) and will support a change in communities – leading not to disappearance, but permanence.

The optimistic statement expressed by the Secretary-General of AMAN about good things coming to those who wait (Nababan and Sihombing 2013) was followed by comments acknowledging that the struggle for these rights to be realised on the ground for Indigenous communities would continue following the court decision. The US Law Library of Congress noted that the decision may result in an increase in disputes over land rights, quoting an official of the Ministry of Forestry who said it would be some time before the decision could be implemented through local government decrees (Johnson 2019). Before that, however, Indigenous communities would have to determine how to evidence their existence and governance in ancestral lands.

Social research and serendipity

This research did not arise deliberately; rather, it sprung from a series of events and encounters that triggered discussions about the Indonesian Constitutional Court case and the realisation of shared interests and disciplinary intersections between the authors. However, following the initiation of the research, and consistent with the legal geography conceptual framework for this chapter (about how the law can empower, vocalise, materialise and enable the reimagining of local communities), it was intentionally designed to be participatory, emancipatory and to draw upon multiple disciplines, perceptions and community experiences. The research reflects and borrows from participatory action research methodologies, social research ideas and reflects a form of serendipitous research. It is additionally "geographic" in that it has been done by multiple authors across spaces and scales with overlaps in Chiang Mai, Thailand and Melbourne, Australia.

The origins of the research can be traced to the time of the court ruling when one of the authors, Cobi Calyx, was based with the Asia Indigenous Peoples Pact (AIPP) in northern Thailand. She learnt of the Indonesian court ruling from the Climate Change Monitoring and Information Network (CCMIN), a project managed by colleagues at AIPP. Calyx had witnessed and heard communities on the Thai–Myanmar border immersed in a struggle to substantiate their land rights and forest management credentials, particularly in landscapes deemed protected by the Thai government. She approached another author, Brad Jessup, to discuss the application and relevance of international law instruments, particularly the Declaration on the Rights of Indigenous Peoples (UN 2007) to that struggle. Informed by Jessup's research and interest in environmental justice (Jessup 2012) and environmental citizenship (Smith and Pansapa 2008), the authors began to explore what an "eco-citizenry" in the forests consistent with the Rio Declaration (UN 1992, Principles 10 and 22) might look like, especially for Indigenous people resident in Thailand who were not citizens of that nation because of their ethnicity (Vandergeest 2003, Roberts 2016). As explained by Smith and Pansapa (2008: 230), environmental citizenship is about having capacity to collaborate with the state about the management of resources and extends to having capacity to influence and contribute to research about their livelihoods. Mindful of this, and drawing on Calyx's own research on deliberative decision making and the use of technologies (Smith 2014, 2017; Calyx 2017), Calyx arranged for Jessup to visit a community that was being assisted by the non-government organisation (NGO) Inter Mountain Peoples Education and Culture in Thailand Association, an AIPP member and part of the Forest Peoples Programme. The community that Jessup visited was using global positioning systems (GPS) technologies to map areas of forest in which it dwelled, farmed and managed, in order to both resist dispossession and counter claims that the community was degrading the protected landscape. The community was doing what groups from around the world have been doing to support land claims for at least seven decades (Chapin et al. 2005: 623). The history of such land mapping is noted in Part 6 of this chapter.

Concurrently with Jessup's post-field desk-based theoretical inquiries into the effect of what he observed in northeast Thailand on the concept of "citizenship" (Jessup and Calyx, forthcoming), Calyx was following the outcomes of the case brought by AMAN in Indonesia, as detailed in Part 2 of this chapter, from her base with AIPP in Thailand. Another author, Mona Sihombing, was more intimately and locally connected with the case. As a development specialist in Indonesia among other places in South East Asia, Sihombing has worked with Indigenous communities on their land claims for over seven years, including for AMAN, the lead petitioner in the court case. Through this work, Sihombing has observed the confrontation with the law by communities which governments have deliberately made distant from, and disempowered by, the law. It was during the after-events of the court case that Sihombing and Calyx were brought together at a workshop in Bangkok, Thailand, about Indigenous peoples' media rights. Both were presenters at the workshop (Smith 2013), which included networking activities where they joined in discussion. Separately, Sihombing facilitated Indigenous community responses to the court case, while Calyx recorded community expressions of land claims through new technologies, including digital media platforms. Calyx saw the value in bringing the authors together to capture what they collectively thought leading up to, during and following the AMAN case in the Indonesian Constitutional Court.

Explained this way, what the authors were doing was trying to make sense of experiences with law and technology using a number of social research methodological approaches that they were individually most familiar with. These approaches included critical legal ethnography, which reflects upon culture, knowledge and power in the law (Thomas 1993), and interdisciplinarity as a way to transform or translate legal information (Vick 2004: 173). With Calyx and Sihombing both directly involved in the research and its outcomes, an action research approach was also employed and, as Kemmis et al. (2013: 5) note, Sihombing in particular was an "insider" of the research. Through assisting and guiding the community groups involved in the research, the authors supported those groups to become active contributors to the changing legal geography of the forests of Asia's Indigenous peoples following the court decision. The authors were also engaged with aspects of "participatory environmental research" (Berkes et al. 2001). While this kind of research is typically associated with environmental volunteerism, it is also pertinent here. Calyx's role with AIPP was sponsored by the Australian government under a programme to deploy young Australians' skills in areas of government aid priority (Georgeou and Engel 2011). The research that followed, including reporting on and analysing the court case and its wider consequences (including beyond Indonesia), was directed towards addressing human rights and environmental concerns while building community and movements around these issues. There were aspects of education, the use of local knowledge and assistance to the Indigenous communities asserting land rights over forests – all of which are features of participatory environmental research.

The authors' grouping was serendipitous: certainly not by chance, but also not by design. As Foster and Ford (2003) explain, research can come about by discussing ideas, while being open to the views of other researchers. Happenstance, however, is much more likely to give rise to research where there is a shared desire to realise an outcome and where researchers are creative, and not set in their approaches to the processes of inquiry. For the authors, the research would never have occurred if one of the authors viewed their ideas as privileged over those of others, or if they had preferred to approach a research problem as more narrowly, disciplinarily or methodologically confined. Serendipity in social research (Fine and Deegan 1996) is marked by the transformation of chance and inquiry into a discovery, and researchers thinking less about what they do and more about what they want to achieve with the research.

Legal geographies are made or conceived. They emerge in response to legal change, reflecting lay legal understandings (Layard and Milling 2015), and the need to adapt (O'Donnell 2016) or exist in resistance to law (Blomley 2003). While not always or necessarily serendipitous, legal geographies are reflexive (Braverman 2014), they are iterative (Bennett and Layard 2015) and they are comprised of multiple perspectives (Braverman 2011). In all these ways, the authors view this research as a reflection both of and on legal geography, and are especially interested in how the use of technology by communities can inform or curate a distinctive legal geography.

Consistent with the main legal geography ideas that inform this chapter, which are visibility, reimagination and materialisation (Scott 2013; Braverman 2011; Barry 2013), the methods that the authors have adopted constitute a form of critical participatory action research (Kemmis et al. 2013). This kind of research attempts to bring together practitioners with scholars who, while working collectively, bring different constructions and experiences together in order to better inform each other about what has been observed or understood. The goal of this research is both educational for the authors but directed towards social improvement: a common vision for both researchers and the researched. The way the research was done, who it was done by and how it is presented, using multiple perspectives and grounding both the doctrinal and the conceptual with the situational, mean that this chapter is providing another voice to the communities whose legal geographies have been intruded and framed by others. It invites legal geographers to think about the degree to which legal geography research can be different when participants are represented in the authorship of the scholarship, and whether personal experience in and of a place enlivens a legal geography differently to one that might otherwise be recorded anthropologically or archivally.

The technological mobilisation in Indonesia: triggered by the Indigenous Peoples Alliance of the Archipelago v Indonesia case

Following the Indonesian Constitutional Court decision, AMAN's member communities started erecting *plang* or placards in their respective customary forests. The community's members then took photos with the *plang* and began sharing them on the Facebook page "Hutan Adat Kita – HAK" (in English: Our Customary Forests administered by Barisan Pemuda Adat Nusantara [BPAN – The Archipelago's Indigenous Youth Front]). The *plang* typically read "This is our customary forest, not the State forest. Indigenous Peoples are implementing the Constitutional Court Decision No. 35/PUU-X/2012". Up until 17 May 2016, the Facebook page had about 50 pictures of communities with their *plang*. This was a process of communities raising awareness of their rights, of their existence and demonstrating a determination not to be excluded from their ancestral lands. Leader of the Hutan Adat Kita group, Simon Pabaras, told the Jakarta Post that their social media activity was aimed at legal and political change:

> The photos are there to affirm that customary forests are forests on customary land and publicly show that indigenous people are eager to reclaim their forests and have the forests returned to them. (Pham 2013)

One of the photos shared in the Facebook page is of the Pandumaan and Sipituhuta communities, which are both AMAN community members in the North

FIGURE 3.1 "Hutan Adat Kita". Source: Mona Sihombing.

Sumatera province (see Figure 3.1). Over decades, these neighbouring communities have jointly fought the conversion of their ancestral forests to a eucalyptus plantation managed by PT Toba Pulp Lestari (PT TPL) (Hadinaryanto 2014). The Pandumaan and Sipituhuta communities saw their lands, comprising 4,100 hectares of myrrh forest, declared state forests for the purposes of the Forestry Law (AMAN 2013a).

The Pandumaan and Sipituhuta communities also started conducting participatory mapping of their ancestral lands, including the myrrh forest that had been taken from them. AMAN, through its regional chapter in Tano Batak, conducted training on participatory mapping on 28 July 2013 for representatives of Pandumaan, Sipituhuta and other communities (AMAN 2013b).

According to Rukka Sombolinggi, a representative of AMAN, "community mapmaking has been a successful tool to show the government that we are here, and that we want to protect our lands" (IRIN 2014). Participatory mapping was discussed in the facts of the Indonesian Constitutional Court case, as evidence to support the legal standing of the Indigenous peoples of Kenegerian Kuntu in the Kampar District Riau Province:

> based on participatory mapping conducted by the Petitioner II, out of (a) total of 280,500 (two hundred eighty thousand and five hundred) Ha of Production Forest Area and Industrial Plantation Forest owned by PT. RAPP, an estimated area of 1,700 (one thousand seven hundred) Ha, is located in customary forest area of Indigenous peoples of Kenegerian Kuntu. (Tebtebba 2013: 18)

Mapping was also presented as evidence to assert the legal standing of the Indigenous peoples of Kasepuhan Cisitu in the Lebak District Banten Province:

> based on participatory mapping (in January 2010), which was facilitated by AMAN, JKPP and FWI, wewengkon (territory) of Kasepuhan Cisitu covers an area of 7,200 acres. Previously, kaolotan (the elders) only estimate area of 5,000 hectares. Mapping was conducted by using a Global Position System (GPS) and Land Sat Imagery. (Tebtebba 2013: 19)

In August 2013, three months after the court decision, AMAN hosted the "Global Conference on Community Participatory Mapping on Indigenous Peoples' Territories" in the North Sumatera province. Attended by participants from 17 countries (Forest Peoples Programme 2013), participants visited the Pandumaan and Sipituhuta communities to learn about their participatory mapping endeavours (see Figure 3.2), and their fight for rights to their customary forests through the use of a geographical information systems (GIS)-based map.

The Sipituhuta-Pandumaan map was registered with a community support agency, Badan Registrasi Wilayah Adat (BRWA) (in English: Ancestral Domains Registration Agency), which was established in part by AMAN in 2010 (BRWA 2019). After verifying ancestral domain maps, BRWA hands them over to government institutions. The first handover after the Indonesian Constitutional Court decision was on 15 July 2013, when AMAN and BRWA submitted 324 verified ancestral domain maps, covering 2,643,261 hectares, to the Ministry of Environment (AMAN 2013c). More than three years later, President Joko

FIGURE 3.2 Photo of Sipituhuta-Pandumaan showing the map of their ancestral land. Source: Mona Sihombing.

Widodo issued a decision recognising the customary forests of nine Indigenous communities in Indonesia, including the Pandumaan and Sipitihuta communities (Mongabay.com 2017). From 2012 until August 2018, it was reported that BRWA and AMAN has submitted maps covering of 9.65 million hectares of ancestral domains in 785 communities (Nugraha 2018).

President Joko Widodo's predecessor, President Susilo Bambang Yudhoyono had issued the "One Map Policy" in 2010 instructing the Presidential Working Unit for the Supervision and Management of Development (UKP4) and related ministers to develop one national map (Indonesia at Melbourne 2016). In February 2016, President Joko Widodo issued a regulation instructing the acceleration of the One Map Policy (ibid). On 11 December 2018, President Joko Widodo finally launched the One Map Policy portal, yet it does not include mapping of ancestral domains, even though 1.2 million hectares of ancestral domains have been officially recognised by local regulations (AMAN 2018a).

As well as social media, an online petition written in both Bahasa Indonesia and English, known as Petisi 35, and referring to the court decision was created to raise awareness and drive change (Pham 2013; Yanto 2013). Petisi 35 had an ambitious goal of gathering 35 million signatures, both online through the change.org platform and offline. With these signatures, AMAN wished to place political pressure during the then-upcoming 2014 general elections:

> Tahun 2014 mendatang, Indonesia akan menghadapi Pemilihan Umum. Semua orang mulai berbondong-bondong mempersiapkan agenda politik 2014 khususnya partai politik. Tidak banyak waktu lagi yang dimiliki. Oleh karena itu, dibutuhkan desakan yang kuat dari berbagai elemen agar pemerintah dan DPR, baik di pusat maupun di daerah segera menindaklanjuti Putusan MK No.35 tersebut. Salah satunya adalah segera mempercepat proses pengesahan Rancangan Undang-Undang (RUU) tentang Pengakuan dan Perlindungan Hak-Hak Masyarakat Adat (PPHMA).

> In the coming 2014, Indonesia will be conducting the General Election. Everybody, especially political parties, has started preparing political agendas for 2014. We don't have much time left. Therefore, [we] need strong pressures from various elements so the Government and House of Representatives on national and local levels follow up the MK 35 Decision immediately. One of the ways is to accelerate the process to legalize the Draft Act on the Recognition and Protection of the Rights of Indigenous Peoples. (AMAN 2013)

Other petitions emerged to demand legal change and to support AMAN in its collaborative work with communities to evidence land claims (Yanto 2013). However, a lack of internet access among Indigenous communities was reported as a barrier to the campaign (Pham 2013; AMAN 2017b), with those Indonesian communities without internet access sharing information instead via voice Short Message Service (SMS) using a local version of FrontlineSMS technology (Weiss

2017). A benefit of using voice technology in particular is that it accommodates Indigenous languages and can include people who lack script or literacy. After the Indonesian Constitutional Court decision, AMAN's community members were mobilised to follow up the decision by means of SMS Adat (AMAN 2017b), which is the SMS Gateway developed and managed by AMAN. It had about 1,800 active subscribers by the end of 2016 (AMAN 2017a). AMAN opted for SMS Adat because it needed effective and efficient communication in making decisions based on informed participation. It also needed to update its members on the critical situations faced by Indigenous activists/communities and inform its members on political issues or policy process taking place at national level (AMAN 2017b).

Since early 2017, AMAN has not been able to rely only on SMS Adat due to changes in the regulations of the provider (AMAN 2017a). This has led to AMAN exploring other technology-based communication tools (AMAN 2018). Among them is the community-based monitoring and information system (CBMIS) and associated information centre (AMAN Maluku Utara 2017). The information centre is a website with the domain name: www.adat.id. The CBMIS runs on an Android-based mobile application (app). The app includes features that allow for data collection and viewing on Android mobile phones and desktop internet browsers, data visualisation and presentation in the form of tables and charts and publication of community information on blogs (AMAN 2017).

Using technology to support land claims

The aftereffects of the AMAN case can be understood as representing an evolution of the use of digital mapping and technologic deployment to enhance participation in forest governance and to support claims to ancestral lands. Technologies have been used to evidence claims in courts (Tebtebba 2013: 18), assert rights against other land-users (Paneque-Gálvez et al. 2017), document site-specific traditional knowledge and practices (Hunt and Stevenson 2017: 7) and manage land-use referrals (Ecotrust Canada 2017). The use of standardised GIS mapping tools is particularly important in land rights contexts given the high evidentiary thresholds that Indigenous communities are required to meet in order to have customary claims to land recognised, often requiring intensive resourcing to document unwritten law and practice (Gilbert and Begbie-Clench 2018: 6–7). Overlaying volunteered geographic information (VGI) (Haworth et al. 2015; Adams 2013; Yudono 2017) onto standardised GIS data can give this volunteered knowledge greater recognition, combining widely known values with customary knowledge. Participatory mapping processes have been acknowledged as sites of empowerment, functioning to share knowledge and strengthen ties within communities (Olson et al. 2016: 349). Mapping projects generally involve the documentation and visualisation of land use and occupancy data, at times in opposition to state endorsed maps, as well as highlighting traditional law and custom. The process and the output of the mapping exercise materialises place, reimagines it and offers an information geography that can be used within the law.

The use of digital technologies, and mapping especially, is an inherently political process (Harley 1988: 129), and this can be seen through the reluctance of the Indonesian government to include ancestral domains on its official mapping project. The validity of VGI is often questioned by authorities, despite research suggesting its value, and this questioning is not limited to VGI from the participatory mapping of Indigenous communities (Smith 2017). Approved maps constitute "authoritative resources that the state mobilises to consolidate their own power" within the colonial landscape (Peluso 1995: 385), and which solidify imagined geographies and exclude others (Hunt and Stevenson 2017: 375). Such maps have operated as "an instrument of colonisation and administrative control" (Gilbert and Begbie-Clench 2018: 8) by presenting as scientific truth certain cultural cartographies which erase Indigenous realities in relation to a specific place.

In this context, participatory mapping by Indigenous communities to document and highlight relationships to land has decolonising potential (Radjawali et al. 2017: 819). Peluso coined the term "counter-mapping" to describe these activities, referring to practices of Indigenous communities in Kalimantan, Indonesia in response to the exploitation of forest resources in the 1990s. She explains that sketch maps were used by Indigenous communities in partnerships with NGOs to outline customary claims to forest resources (Peluso 1995: 384–5). This process reinforced the legitimacy of claims, denaturalising the narrative presented in state maps by presenting alternate representations that privileged Indigenous worldviews.

Where communities are in a position to access new technologies, the effectiveness of counter-mapping practices may be enhanced. VGI can be combined with text, sound and images, which can be shared through social media, generating further engagement. Visual and voice information, such as the information shared through Frontline SMS technology discussed in Part 6, allows a diversity of languages and cultures to engage and express themselves. GIS technology allows the sharing of large data sets with relative ease, as well as allowing for presentation of layered and interactive mapping which better reflects the complex use of space. VGI can be combined with standardised GIS layers to provide richer maps. The use of drones to capture high-resolution photographs has also been adopted to bolster the accuracy, and therefore the legitimacy, of community mapping projects (Radjawali et al. 2017: 817). Rather than freezing cultural practice at a moment in time, digital mapping presents changing community practices within living and evolving Indigenous cultures. As mapping technologies are evolving from specialised to more accessible platforms (ibid: 819), the participatory potential of projects to support environmental citizenship is being enhanced.

Despite the possibilities created by new technology, participatory mapping is not a panacea to the challenges faced by Indigenous communities in seeking to have their rights recognised. Counter-mapping is a powerful political tool in that it communicates Indigenous expressions of rights to land through the language of government and industry (Cullen 2015: 476). However, scholars have also indicated that such mapping risks further erasing Indigenous epistemologies governing rights to land (Bryan 2011: 40; Gilbert and Begbie-Clench 2018: 7). As in the AMAN case (Tebtebba 2013: 6), mapping Indigenous land use and occupancy also involves negotiating degrees of disclosure of restricted and sacred knowledge to third parties, including governments. Social media and other technologies offer communities the capacity to connect, share and support each other, and combining the use of technologies, for example sharing mapping data via social media, can allow for greater engagement within and beyond communities. While social media technologies might not have the clear connection with land rights and claims that GIS and VGI have, they do facilitate the expression of other civil and political rights, and in doing so better position and empower local communities with networked affiliations to advance claims to land. Mapping technologies allow communities to evidence claims more readily and convincingly, while other communication technologies can support suppressed or hidden communities to express and reimagine their geographies.

Conclusion

This chapter has highlighted how a legal decision upholding the constitutional rights to property of Indigenous peoples, handed down at a time of increased awareness and use of new and adapted digital technologies, has spawned a revived interest in and capacity to achieve customary tenure over land in Indonesia. As one of the most populous nations in the world with a richness of languages and cultures, Indonesia provides exceptional case studies on how digital technologies can allow people to assert and express their claims to land and resources. Even on the most far-flung islands of the archipelago, communities are beginning to use digital technologies to engage with each other and the world, allowing their unique knowledge of place and environment to be shared in unprecedented ways.

By endorsing the use of digital and participatory mapping technology by communities, Indonesia's Constitutional Court has brought recognition of rights within reach to hundreds of groups deprived of their forests by the operation of the 1999 Forestry Law. Two such communities are the Pandumaan and Sipituhuta communities, whose participatory mapping exercise supported by the lead applicant in the *Indigenous Peoples Alliance of the Archipelago v Indonesia* case saw them reclaim their ancestral forests. Their efforts reflected the statement of the outgoing Indonesian Prime Minister who spoke about the overlap between environmental and human rights:

> Forest management is a cross-cutting issue, and not only about keeping the trees. It is about striking a balance between the need for conserving the environment, and guaranteeing the rights of local communities over their customary forests. (Yudhoyono 2014)

Other communities, however, continue to wait for the Indonesian government to confirm their land claims, as their reimagining of their ancestral geography has been completed but not sanctioned by those charged with responsibility for an official map. The terrain those communities now enter is another information geography (Barry 2013) controlled by the Indonesian government. Yet their voices and claims are still heard, shared and promoted as they are through social media and other forms of telecommunication, in Indigenous landscapes across an increasingly technologically just (Meikle and Sugden 2015) globe that is alive to the potential of the goals of environmental citizenship. The support of compatriot communities and researchers, such as the authors, who share an imagination of a legal geography of the forests controlled by those with long and deeply held rights over it means their stories, like their existence, will not go away.

Note

1 The case was presented and judged in Bahasa Indonesia, however this paper quotes the English translation published by Tebtebba (2013), the Indigenous Peoples' International Centre for Policy Research and Education, which was checked against a translation by the United Nations Office for REDD+ Coordination in Indonesia (2013) for consistency.

References

Adams, D. "Volunteered Geographic Information: Potential Implications for Participatory Planning", *Planning Practice & Research* 28, no. 4 (2013): 464–469.

Agence, France-Presse. 2013. "Indonesia Court Ruling Boosts Indigenous Land Rights". *Jakarta Globe*, 17 May 2013, last accessed 17 March 2019. http://ccmin.aippnet.org/index.php?option=com_content&view=article&id=1131:indonesia-court-ruling-boosts-indigenous-land-rights&catid=28:indonesia&Itemid=8.

Aliansi Masyarakat Adat Nusantara (AMAN) and Asia Indigenous Peoples Pact (AIPP). 2016. "The Situation of Human Rights of Indigenous Peoples in Indonesia," Aliansi Masyarakat Adat Nusantara and Asia Indigenous Peoples Pact, last accessed December 18, 2018, http://www.aman.or.id/wpcontent/uploads/2016/09/INDONESIA_AMAN_AIPP_UPR_3rdCycle.pdf.

Aliansi Masyarakat Adat Nusantara (AMAN). 2013. "Constitutional Court Agrees on Judicial Review of UKK," Aliansi Masyarakat Adat Nusantara (AMAN), last accessed January 27, 2014, http://www.aman.or.id/2013/05/16/constitutional-court-agrees-on-judicial-review-of-uuk/.

Aliansi Masyarakat Adat Nusantara (AMAN). 2013a. "Eucalyptus Plantation Case in Pandumaan and Sipituhuta Indigenous Communities, North Sumatera, Indonesia". Prepared for the "Global Conference on Community Participatory Mapping on Indigenous Peoples' Territories" in Samosir, North Sumatera, 25–31 August 2013 organised by AMAN and Tebtebba.

Aliansi Masyarakat Adat Nusantara (AMAN). 2013b. "Pelatihan pemetaan partisipatif di Pandumaan Sipituhuta". Accessed 16 January 2018. http://www.aman.or.id/pelatihan-pemetaan-partisipatif-di-pandumaan-sipituhuta/.

Aliansi Masyarakat Adat Nusantara (AMAN). 2013c. "Penyerahan Peta Wilayah Adat oleh AMAN dan BRWA kepada Kementerian Lingkungan Hidup", Last accessed

16 January 2019. http://www.aman.or.id/penyerahan-peta-wilayah-adat-oleh-aman-dan-brwa-kepada-kementerian-lingkungan-hidup/.

Aliansi Masyarakat Adat Nusantara (AMAN). 2017a. "Report of Directorate of Information and Communication 2012–2017 to be included into AMAN Secretary General's report to the 5th Congress of Indigenous Peoples of the Achipelago, March 2017".

Aliansi Masyarakat Adat Nusantara (AMAN). 2017b. "Adat SMS". Document of Directorate of Information and Communication.

Aliansi Masyarakat Adat Nusantara (AMAN). 2018. "BPAN akan mengembangkan Aplikasi Perangkat Mobile dan Teknologi Data Center untuk Pemuda Adat", last accessed 16 January 2018. http://www.aman.or.id/bpan_akan_mengembangkan_teknologi_berbasis_aplikasi_mobile_dan_data_center_untuk_pemuda_adat/.

Aliansi Masyarakat Adat Nusantara (AMAN). 2018a. "Geoportal Kebijakan Satu Peta Diluncurkan untuk Siapa?", last accessed 16 January 2019. http://www.aman.or.id/geoportal-kebijakan-satu-peta-diluncurkan-untuk-siapa/.

AMAN Maluku Utara. 2017. "Adakan Training, AMAN Memperkenalkan CBMIS Kepada Kadernya di Maluku Utara", last accessed 16 January 2018. http://malut.aman.or.id/2017/02/06/adakan-training-aman-memperkenalkan-cbmis-kepada-kadernya-di-maluku-utara/.

Badan Registrasi Wilayah Adat (BRWA). 2019. "Profil", last accessed 16 January 2018. http://www.brwa.or.id/pages/about.

Barry, Andrew. *Material Politics: Disputes Along the Pipeline*. Chichester: John Wiley & Sons, 2013.

Bartel, Robyn, Nicole Graham, Sue Jackson, Jason Hugh Prior, Daniel Francis Robinson, Meg Sherval and Stewart Williams. "Legal geography: An Australian perspective," *Geographical Research* 51, no. 4 (2013): 339–353.

Bennett, Luke and Antonia Lanyard. "Legal geography: Becoming spatial detectives," *Geography Compass* 9, no. 7 (2015): 406–422.

Berkes, Fikret, Jack Mathias, Mina Kislalioglu and Helen Fast. "The Canadian Arctic and the Oceans Act: The development of participatory environmental research and management". *Ocean & Coastal Management* 44, no. 7–8 (2001): 451–469.

Blomley, Nicholas. "Law, property, and the geography of violence: The frontier, the survey, and the grid," *Annals of the Association of American Geographers* 93, no. 1 (2003): 121–141.

Boyd, William. "Climate change, fragmentation, and the challenges of global environmental law: Elements of a post-Copenhagen assemblage," *University of Pennsylvania Journal of International Law* 32, no. 2 (2010): 457–550.

Braverman, Irus. "Hidden in plain view: Legal geography from a visual perspective," *Law, Culture and the Humanities* 7, no. 2 (2011): 173–186.

Braverman, Irus. "Whose afraid of methodology: Advocating a methodological turn in legal geography," in *The Expanding Spaces of Law: A Timely Legal Geography*, edited by Braverman, Irus, Nicholas Blomley, David Delaney, and Alexandre Kedar. 120–141. Stanford: Stanford University Press, 2014.

Bryan, Joe. "Walking the line: Participatory mapping, indigenous rights, and neoliberalism," *Geoforum* 42, no. 1 (2011): 40–50.

Calyx, Cobi. "Tradeoffs in deliberative public engagement with science. PhD thesis, Australian National University," (2017).

Carmalt, Jean. "For critical geographies of human rights," *Progress in Human Geography* 42, no. 6 (2018): 847–861.

Chapin, Mac, Zachary Lamb and Bill Threlkeld. "Mapping indigenous lands," *The Annual Review of Anthropology* 34, no. 1 (2005): 619–638.

Climate Change Monitoring and Information Network (CCMIN). (2010). Aliansi Masyarakat Adat Nusantara (AMAN), last accessed 27 January 2014. http://ccmin.ai ppnet.org/index.php/aman-indonesia.

Cullen, Alexander. "Making sense of claims across institutional divides: Critical PGIS and mapping customary land in Timor-Leste," *Australian Geographer* 46, no. 4 (2015): 473–490.

Danielsen, Finn, Margaret Skutsch, Neil Burgess, Moestrup Jensen, Herizo Andrianandrasana, Bhaskar Karky, Richard Lewis, Jon Lovett, John Massao, Yonika Ngaga, Pushkin Phartiyal, Michael Poulsen, Singh Soi, Silvia Solis, Marten Sørensen, Ashish Tewari, Richard Young and Eliakimu Zahabu. "At the heart of REDD+: A role for local people in monitoring forests?," *Conservation Letters* 4, no. 2 (2011): 158–167.

Ecotrust Canada. *Referrals Software: An Analysis of Options.* 2017.

Fine, Gary A. and James G. Deegan. "Three principles of serendip: Insight, chance, and discovery in qualitative research," *International Journal of Qualitative Studies in Education* 9, no. 4 (1996): 434–447.

Forest Peoples Programme. 2013. "Global conference in Lake Toba, Indonesia, highlights multiple benefits of community mapping for indigenous peoples", last accessed 16 January 2018. https://www.forestpeoples.org/en/topics/participatory-resource-ma pping/news/2013/10/global-conference-lake-toba-indonesia-highlights-.

Foster, Allen and Nigel Ford. "Serendipity and information seeking: An empirical study," *Journal of Documentation* 59, no. 3 (2003): 321–340.

Georgeou, Nichole and Susan Engel. "The impact of neoliberalism and new managerialism on development volunteering: An Australian case study," *Australian Journal of Political Science* 46, no. 2 (2011): 297–311.

Gilbert, Jeremie and Ben Begbie-Clench. "'Mapping for rights': Indigenous peoples, litigation and legal empowerment," *Eramus Law Review* 1, (2018): 6–13.

Gillespie, Josephine, "A legal geography of property, tenure, exclusion, and rights in Cambodia: exposing an incongruous property narrative for non-Western settings," *Geographical Research* 54, no. 3 (2016): 256–266.

Gillespie, Josephine, "Engaged enquiry in environmental law: Understanding people/ place connections through a geographically informed human rights lens," in *Research Methods in Environmental Law: A Handbook*, edited by Andreas Mihalopoulos-Philippopoulos and Victoria Brooks. 265–285. Cheltenham: Edward Elgar, 2017.

Hadinaryanto, Satria Eka. 2014. "Special report: Lake Toba indigenous people fight for their frankincense forest", last accessed 16 January 2019. https://news.mongabay.com /2014/05/special-report-lake-toba-indigenous-people-fight-for-their-frankincense -forest/.

Harley John B. "Maps, knowledge, and power" in *The Iconography of Landscape*, edited by Denis Cosgrove and Stephen Daniels. 277–312. Cambridge: Cambridge University Press, 1988.

Haworth, B., E. Bruce and P. Middleton. "Emerging technologies for risk reduction: assessing the potential use of social media and VGI for increasing community engagement", *The Australian Journal of Emergency Management* 30, no. 3 (2015): 36–41.

Hunt, Dallas and Shaun Stevenson. "Decolonising geographies of power: Indigenous digital counter-mapping practices on Turtle Island," *Settler Colonial Studies* 7, no. 3 (2017): 372–392.

Indonesia at Melbourne. 2016. "Getting the One Map policy right", last accessed 16 January 2018. http://indonesiaatmelbourne.unimelb.edu.au/getting-one-map-policy -right/.

Interamerican Court of Human Rights. *Maya Indigenous Community of the Toledo District v Belize*. 2004.

Integrated Regional Information Networks (IRIN). 2014. "Indonesian indigenous groups fight climate change with GPS mapping," *Integrated Regional Information Networks (IRIN)*, 8 January 2014, last accessed 17 March 2019. http://www.irinnews.org/feature/2014/01/08.

Jessup, Brad. "Environmental justice as spatial and scalar justice: A regional waste facility or a local rubbish dump out of place?," *McGill International Journal of Sustainable Development Law and Policy* 9, no. 2 (2014): 69–107.

Jessup, Brad. "The journey of environmental justice through public and international law," in *Environmental Discourses in Public and International Law*, edited by Brad Jessup and Kim Rubenstein. 47–70. Cambridge: Cambridge University Press, 2012.

Jessup, Brad and Cobi Calyx. Forthcoming. "Human Rights and Environmental Citizenship in the Forest" (draft/working paper).

Johnson, Constance. 2019. "Indonesia: Forest Rights of Indigenous Peoples Affirmed". *The Law Library of Congress*, 4 January 2019.

Kemmis, Stephen, Robin McTaggart and Rhonda Nixon. *The Action Research Planner: Doing Critical Participatory Action Research*. New York: Springer Publications, 2013.

Kleden, E. O., L. Chidley and Y. Indradi, *Forests for the Future: Indigenous Forest Management in a Changing World*. The Indigenous Peoples Alliance of the Archipelago and Down to Earth, 2009.

Layard, Antonia and Jane Milling. "Creative place-making: Where legal geography meets legal consciousness", in *Creative Economies, Creative Communities: Rethinking Place, Policy and Practice*, edited by Saskia Warren and Phil Jones. 79–100. Abingdon: Routledge, 2015.

Meikle, Amber and Jodi Sugden, *Introducing Technology Justice: A New Paradigm for the SDGs*. Practical Action Publishing, 2015.

Mihalopoulos-Philippopoulos, Andreas and Sharron FitzGerald. "From space immaterial: The invisibility of the lawscape", *Griffith Law Review* 17, no. 438 (2008): 438–453.

Mihalopoulos-Philippopoulos, Andreas. *Spatial Justice: Body, Lawscape, Atmosphere*. London: Routledge, 2015.

Mongabay.com. 2017. "Jokowi grants first-ever indigenous land rights to 9 communities", *Mongabay.com*, 4 January 2017, last accessed 16 January 2018. https://news.mongabay.com/2017/01/jokowi-grants-first-ever-indigenous-land-rights-to-9-communities/.

Nababan, A. and Mona Sihombing. 2013. "Constitutional Court agrees on judicial review of UUK" AMAN, 16 May 2013, last accessed 3 June 2019. http://www.ccmin.aippnet.org/index.php?option=com_content&view=article&id=1127:constitutional-court-agrees-on-judicial-review-of-uuk&catid=110:news&Itemid=48.

Nugraha, Indra. 2018. "Menanti Peta Wilayah Adat Masuk Kebijakan Satu Peta", *Mongabay.com*, 11 August 2018, last accessed 16 January 2019. https://www.mongabay.co.id/2018/08/11/menanti-peta-wilayah-adat-masuk-kebijakan-satu-peta/.

O'Donnell, Tayanah. "Legal geography and coastal climate change adaptation: The Vaughan litigation", *Geographical Research* 54, no. 3 (2016): 301–312.

Olson, Rachel, Jeffrey Hackett and Steven DeRoy. "Mapping the digital terrain: Toward indigenous geographic information and spatial data quality indicators for indigenous knowledge and traditional land-use data collection", *The Cartographic Journal* 53, no. 4 (2016): 348–355.

Palangka Raya Declaration on Deforestation and the Rights of Forest Peoples. 2014.

Paneque-Gálvez, Jaime, Nicholas Varga-Ramirez, Brian Napoletano and Anthony Cummings. "Grassroots innovation using drones for indigenous mapping and monitoring", *Molecular Diversity Preservation International Institute* 6, no. 4 (2017): 1–25.

Parker, Diana. 2014 "Indigenous communities demand forest rights, blame land grabs for failure to curb deforestation". *Monga Bay.com*, 24 March 2014, last accessed 17 March 2019. https://news.mongabay.com/2014/03/indigenous-communities-dem and-forest-rights-blame-land-grabs-for-failure-to-curb-deforestation/.

Peluso, Nancy L. "Whose woods are these? Counter-mapping forest territories in Kalimantan, Indonesia". *Antipode* 27, no. 4 (1995): 383–406.

Pham, Jacqueline. 2013. "Indigenous people fight for rights online". Jakarta Globe, 18 November 2013, last accessed 17 March 2013. https://iva.aippnet.org/indoensia-i ndigenous-people-fight-for-rights-online/.

Radjawali, Irendra, Oliver Pye and Michael Flinter. "Recognition through reconnais-sance? Using drones for counter-mapping in Indonesia". *The Journal of Peasant Studies* 44, no. 4 (2017): 817–833.

Roberts, Kimberly. "It takes a rooted village: Networked resistance, connected communities, and adaptive responses to forest tenure reform in Northern Thailand", *ASEAS-Österreichische Zeitschrift für Südostasienwissenschaften* 9, no. 1 (2016): 53–68.

Scott, Dayna. "Situating Sarnia: Unimagined communities in the new national energy debate", *Journal of Environmental Law and Practice* 25, no. 1 (2013): 81–112.

Smith, Cobi. "A case study of crowdsourcing imagery coding in natural disasters", in *Data Analytics in Digital Humanities*, edited by Shalin Hai-Jew. 217–230. Cham: Springer, 2017.

Smith, Cobi. 2013. "Indigenous Representation in Wikipedia". Conference presentation, Indigenous Voices in Asia, 7 July 2013, last accessed 17 March 2019. https://www.sli deshare.net/cobismith/iv-asiawikipedia.

Smith, Cobi. "Public engagement in prioritizing research proposals: A case study", *SAGE Open* 4, no. 1 (2014) doi: 2158244014523791.

Smith, Mark J. and Piya Pangsapa. *Environment and Citizenship: Integrating Justice, Responsibility and Civic Engagement.* London: Zed Books, 2008.

Tebtebba. 2013. "Indonesian Constitutional Court's Decision regarding the 1999 Forestry Law", last accessed online 8 January 2014. http://www.tebtebba.org/index.php/all-re sources2/all-resources-menu-2?download=885:indonesian-constitutional-courts -decision-regarding-the-1999-forestry-law.

Thomas, Jim. *Doing Critical Ethnography.* Newbury Park, CA: SAGE Publications, 1993.

United Nations (UN). *Rio Declaration on Environment and Development.* 1992.

United Nations Office for REDD+ Coordination in Indonesia. 2013. "DEMI KEADILAN BERDASARKAN KETUHANAN YANG MAHA ESA", *United Nations Office for REDD+ Coordination in Indonesia*, last accessed 10 June 2013. http: //www.UNORCID.org/index.php/events-menu/past-events/181-national-works hop-on-the-forestry-law-constitutional-court-ruling-no-35-puu-x-2012.

United Nations (UN). *Declaration on the Rights of Indigenous Peoples.* 2007.

Vandergeest, Peter. "Racialization and citizenship in Thai forest politics", *Society & Natural Resources* 16, no. 1 (2003): 19–37.

Vick, Douglas W. 'Interdisciplinarity and the discipline of law", *Journal of Law and Society* 31, no. 2 (2004): 163–193.

Weiss, Jessica. 2017. "How citizens use mobile news to inform their communities in Indonesia", *International Center for Journalists*, 27 December 2017, last accessed 17

March 2019. https://ijnet.org/en/story/how-citizens-use-mobile-news-inform-th eir-communities-indonesia.

White, Andy and Ajejhandra Martin. *Who Owns the World's Forests? Forest Tenure and Public Forests in Transition*. Washington DC: Forest Trends and Center for International Environmental Law, 2002.

Wright, Glen. "Indigenous people and customary land ownership under domestic REDD frameworks: A case study of Indonesia", *Law, Environment and Development Journal* 7, no. 2 (2011): 117.

Yanto, Hari. 2013. "Indonesia government: Implement the Constitutional Court's decision no. 35/PUU-X/2012 and immediately adopt the Bill on Indigenous Peoples", *Change.org*, last accessed 28 May 2014. http://www.change.org/id/petisi/indonesi a-government-implement-the-constitutional-court-s-decision-no-35-puu-x-2012 -and-immediately-adopt-the-bill-on-indigenous-peoples.

Yudhoyono, S. B. 2014. Speech presented at the Forests Asia Summit, *Jakarta*. 5 May 2014, last accessed 6 May 2014. http://blog.cifor.org/22330/indonesian-president-c alls-on-successor-to-continue-moratorium-on-forest-concessions.

Yudono, Adipandang. "Towards democracy in spatial planning through spatial information built by communities: The investigation of spatial information built by citizens from participatory mapping to volunteered geographic information in Indonesia", *IOP Conference Series: Earth and Environmental Science* 70, no. 1 (2017) doi: 10.1088/1755-1315/70/1/012002.

4

ISLAM, LEGAL GEOGRAPHY AND METHODOLOGICAL CHALLENGES IN INDONESIA

Christine Schenk

Introduction

For much of Indonesia's post-war history, religion and especially Islam has been both a sensitive and a politically relevant topic. In the 1945 *Constitution of the Republic of Indonesia* ("Indonesian Constitution"), Islam received a *primus inter pares* status among the six officially recognised religious affiliations[1] (Effendy 2003), which served to influence the shaping of a strong Muslim judiciary in a country hosting the largest Muslim population worldwide.[2] Thus, scholars working on the juncture of Islam and state bureaucracy have highlighted the prominent role of the Muslim judiciary, the religious courts and their legal professionals. Once legal professionals are trained and have taken up positions, they become part of an administrative elite (Gade & Feener 2004; Feener 2007; Hooker 2008) with influence on administrative structures. At the same time, local, and often religious, elites seize opportunities to shape more place-based regulation of the social context vis-à-vis a national level that shapes the "politics of Islamic law" (Hussin 2016, p.15). Socio-legal matters, such as marriage, divorce, inheritance, child custody, charitable endowments under Islamic law (*waqf*) and – as in the Indonesian province Aceh – criminal law, are regulated according to Muslim precepts, both in terms of their administration and bureaucracy. In addition, there are many diverse active Muslim organisations which stimulate vibrant public debate about how to live in accordance with Islam. In consequence, any inquiry into law and its implementation will raise the question of the role of Islam and how to include it in research on socio-legal matters.

Braverman (2014, p.120) suggests in her chapter on methodology that legal geographers are by training and interest familiar with administrative and bureaucratic reasoning, government schemes and involved experts. While legal geographers are often practising experts who have become academics, their positionality is (still) often influenced by their previous experiences in administrative and bureaucratic work. Positionality relates to the researcher's social, cultural and subjective positions (and other psychological processes) which are reflected in the kinds of questions asked, theoretical framings, reading of texts and relations between different persons involved in the research (England 2016). Building on the requests by Braverman (2014, p.120f) and Bennett and Layard (2015) to situate the legal geography project when crafting a methodology, this chapter focuses on *why* it is important to pay attention to Indonesia's main religious affiliation – Islam and the Muslim judiciary – and the implications of such a focus for developing a research design.

To respond to this request, this chapter argues that a research design to study legal geographies in Indonesia needs to engage specifically with the Muslim judiciary and the "politics of Islamic law" (Hussin 2016, p.15). Indonesia's "politics of Islamic law" is grounded in its inter-legality "constituted by different legal spaces operating simultaneously on different scales and from different interpretive standpoints" (De Sousa Santos 1987, p.288; see also Benda-Beckmann & Benda-Beckmann 2014). In Indonesia, inter-legality emerges from the co-existence of customary law (*adat*) and formalised religious law. While customary law is often considered as a kind of place-based regulation of social matters (Bowen 2003), religious law or Islamic law is subject to negotiation between colonial, post-colonial, religious and local elites as they struggle for the codification of particular readings of Islam into law and its implementation. I examine which readings of Islamic law based on different schools of religious thought are applied in practice. My focus is on how such considerations are reflected in publicly accessible documentation and legal expertise, in order to understand the socio-politically spatialised struggles in the legal arena that aim to regulate communities and its inhabitants.

In what follows in the first section, I highlight the historical roots of Indonesia's legal pluralism. From roughly the tenth to the twentieth century CE, the dominant denomination in Indonesia, Sunni Muslims, in principal agreed upon the sources and methods for formulating Islamic law (*Sharia*). Since then, Islam itself has increasingly provided concepts and guidance for processes of legal drafting and has gradually come to inform or even replace important *adat* and, as a result, has created "alternative forms of ordering, legal "grey zones", silences, areas of "non-law" … where laws fail to keep pace with life" (Robinson & Graham 2018, p.4). Such spatial-legal arrangements shaping a territorial project are particularly important with regard to geo-legality, where the security of bodies, most often female bodies, is at stake (Brickell & Cuomo 2019).

In the second section, I highlight how Islam has been institutionalised in the governance of Indonesia in various guises. Religious courts are responsible for regulating Muslim affairs nationally. Muslim leaders should be consulted by politicians and law-makers, either informally or formally, during or after legal drafting processes. Legal professionals in the judiciary have been trained in institutions and courses that focus (solely) on the application of *Sharia*. Even though *adat* is the traditional local source of law, increasingly Muslim leaders are legitimised by *adat* itself to influence or even shape the regulation of community affairs.

In the third section, I present possibilities for ways of carrying out research where legal geography and Islam are at stake. In so doing, I build on the argument that positionality is crucial in ethically sensitive contexts and requires a collaborative research approach that includes different positionalities and thus responds to different expectations and opinions (Schenk 2013). Here I extend this argument by highlighting how engagement with the Muslim judiciary influences, and is influenced by, the researchers' and the participants' positionalities, and take account of the different positionalities of the researchers involved.

Islam and the background to Indonesia's legal pluralism

This section highlights the historical importance of Islam in relationship to Indonesia's legal pluralism based on state law, Islamic law and *adat*. These laws often exist in parallel and can result in "alternative forms of ordering, legal "grey zones", silences, areas of "non-law" … where laws fail to keep pace with life" (Benda-Beckmann & Benda-Beckmann 2014; Robinson & Graham 2018, p.4). In this chapter, the implications of this co-existent legal pluralism are examined in relation to a feminist legal geography (Brickell 2016; Brickell & Cuomo 2019) that emphasises the role of the well-being of the (gendered) body vis-à-vis heteronormative societal patterns.

In Western societies, law-making is often understood as a process of defining a "territorial project" (Murphy 2012, p.168) in the sense that it legitimates epistemic sovereignty and the spatial diffusion of administrative control in social fields across a bounded territory. This kind of top-down idealistic view of the state as container with clear boundaries defined by law is confronted with the messy social realities of a social space. Power is by no means concentrated in the state, particularly not in post-colonial contexts. Instead, this social space is shaped by the materialisation of power and discourse, with all their indeterminate cultural, political and historical meanings (Delaney, Ford & Blomley 2001, p.xviii). In post-colonial contexts such social space is not only shaped by fragmented administrative set-ups that diffuse aspirations to epistemic sovereignty, but also by a problematic inter-legality.

Inter-legality can be informed by the interaction of place, nationalist ideas of state rule and cultural and/or religious norms; such inter-legality is visibly present and researched in contexts with colonial legacies, but has tended to be underestimated in relation to more general forms of place-based regulation (Robinson & Graham 2018). While positive law derived from an authority such as a government often results in codified state law, it does not necessarily address all existent non-state rules regulating the social context of a place (Berman 2012).

The importance of place-based forms of regulation has been discussed in recent publications. For example, Bartel (2018) analyses the political case of

Silent Spring (drawing on Rachel Carson's book, *Silent Spring*), highlighting the importance of place-based knowledge to develop legislation on pest control. In Cambodia, law-making for the protection of wetlands lacks not only place-based knowledge, but often excludes female perspectives, because law-making is inherently male-driven (Gillespie 2018). Both papers point to the importance of the nexus between place-based regulations and female (non-)representation. This is central to understanding the operation and interaction of legal spaces in contexts where place-based regulations not only operate informally, but also represent a parallel law legitimised though its colonial legacy and represented by (predominantly male) actors.

The parallel operation of such legal spaces becomes particularly pertinent in contexts with colonial legacies such as Indonesia, where state law, *adat* and Islamic law operate side-by-side as overlapping legal constructions. In Indonesia, *adat* occupies a unique position as legal space, often in contrast to state law on the national level. At the community level, *adat* regulates day-to-day matters, at times in opposition to the state. Bowen (2003, p.59) considers "*adat* as 'not-the-state'"; although Cammack (2007, p.148) points out that in colonial times, *adat* was considered as Dutch "*adat* law" when the regulation of inheritance was delegated to civil courts. The practice and rationale of *adat* slowly changed over the centuries to become more a matter of consensus-making situated in opposition to jurisdictions based on written texts. In more recent times, *adat* has become an instrument for proposing and developing norms and values in conformity with *Sharia* (Feener 2013). Thus, both *adat* and Islamic law could be considered as place-based regulations, because they are both legal orders that regulate community matters in relation to an understanding of "Muslimness" separate and distant from state-based constructions. The coexistence of multiple legal constructions entails different perceptions of the meanings of different laws as well as of social relations and, in this sense, "legal phenomena can be seen as constitutive of social relations, social consciousness and experiential reality" (Blomley, Ford & Delaney 2001, p.xvi).

Islamic law in Indonesia, drawing on the dominant *Shafi'i* school, substantially differs from Islamic traditions in the Middle East since it has embraced the forms of social relations, social consciousness and experiential reality prevailing in Indonesia (Feener 2007, p.2f). Approaches to Islamic law itself are embedded in a framework of "discursive traditions" (Asad 1993, p.29) that considers Islamic law as dynamic and not fixed in one singular approach.[3] Drawing on comparative research in India, Malaysia and Egypt, Hussin (2016) challenges the point of view that colonial regimes have imposed their vision of law and instead argues for a focus on the role of politics in shaping Islamic law. Such politics are shaped by local and colonial elites, where local elites often allied with colonial power-holders to realise their vision of society and state (Hussin 2016, p.15). In relation to Indonesia, law-making needs to be considered as a phenomenon in which Muslim representation on all administrative levels is important.

Scholars of Islamic studies focusing on Indonesia (e.g. Feener 2007; Van Bruinessen 2013) differentiate between two main streams or schools of religious thought, namely the traditionalist and modernist (or reformist). A rather traditionalist interpretation of Islam is called *dayah*. Representatives of this school are predominantly represented in traditional religious education systems such as boarding schools (*pesantren*) focusing on Islamic legal texts in line with the reinterpretation of two main sources of Islam (*Qu'ran and Sunnah*). This traditionalist interpretation also involves tolerance towards cultural expressions, such as Sufism. The other influential school of religious thought has been identified as the "modernist". This school subscribes to modern learning as a way of enhancing the power of reason. However, it is also related to reform-minded ideas in line with Indonesia's state-based Islam that focuses on the institutionalisation of Islam in government institutions such as the judiciary, the civil service and schools.

Two developments supported the role of Islamic law in the state bureaucracy. First, the *Religious Judicature Act 1989* (Indonesia) strengthened the jurisdiction of the Islamic judiciary while making civil courts responsible for non-Muslim family management. Second, in 1991 Haji Mohamed Suharto, then President of Indonesia, promulgated the *Compilation of Islamic Laws* (*Kompilasi Hukum Islam*) (KHI) "an Indonesian-language Manual of rules covering marriage, divorce, inheritance, and charitable foundations" (Cammack 1999, p.15) that had been enforced through presidential decree. Through this compilation, and despite counter drafts to amend it (Mulia & Cammack 2007), Islamic law was adopted into state law. The KHI can be considered as a form of Islamic modernism (Esposito 1998): "Modernism's invitation to reinterpret the revealed sources in light of contemporary needs opens the possibility of clothing the requirements of the statute with the authority of the divine law" (Cammack 1999, p.18). The need for this modernisation arose because Islamic texts are so diverse and there was no systematic reference to indicate which text applies to contemporary needs.

The implications of the legal pluralism resulting from entwinement of *adat*, different interpretations of Islamic law and state law (Geertz 1971; Brickell & Platt 2015), resonate with feminist legal geographies (Brickell & Cuomo 2019) that investigate the bodily connotations of law. Researchers working on Indonesia have analysed the consequences of such legal pluralism and its potential injustices in relation to children and women in legal regimes relating to marriage and divorce (Butt 2008; O'Shaughnessy 2008; Bedner & Van Huis 2010), sexual identities (Blackwood 2005) and the counter-agency of women's groups (Aisyah & Parker 2014; Eidhamar 2017) resisting patriarchal readings of Islamic texts (Van Doorn-Harder 2006). Questions around maintenance, child custody and domestic violence, often related to unregistered marriages (Nurlaelawati 2010; Nurlaelawati 2016), remain unresolved, and the consequences are often inscribed on female bodies – physically, mentally and economically (Schenk 2019).

In these examples, legal pluralism provides the ground for a "normative vacuum" (Griffiths 1986, p.34) in which legal interpretation is derived from diverse sources, but is far from being definite and might not provide security to all "bodies that make territory" (Smith 2012). Farries and Sturm (2018) showed that cyber-misogyny on the internet can prevail notwithstanding that safety for users is guaranteed in writing, but not in the everyday digital sphere where they can become objects of oppression. Such oppression can be also understood as "rule *by* law", while "rule *of* law" refers to a system of rights and responsibilities (Brickell 2016, emphasis in original). Butler (2002, p.183ff) challenges in her book *Gender Trouble* (2002, p.183ff) challenges the notion of a law that intends to establish security for everyone. Rather, Butler asks what kind of security law any law establishes, along with who might benefit from this notion of security, who might be excluded and who was involved in crafting the law.

In this section, it can thus be seen that Indonesia's legal pluralism is far from being a coherent territorial project. *Adat* as customary law regulating community matters often functions in parallel with state law and increasingly incorporates Islamic law with its inherent heteronormativity, often along the aim of securing the biological reproduction of the population. On the level of national law, both civil law and the KHI add to the plethora of legal orders. The co-existence of these plural legal orders creates a normative vacuum that often has dire consequences for those who are most vulnerable. In the next section I analyse the institutionalisation of Islam not only in laws, but also in the state administration and bureaucracy, including the role of the Muslim judiciary, and the inherent heteronormativity and masculinity of law-making.

Role of the Muslim judiciary in shaping the legal geography of Indonesia

Islam-inspired statecraft in Indonesia became particularly controversial in Indonesia's move away from colonial Netherlands. The draft preamble of the *Jakarta Charter* (1945) included one of its purposes as being: "to carry out Islamic law for adherents of Islam". Although this clause was removed from the final draft of the Indonesian Constitution (18 August 1945), "the Islamic and nationalist leaders agreed on a secular constitution based on five principles" (*Pancasila*), the first of which was "the belief in one God" (2008, p.6). This principle was seen as a reference to Islamic law, but not to the creation of an Islamic state (Hooker 2003, p.38). This struggle around the role of Islamic law set the scene for the legitimacy of Islam in Indonesia's jurisdiction and has continued to shape the question of how to institutionalise Islam in the state management of Indonesia and thus increasing the sovereignty of Islamic law in Indonesia's inter-legality. This question is examined here by critically discussing the role of the Muslim judiciary and their increasing influence on law-making and examining the judiciary's importance in matters related to the strong role of Islam in regulating the communities of Indonesia. In so doing, this examination will also highlight the role of legal professionalism and its inherent masculinity, and emphasise the importance of influences from neighbouring countries such as Malaysia.

The Islamic courts constitute the backbone of the Muslim judiciary (Cammack 2007; Nurlaelawati 2010). In colonial times the small number of these courts was alleged to be due to Dutch hostility to Islam. However, their number, spatial distribution across the islands and the extent of their juridical power slowly expanded and then increased rapidly after Independence (Cammack 2007). In response to clashes over the Indonesian Constitution, nationalist-minded groups advocated a largely secular Indonesian Constitution, while Muslim voices aimed for Islamic law to be included in political rule and statecraft (Hooker 2003). The Ministry for Religion, created in 1946 (Hooker 2003; Hooker 2008), was entrusted with the supervision of the Islamic courts (responsible for issuing divorces and their registration). The Office for Religious Affairs (*Kantor Urusan Agama*) (KUA) was responsible for the registration of marriages. The Ministry of Home Affairs and its subsidiaries were given the task of registering non-Muslims and managing their family matters (Nurlaelawati 2010; Schenk 2018; Schenk 2019).

The institutionalisation of Islam in law is supported by a pluralistic group of contributors including Muslim political parties and Muslim mass organisations, of which two groups, the *Muhammadiya* and the *Nahdlatul Ulama* (NU), are the best known (Van Bruinessen 2013). The shorthand dichotomy between "modernist" and "reformist" schools of religious thought can be applied here: *ulama* (or the established scholars) that are members of *Muhammadiya*, are considered modernist, while the *ulama* of the NU are characterised as subscribing to traditionalist ideas (Van Bruinessen 2013). Fealy and Bush (2014) argue that the NU *ulama* have significantly influenced national public debates since their inception in 1926, and especially after Indonesia's Independence. The NU and *Muhammadiyah* are only two organisations in a vibrant field of civil society organisations that are members of the *Majelis Ulama Indonesia* (MUI) or the Council of Islamic Scholars. The MUI is an umbrella organisation that functions as an interface between the government and the Islamic communities of Indonesia. While it has been given the task of developing guidelines for the government regarding Islamic life, it also informs the public on Islamic life through the provision of legal advice (*fatwa*) (Hasyim 2015). These forms of legal advice have impressive outreach because they are issued by influential Muslim leaders, and those by the MUI are particularly significant (Gillespie 2007). Depending on the educational background of the elected members of the MUI, opinions can be controversial because they can range from liberal to conservative readings of Islam (Hasyim 2015).

The training of legal professionals is another field that highlights the institutionalisation of Islam in the Muslim judiciary. The manner in which jurists who are specialised in Islamic law are trained often differs significantly from the training of classical jurists (Lombardi & Feener 2012). These jurists, especially those representing litigants, are often trained to provide legal advice in line with *Sharia* interpretations rather than positive civil law. In addition, judges in *Sharia* courts might have received a specific form of training in *Sharia*, and might adopt such specific interpretations of *Sharia* in their judgments. Of the manifold options possible in *Sharia*, the challenge is to decide which reading applies in any given case. The interpretation of *Sharia* also depends on the institution where jurists specialised in Islamic law, especially judges, obtain training (Lombardi & Feener 2012, p.12).

Many legal professionals working in Indonesia have been trained in Malaysia, for instance, at the International Islamic University Malaysia. Mohamad (2013) argues that the bureaucratisation of Islam has triggered a homogenisation of Islam, based on an official and codified definition of specific behavioural norms for Muslims. The influence of such training has thus served to provide a template for the spread of a more homogenised form of Islam (that is, Muslim rule *by* law), which is observable in Malaysia's neighbouring countries such as Indonesia. Similarly, the application of Islamic law by Indonesia's state administration and bureaucracy is heavily influenced by both the curricula and political intentions of the training centres, where legal professionalism is shaped (Feener 2007; Feener 2012). From this it can be seen that it is important to engage with the background, training, intellectual assumptions and work experience of judges and lawyers in order to understand the "politics of Islamic law" (Hussin 2016, p.15). Consequently, it is also important to acknowledge that administrative levels don't necessarily correlate with spatial scales.

This discussion has analysed the important role of Islam in the Indonesian Constitution and how it translates into Indonesia's administration and bureaucracy which, although officially referring to a secular constitution, is informally influenced by a form of Islam-inspired state management. This aspect of Indonesian constitutional law thus strengthens the role of Islamic law in Indonesia's inter-legality. Furthermore, the influential role of the legal professional training of Indonesia's administration and bureaucracy in neighbouring Muslim-majority countries (and other such scholarly exchanges on the interpretation of Islamic law) places Indonesia within a wider transnational Islamic legal realm. Some of these approaches to Islamic law may serve to challenge a focus on individual rights and responsibilities, regardless of any religious affiliation, instead of emphasising the sovereignty of the *Sharia* that regulates individuals of the Muslim community (*ummah*). The next section will consider some of the methodological challenges arising from these conditions.

The "hows" and "whats" of Islam in designing research on legal geography in Indonesia: the role of positionality and collaborative research

In this section, suggestions are made for how to consider the role of Islam and its proponents in research design drawing on the concept of positionality. Positionality is important in any fieldwork because it frames the social, cultural and political ways of approaching, conducting and writing about research (Rose 1997; Mohammad 2001; Franks 2002; Sultana 2007). In settings where social and cultural differences are at stake, and where fieldwork is informed by "ask[ing] how people live, constitute, and imagine social space, place and landscape as well as how people understand themselves as living, doing and imagining the legal"

(Braverman 2014, p.124), the positionality of the researcher plays a crucial part in becoming a "spatial detective" (Bennett & Layard 2015) in order to understand the materialisation of law. In so doing, I reflect on insights from ethnographic fieldwork in Indonesia on the registration of marriages, and include the analysis of an expert hearing on legal reform of the registration of marriages and divorces (Schenk 2018; Schenk 2019). My research has also been subject to my own positionality as a white, non-Muslim and female academic, who is fluent in Indonesian. In the following, I situate positionality in the context of cultural and social protocols and present how I dealt with it through my organisation of the fieldwork, which included a combination of different sources or "multi-sited fieldwork" (Braverman 2014; Bennett & Layard 2015). Such multi-sited fieldwork aims to balance any positionality in a non-Western setting (Gillespie 2016, and see Chapter 2 of this collection) with different avenues of research, and seeks to both avoid relying one-sidedly on known and like-minded contacts alone, and respond to the "politics of Islamic law" (Hussin 2016, p.15)

Positionality in religious contexts has long been, and continues to be, both a sensitive and politically significant topic in Indonesia (Schenk 2013). Consequently, research on the Muslim judiciary and on the role of religion in political processes or how religion influences the course of everyday life, benefits from paying attention to social and cultural expectations and protocols (e.g. the central role of the family unit in society). It raises the question of who can legitimately ask questions and carry out research on such politically sensitive topics. "Critical" questions challenging different normative expectations and protocols might upset research participants and lead to heightened emotions. These heightened emotions or even hostility can result in the researcher not being able to continue with the interview or, indeed, the whole research project. Even the fact that a researcher comes from the West can lead to their being perceived as meddling in national affairs and evoke hostile reactions (Jones & Ficklin 2012; Gillespie 2016). There are two issues relating to positionality that I would like to explore here: the dual identity of a researcher, and the cultural and political restrictions on the research itself.

The first is the dual identity of being both "an expert" with, for example, expertise in legal issues, and being an academic and/or an activist. Being both can allow access to the different perspectives of those working in different positions and organisational backgrounds in "living, doing and imagining the legal" (Braverman 2014, p.124) and might also produce discussions in which the researcher turns into an expert and vice-versa. If the researcher is considered as an expert s/he turns into a respondent, or potential source for, providing expertise on the research participant's inquiry. The researcher can turn into an authority providing advice instead of researching the social context informing legal problems.

The second aspect of the issue of positionality refers to the potential cultural and political restrictions in the field, especially if research participants consider the researcher as an intruder into their affairs and are emotionally opposed to the research. Jones and Ficklin (2012) argue that trust is essential in establishing types of relational fieldwork. To address these restrictions, the researcher and research participants must develop some form of social relationship. Building on such relational fieldwork, Brigg and Bleiker (2010) examine the importance of the self as producer of knowledge in interaction with knowledge communities. Relational fieldwork does not just start with empathy, for instance by saying "I like Acehnese food" or any other appreciative, but largely superficial, comments. Rather, it starts with "a recognition of the importance of the 'position' or 'positionality' of the researcher: that we see the world from specific embodied locations" (Rose 1997; Valentine 2002, pp.116–17). Thus, recognising one's own educational background, origin, gender difference and being aware of different perceptions of the role of religion in shaping (or not shaping) everyday life has proven important to me when carrying out research. Such recognition avoids reproducing "the illusion of sameness" (Valentine 2002, p.123) between researcher and participants and, instead, incorporates one's positionality into fieldwork.

In my research on the registration of marriages and divorces, I "studied up" power relations in bureaucratic organisations and legislative processes. To "studying up" involves an ethnographic approach that focuses not only on disadvantaged actor groups, but connects the inquiry to the power structures of elites (Mukherjee 2017). Local and national elites often are key in shaping legislative processes. Most of my fieldwork starts with the following questions: "What genuine interests can I share with my research respondent/s?"; and "How does this help me to develop positive relationships or even empathy with research participants?". Empathy "refers to the capacity to understand the experiential frame of reference of another without losing an awareness of (its difference from) one's own" (Bondi 2003, p.65). From these questions, various further avenues can develop such as exploring the self as a source of knowledge (Brigg & Bleiker 2010), which might involve articulating one's own emotions within the research setting. This notion is relevant to a situation during my research where I used my role as a mother to establish trust between myself as researcher and some of my participants. At the same time, I pondered the issues of family management with my colleagues, which aims to link the personal to research work (Moser 2008). Moser (2008, p.383) describes how one's personality responds to aspects of research, "such as my social skills, my emotional responses to, and interest in, local events, how I conducted myself and the manner in which I navigated the personalities of others". These social skills support the research process and have a positive impact on the insights being shared, but also assign more weight to the researcher's interaction with the research participant.

However, during my fieldwork I faced situations that required me to think of and craft alternative avenues for research. One avenue involved swapping roles with Muslim co-researcher(s), who were able to contribute Islamic idioms and ideas to the discussions with participants, which served as icebreakers for establishing research relationships. Such practice proved particularly important when engaging with persons with an elite background. However, the practice of "studying-up" elites can uncover questions other than the research question, as persons with such a background might feel inspired by hearing themselves talk (Braverman 2014, p.125). I often experienced this kind of self-indulgence with experts who used me and my co-researcher(s) as "good listeners". Quite often, the resulting flow of words could hardly be interrupted, and this made me think about the types of multi-sited fieldwork that take place in the archives of parliamentary services, court rooms, offices of lawyers, mosques, help centres for vulnerable persons, administrative offices and private houses. The ability to research a diversity of locations and places and forms of regulations tailored to different local contexts is important in addressing the conundrum of "politics of Islamic law" (Hussin 2016, p.15).

In my interviews, I inquired about the most contentious issues such as the registration of marriages and the codification of polygamy in law-making and how they relate to particular justifications present in Islamic and religious perspectives. I identified "signifiers" (Dixon 2010, p.393): for example, proposing to issue a *fatwa* can be a signifier for ambitions to actively shape the regulation of a Muslim society. In order to understand how a research participant framed her/his ideas on a *fatwa*, I asked questions to uncover where the research participant had received legal training and whether the research participant subscribes to a modernist or traditionalist school of religious thought. These lines of questioning enabled me to find out their points of view on drafting law and the role of Islamic teachings within the law. In parallel, I conducted interviews with Muslim leaders to help me understand religious justifications and underlying rationales for the exercise of Muslim leadership.

One way I balanced the like-mindedness of research participants was to analyse "expert hearings" of legal processes. In Indonesia, these sources can be accessed upon application to the Parliamentary Information Service or similar public services. In order to analyse the different positions on the registration of marriages and divorces, I examined the *Expert Hearing on the by-law on population administration of the Province of Aceh*, conducted between March and June 2008 and recorded in verbatim reports in Indonesian (Schenk 2019). These hearings documented the statements made during the legal reform comprising civil servants from different governmental institutions (religious courts, civil registration, Office for Religious Affairs), legal experts, lawyers representing women's organisations and Muslim leaders. Since most of the participants in the hearings were important figures in Acehnese society, their backgrounds and importance could serve as bases for more detailed interviews. But such research is not adequate on its own and needs to be complemented with accounts of persons experiencing the lack of individual rights or its implementation.

This section has discussed how selected methods and diverse research sites relate to furthering an understanding of the role of the judiciary. A researcher's (Western) positionality will require taking into account cultural and political expectations and protocols. Besides a collaborative research approach, I have presented various techniques of multi-sited fieldwork to deal with these expectations and protocols, while paying particular attention to the Muslim judiciary, for example through research methods including the analysis of verbatim reports such as expert hearings. Developing signifiers and asking where research partici-pants have received their legal training can clarify how state law is informed by particular schools of religious thought.

Conclusion

This chapter has examined the role of Islam in shaping Indonesia's legal geog-raphy and has considered ways of studying such complex conditions. Due to the Indonesian Constitution, Islamic law has a strategic role in shaping a judi-ciary divided by religious affiliations (Muslim and non-Muslim administration and bureaucracy). Besides Islamic law, Indonesia's legal pluralism comprises both customary law (*adat*) and forms of positive law that are not constitutive of the place-based regulations underpinning social practices prevailing in villages across Indonesia – and vice versa. The gap between place-based regulations and codified law particularly affects the regulation of family matters such as divorces, main-tenance and the payment of alimony. Women and children often depend on the legal interpretation and jurisdiction of the judge at the religious courts aiming to regulate rights and responsibilities of "bodies that make territory" (Smith 2012). In turn, one can say that the role of the Muslim judiciary, its local and national elites, is important to any understanding of the rationale underlying social pro-cesses, their legislation and jurisdictions in shaping the politics of Islamic law that forms part of the multi-legal territorial project of post-colonial Indonesia.

In order to understand the rationale of the Muslim judiciary, this chapter has argued for a comprehensive and explorative engagement with the politics of Islam. In so doing, I have highlighted the role of the researcher's positional-ity, which may confront boundaries if some topics cannot be voiced due to the researcher's gender and, possibly, Western origin. To address positionality, I have highlighted the importance of a collaborative research team involving different origins and genders. However, to approach groups facing severe injustices, mul-tiple research approaches are necessary involving fieldwork in different places and through mixed methods. In addition, the analysis of public records, such as those of expert hearings, can provide insights into important societal discourses. I have briefly explained the application of signifiers that can help to identify which reading of Islamic law, including which school of religious thought and its sub-denominations, has most influenced members of the Muslim judiciary. These signifiers draw on questions asking where research participants received their training because legal professionalism both influences the exercise of juris-diction and shapes the translation of social processes into codified law.

Notes

1 Islam, Protestantism, Catholicism, Hinduism, Buddhism *and* Confucianism. The author is aware that Protestantism and Catholicism are different branches of Christianity.

2 Approximately 88% (Government of Indonesia 2010) of Indonesia's population (app. 264 Mio., Worldbank 2017) are Muslim adherents.
3 Islamic law is here referred to as *fiqh* or Islamic jurisprudence, while being aware that the term *Sharia* comprises both *Sharia* as way of life and *fiqh* (Hallaq 2001; Hallaq 2009).

References

Aisyah, S., and Parker, L. 2014, 'Problematic conjugations: Women's agency, marriage and domestic violence in Indonesia', *Asian Studies Review*, vol. 38, pp.205–223.

Asad, T. 1993, *Genealogies of Religion: Discipline and Reasons of Power in Christianity and Islam*, John Hopkins University Press, Baltimore, London.

Bartel, R. 2018, 'Place-speaking: Attending to the relational, material and governance messages of Silent Spring', *The Geographical Journal*, vol. 184, no. 1, pp.64–74.

Bedner, A., & Van Huis, S. 2010, 'Plurality of marriage law and marriage registration for Muslims in Indonesia: A plea for pragmatism', *Utrecht Law Review*, vol. 6, pp.175–191.

Benda-Beckmann, F., & Benda-Beckmann, K. 2014, 'Places that come and go: A legal anthropological perspective on the temporalities of space in plural legal orders', in *The Expanding Spaces of Law: A Timely Legal Geography*, eds I. Braverman, N. Blomley, D. Delaney & A. Kedar, Stanford University Press, Stanford, pp.30–52.

Bennett, L., & Layard, A. 2015, 'Legal geography: Becoming spatial detectives", *Geography Compass*, vol. 9, pp.406–422.

Berman, P. S. 2012, *Global Legal Pluralism: A Jurisprudence of Law beyond Borders*, Cambridge University Press, New York.

Blackwood, E. 2005, 'Transnational sexualities in one place – Indonesian readings', *Gender & Society*, vol. 19, pp.221–242.

Blomley, N., Ford, R. T., & Delaney, D. 2001, *Legal Geographies Reader*, Blackwell, Oxford.

Bondi, L. 2003, 'Empathy and identification: Conceptual resources for feminist fieldwork' *ACME: An International E-journal for Critical Geographies*, vol. 2, pp.64–76.

Bowen, J. R. 2003, *Islam, Law, and Equality in Indonesia: An Anthropology of Public Reasoning*, Cambridge University Press, Cambridge.

Braverman, I. 2014, 'Who's afraid of methodology', in *The Expanding Spaces of Law*, eds I. Braverman, N. Blomley, D. Delaney & A. Kedar, Stanford University Press, Stanford.

Brickell, K. 2016, 'Gendered violences and rule of/by law in Cambodia', *Dialogues in Human Geography*, vol. 6, pp.182–185.

Brickell, K., & Cuomo, D. 2019, 'Feminist geolegality', *Progress in Human Geography*, vol. 43, pp.104–122.

Brickell, K., & Platt, M. 2015, 'Everyday politics of (in)formal marital dissolution in Cambodia and Indonesia', *Ethnos*, vol. 80, pp.293–319.

Brigg, M., & Bleiker, R. 2010, 'Autoethnographic International Relations: Exploring the self as a source of knowledge', *Review of International Studies*, vol. 36, pp.779–798.

Butler, J. 2002, *Gender Trouble*, Routledge, London, New York.

Butt, S. 2008, 'Polygamy and mixed marriage in Indonesia: The application of the marriage law in the Courts', in *Indonesia: Law and Society*, ed. T. Lindsey, The Federation Press, Annandale, pp.122–144.

Cammack, M. 1999, 'Islam, nationalism, and the state in Suharto's Indonesia', *Wisconsin International Law Journal*, vol. 17, pp.1–42.

Cammack, M. 2007, 'The Indonesian Islamic judiciary', in *Islamic Law in Contemporary Indonesia*, eds R. M. Feener & M. Cammack, Harvard University Press, Cambridge, MA, pp.146–169.

Constitution of the Republic of Indonesia. 1945.

De Sousa Santos, B. 1987, 'Law: A map of misreading. Toward a postmodern conception of law', *Journal of Law and Society*, pp.279–302.

Delaney, D., Ford, R. T., & Blomley, N. 2001, 'Preface: Where is law?', in *Legal Geographies Reader*, eds N. Blomley, D. Delaney & T.R. Ford, Blackwell, Oxford, pp.xiii–xxii.

Dixon, D. 2010, 'Analyzing meaning', in *Research Methods in Geography*, eds B. Gomez & J.P. Jones III, Wiley-Blackwell, Malden, Oxford, pp.392–408.

Effendy, B. 2003, *Islam and the State in Indonesia*, Institute of Southeast Asian Studies, Singapore.

Eidhamar, L. G. 2017, '"My husband is my key to paradise" – Attitudes of Muslims in Indonesia and Norway to spousal roles and wife-beating', *Islam and Christian–Muslim Relations*, vol. 29, pp.1–24.

England, K. 2016, 'Positionality', in *International Encyclopedia of Geography: People, the Earth, Environment and Technology*, eds D. Richardson, N. Castree, M. F. Goodchild, A. Kobayashi, W. Liu & R. A. Marston, doi:10.1002/9781118786352.wbieg0779.

Esposito, J. L. 1998, *Islam: The Straight Path*, Oxford University Press Oxford, New York.

Farries, E., & Sturm, T. 2018, 'Feminist legal geographies of intimate-image sexual abuse: Using copyright logic to combat the unauthorized distribution of celebrity intimate images in cyberspaces', *Environment and Planning A: Economy and Space*, vol. 0, no. 0, pp.1–21.

Fealy, G., & Bush, R. 2014, 'The political decline of traditional Ulama in Indonesia: The state, Umma and Nahdlatul Ulama', *Asian Journal of Social Science*, vol. 42, pp.536–560.

Feener, M. 2007, *Muslim Legal Thought in Modern Indonesia*, Cambridge University Press, Cambridge.

Feener, M. 2012, 'Social engineering through Shari'a: Islamic law and state-directed da'wa in contemporary Aceh', *Islamic Law and Society*, vol. 19, pp.275–311.

Feener, M. 2013, *Shari'a and Social Engineering. The Implementation of Islamic Law in Contemporary Aceh*, Indonesia, Oxford University Press, Oxford.

Franks, M. 2002, 'Feminisms and cross-ideological feminist social research: Standpoint, situatedness and positionality–developing cross-ideological feminist research', *Journal of International Women's Studies*, vol. 3, pp.40–53.

Gade, A., & Feener, M. 2004, 'Muslim thought and practice in contemporary Indonesia', in *Islam in World Cultures. Comparative Perspectives*, ed. M. Feener, ABC CLIO, Santa Barbara, Denver, Oxford, pp.183–218.

Geertz, C. 1971, *Islam Observed: Religious Development in Morocco and Indonesia*, University of Chicago Press, Chicago.

Gillespie, J. 2016, 'A legal geography of property, tenure, exclusion, and rights in Cambodia: Exposing an incongruous property narrative for non-Western settings', *Geographical Research*, vol. 54, pp.256–266.

Gillespie, J. 2018, 'Wetland conservation and legal layering: Managing Cambodia's great lake', *The Geographical Journal*, vol. 184, pp.31–40.

Gillespie, P. 2007, 'Current issues in Indonesian Islam: Analysing the 2005 Council of Indonesian Ulama Fatwa No. 7 opposing pluralism, liberalism and secularism', *Journal of Islamic Studies*, vol. 18, pp.202–240.

Government of Indonesia, 2010, *Kewarganegaraan, suku bangsa agama dan bahasa sehari-hari penduduk Indonesia. Hasil Sensus Penduduk 2010*, National Statistics Office [Badan Statistik Nasional] Badan Statistik Pusat, Jakarta.

Griffiths, J. 1986, 'What is legal pluralism', *Journal of Legal Pluralism & Unofficial Law*, vol. 24, pp.1–55.

Hallaq, W. B. 2001, *Authority, continuity and change in Islamic law*, Cambridge University Press, Cambridge, New York.

Hallaq, W. B. 2009, *An Introduction to Islamic Law*, Cambridge University Press, Cambridge.

Hasyim, S. 2015, 'Majelis Ulama Indonesia and pluralism in Indonesia', *Philosophy & Social Criticism*, vol. 41, pp.487–495.

Hooker, M. B. 2003, 'The state and Shari'a in Indonesia', in *Shari'a and Politics in Modern Indonesia*, eds A. Salim & A. Azra, Institute of Southeast Asian Studies, Singapore, pp.33–47.

Hooker, M. B. 2008, *Indonesian Syariah. Defining a National School of Islamic Law*, Institute of Southeast Asian Studies, Singapore.

Hussin, I. 2016, *The Politics of Islamic Law. Local Elites, Colonial Authority, and the Making of the Muslim State*, University of Chicago Press, Chicago.

Jakarta Charter 1945.

Jones, B., & Ficklin, L. 2012, 'To walk in their shoes: Recognising the expression of empathy as a research reality', *Emotion, Space and Society*, vol. 5, pp.103–112.

Lombardi, C. B., & Feener, R. M. 2012, 'Why study Islamic legal professionals', *Pacific Rim Law & Policy Journal*, vol. 21, pp.1–12.

Mohamad, M. 2013, 'Legal-bureaucratic Islam in Malaysia: Homogenizing and ring-fencing the Muslim', in *Encountering Islam: The Politics of Religious Identities in Southeast Asia*, ed. H. Yew-Foong, Institute of South-Asean Studies, Singapore, pp.103–132.

Mohammad, R. 2001, "Insiders' and/or 'outsiders': Positionality, theory and praxis', in *Qualitative Methodologies for Geographers: Issues and Debates*, eds M. Limb & C. Dwyer, Arnold, London, pp.101–117.

Moser, S. 2008, 'Personality: A new positionality?', *Area*, vol. 40, pp.383–392.

Mukherjee, S. 2017, 'Troubling positionality: Politics of "studying up" in transnational contexts,' *The Professional Geographer*, vol. 69, pp.291–298.

Mulia, S., & Cammack, M. 2007, 'Toward a just marriage law: Empowering Indonesian women through a Counter Legal Draft to the Indonesian Compilation of Islamic Law', in *Islamic law in contemporary Indonesia. Ideas and institutions*, eds M. Feener & M. Cammack, Harvard University Press, Cambridge, MA, pp.128–145.

Murphy, A. B. 2012, 'Entente territorial: Sack and Raffestin on territoriality', *Environment and Planning-Part D*, vol. 30, pp.159–172.

Nurlaelawati, E. 2010, *Modernization, Tradition and Identity: The Kompilasi Hukum Islam and Legal Practice in the Indonesian Religious Courts*, Amsterdam University Press, Amsterdam.

Nurlaelawati, E. 2016, 'The legal fate of Indonesian Muslim women in court: Divorce and child custody', in *Religion, Law and Intolerance in Indonesia*, eds T. Lindsey & H. Pausacker, Routledge, London, New York, pp.353–368.

O'Shaughnessy, K. 2008, *Gender, State and Social Power in Contemporary Indonesia: Divorce and Marriage Law*, Routledge, Abingdon, New York.

Robinson, D. F., & Graham, N. 2018, 'Legal pluralisms, justice and spatial conflicts: New directions in legal geography', *The Geographical Journal*, vol. 184, no. 1, pp.3–7.

Rose, G. 1997, 'Situating knowledge: Positionality, reflexivities and other tactics', *Progress in Human Geography*, vol. 21, pp.305–320.

Schenk, C. G. 2013, 'Navigating an inconvenient difference in antagonistic contexts: Doing fieldwork in Aceh, Indonesia', Singapore Journal of Tropical Geography, vol. 34, pp.342–356.

Schenk, C. G. 2018, 'Islamic leaders and the legal geography of family law in Aceh, Indonesia', *The Geographical Journal*, vol. 184, no. 1, pp.8–18.

Schenk, C. G. 2019, 'Legal and spatial ordering in Aceh, Indonesia: Inscribing the security of female bodies into law', *Environment and Planning A: Economy and Space*, vol. 51, no. 5, pp.1128–1144.

Smith, S. 2012, 'Intimate geopolitics: Religion, marriage, and reproductive bodies in Leh, Ladakh', *Annals of the Association of American Geographers*, vol. 102, pp.1511–1528.

Sultana, F. 2007, 'Reflexivity, positionality and participatory ethics: Negotiating fieldwork dilemmas in international research', *ACME: An International E-journal for Critical Geographies*, vol. 6, pp.374–385.

Valentine, G. 2002, 'People like us: Negotiating sameness and difference in the research process', in *Feminist Geography in Practice: Research and Methods*, ed. P. Moss, Blackwell, Oxford, pp.116–126.

Van Bruinessen, M. 2013, 'Overview of Muslim organizations, associations and movements in Indonesia', in *Contemporary Developments in Indonesian Islam: Explaining the "Conservative Turn"*, ed. M. Van Bruinessen, Institute of Southeast Asian Studies, Singapore, pp.21–59.

Van Doorn-Harder, P. 2006, *Women Shaping Islam: Indonesian Women Reading the Qur'an*, University of Illinois Press, Champaign.

Worldbank, 2017, *Country Profile Indonesia*, viewed 10 January 2019, <https://databan k.worldbank.org/data/views/reports/reportwidget.aspx?Report_Name=Count ryProfile&Id=b450fd57&tbar=y&dd=y&inf=n&zm=n&country=IDN>.

5

PATENT LANDSCAPING FOR VANUATU

Specific legal geographic methods for Indigenous knowledge protection and promotion

Daniel F. Robinson, Margaret Raven, Donna Kalfatak, Trinison Tari, Hai-Yuean Tualima and Francis Hickey

Introduction

Much of the study in the field of legal geography seeks to understand the co-constitution of both law and place, which is often discussed by legal geographers in regard to planning and environmental laws, through case law analysis and in studying laws in different places and contexts (Bartel & Graham 2016; Bennett & Layard 2015; Delaney 2015a). This chapter uses a legal geography lens to interrogate an intersection of laws – where intellectual property laws meet with environmental (biodiversity) laws and also Indigenous rights. Our chapter here builds upon several years of research relating to the issue of "biopiracy" (see Robinson 2010) and attempts at finding preventative measures to help protect and promote Indigenous knowledge. We provide a short introduction to these legal frameworks and contexts, before explaining a specific methodological approach – patent landscaping – which has been useful for our legal geography research in several countries, jurisdictions and contexts. We then provide a simple case study of a "patent landscape" using species native and endemic to Vanuatu. We use this case study to highlight that there is ongoing potential for "biopiracy" to occur unless there are specific legal interventions.

Some of the main concerns of legal geography scholarship are in working towards revealing and interrogating the countless ways in which "law makes space" and "place makes law". These concerns extend to critical considerations of the political-geographical interrelationships between them, demonstrating how issues of spatial justice are mediated by and through dynamic and diverse "lawscapes" (Graham 2011). Legal geography scholarship thereby offers scholars, and also activists, paradigmatic strategies for approaching the spatial and material dimensions of social and environmental justice (Delaney 2015b; Robinson & Graham 2018).

"More-than-human" bio-geographies, or human-plant geographies, are also of interest to this paper. These scholarly initiatives and methodologies attempt to break down binaries of nature and culture, ecology and society, and to understand the many varying perspectives and worlds of plants, their identity and collectivity, their connections and transformations (as food, timber, commodity, sacred object, part of ritual, as living entity) and their webs of relationality (from Head, Atchison & Gates 2012; Head & Atchison 2009). In this case, a "legal human-bio-geography"[1] is probably the most apt description, because we are interested in the way patent laws and biodiversity laws internationally have shaped the perception of plants as ownable or monopolisable "inventions", genetic resources and commodities. On the other hand, *Kastom* (or customary/traditional laws and practices) in Vanuatu may define relationships with plants in a number of different ways – as part of ceremonies, as having totemic or customary significance or as being something closer to a commodity (see Robinson et al. 2019). The significance of certain species is often embedded in place-based social and customary relationships and laws, and so any attempts at state legal reforms in countries in the Pacific require specific attention to these realities (Robinson & Forsyth 2016).

Legal contexts

This chapter cannot examine in full detail the many relevant international laws, save to provide a brief explanation of the "lawscape" (Graham 2011). A simple starting point is to explain that, in the early 1990s, a discourse of "biopiracy" emerged, which was described around this time by the non-government organisation (NGO), the Rural Advancement Foundation International (RAFI) as follows:

> Biopiracy refers to the appropriation of the knowledge and genetic resources of farming and indigenous communities by individuals or institutions who seek exclusive monopoly control (patents or intellectual property) over these resources and knowledge. (RAFI 1995; cited by Robinson 2010)

The globalisation of intellectual property rights has occurred most significantly through the *Agreement on Trade-Related Aspects of Intellectual Property Rights* (TRIPS Agreement), which is Annex 1C to the 1994 *Marrakesh Agreement Establishing the World Trade Organization* (WTO Agreement). The TRIPS Agreement includes provisions that have generated significant debate in this area, including Article 27.3(b):

> 3. Members may also exclude from patentability: (b) plants and animals other than micro-organisms, and essentially biological processes for the production of plants or animals other than non-biological and microbiological processes. However, Members shall provide for the protection of

plant varieties either by patents or by an effective sui generis system or by any combination thereof. The provisions of this subparagraph shall be reviewed four years after the date of entry into force of the WTO Agreement.[2]

This article has been heavily debated by parties to the World Trade Organization (WTO), and, as such, a broader review has been mandated under Paragraph 19 of the 2001 *Doha Declaration on the TRIPS Agreement and Public Health*. This paragraph requires that the TRIPS Council should also look at the relationship between the TRIPS Agreement and the *United Nations Convention on Biological Diversity* (1992) (CBD), and at the protection of traditional knowledge and folklore.[3] Negotiated at approximately the same time, the CBD has a number of provisions relating to this concern about "biopiracy" and related issues – including the "access and benefit-sharing" provisions described below.

The CBD also recognises state sovereign rights over biological and genetic resources. It specifies that "utilisation of genetic resources" including research and development (R&D) should be undertaken according to terms of "fair and equitable access and benefit-sharing" procedures (ABS) including prior informed consent (PIC) for access and mutually agreed terms (MAT). It was envisaged in the early 1990s that ABS might be a win-win for conservation and for science, providing a funding mechanism back to communities and governments charged with conservation activities. In reality, it has only sometimes generated significant monetary benefits (see Robinson 2014 for a comprehensive overview). While "state sovereignty" is the focus of the CBD, it also acknowledges – in Article 8(j) – the important role of traditional knowledge. However, the CBD has several specific areas of focus that fall short of the rights sought by Indigenous peoples as reflected in the 2007 *United Nations Declaration on the Rights of Indigenous Peoples* (UNDRIP). For many years, countries had difficulties implementing the ABS provisions of the CBD, and Indigenous peoples and local communities also had limited success asserting rights relating to traditional knowledge under the CBD. This led to the negotiation of the *Nagoya Protocol on Access to Genetic Resources and the Fair and Equitable Sharing of Benefits Arising from the Utilization* to the CBD (Nagoya Protocol) in 2010, which was influenced on the one hand by the International Indigenous Forum on Biodiversity and other like-minded NGOs and groups, and on the other had its language dampened by some countries such as Canada and many of the European delegates (Bavikatte & Robinson 2011).

The Nagoya Protocol was gradually negotiated by 2010 and came into force in 2014. It provides a legal framework for the CBD's objective related to the fair and equitable sharing of benefits arising out of the utilisation of genetic resources. The Nagoya Protocol goes further than the CBD in recognising Indigenous rights over genetic resources and traditional knowledge, albeit using some constrained language. The Nagoya Protocol requires PIC for access to genetic resources from Indigenous peoples, "where they have established rights" over those resources (Article 6). The Nagoya Protocol extends access rules to traditional knowledge associated with genetic resources (Article 7) and asks parties to consider the customary laws and community protocols of communities, with respect to traditional knowledge associated with genetic resources (Article 12). However, while access rules pertaining to traditional knowledge include both PIC and MAT, the latter is ambiguously worded for genetic resources (in Article 6.2), to be required only where Indigenous and local communities have "established" rights. In some countries, established rights might be clearly defined by the constitution or by statutes. In countries like Vanuatu, for example, there is widespread recognition of *Kastom* rules and rights relating to land, which would extend to the resources attached to it. However, in countries like Australia, "established rights" might be more complicated for Indigenous peoples under the legal system. For example, rights relating to genetic resources may be recognised in some jurisdictions (e.g. Commonwealth lands under *Environment Protection and Biodiversity Conservation Act 1999* (Cth) s.8), however states and territories do not have a clear statutory indication of such rights.

The geo-political nature of the negotiations leading to the Nagoya Protocol is one of the many examples where a legal geographical lens is useful. By adopting such a lens, we can see international laws not as static, but as evolving due to different commercial, political and Indigenous imperatives and demands over time. In this case, while the Nagoya Protocol still adopts a "sovereign rights" perspective over nature as "genetic resources", it also indicates a broader suite of rights that may be adopted by Indigenous peoples. While the Nagoya Protocol adopts a language that objectifies plants and animals as "genetic resources", it also usefully acknowledges the customary laws of Indigenous peoples and encourages their recognition by governments, which might include recognition of the sacred or spiritual connections with some plants and animals. Head, Atchison & Gates (2012, p.4) encourage us to unpack deeper ontological questions about plants: "how are plant worlds approached in different human framings?" Their human bio-geographies of wheat, for example, help us to think about the "plantiness" of plants as opposed to animals, which are more commonly studied as part of scholarly ethical engagement with nature. Compared to animals, humans are more inclined to think of plants not as individuals but as different types of collectives or assemblages – forest, food, commodities, habitat, biodiversity and for carbon storage (Head, Atchison & Gates 2012, p.4). Humans also tend to think more about the "humane" treatment of animals, whereas plants are more readily decided to be a commodity, disconnected from humans and worthy of less ethical (or legal) engagement. But in thinking about our human/nature relations, Head (2016) urges us to recognise both human power and its embeddedness within material relations, including particularly our relations with plant species.

In addition, the plants in question are easily moved, their seeds may be carried on the wind or by sea, may be found across many borders, and so the provisions for ABS need to also consider the sources and multi-jurisdictionality of the plant bio-geographies. Our study also considers that in describing our patent landscape methods, we focus on specific native and endemic species, while also comparing these to the wider bio-geographies of plants in tropical Asia, Australasia and the Pacific Islands. The human uses of those plants as "genetic resources and associated traditional knowledge" in the Nagoya Protocol also broaden our study to include consideration of the human knowledge vectors where there have been traditional uses for medicines, foods, remedies, skin and hair care, as fibres or building instruments, among other sacred, customary or traditional uses by Indigenous peoples and local communities. Thus, we use the phrase "legal human bio-geographies" to explain this socio-legal-natural complex. In the next section, we explain "patent landscaping", which is a methodological tool we use to examine research relating to the plants and animals in question, and to identify if traditional knowledge might be used towards monopolised inventions.

The methods: patent landscaping

Patent: "the official legal right to make or sell an invention for a particular number of years".[4]
Landscape: "the shape of the land and related features in a particular area".[5]

If legal geography provides a field of study at the intersection of law/the legal and the space/place/scale/nature contexts of geography, then, by definition, our chosen method of patent landscaping (or patent mapping) seems an ideal fit. We are mapping or scoping the landscape of technical and legal information, used to assert monopoly rights for innovations. Bubela et al. (2013, p.202) state that, as a type of methodology, "a landscape is an analysis of the relationships between multiple sets of indicators measured against temporal, technical or spatial dimensions" and can be applied to patents, scientific articles, clinical trials and other indicators. While they vary greatly in scale and scope, the notion of a patent landscape is increasingly used to map trends in science and technology, as industries become more knowledge-intensive or as the "value-added" component of their production expands (also see Robinson & Raven 2017). Patent landscaping examines the filing for legal rights of monopoly by patent-holders and inventors and provides a "map" or an array of the results. Our patent mapping is particularly interested in identifying trends in patenting "nature" – in other words, who is filing patents on extracts or uses of plants or animals. The results provide a snapshot of legal assemblages or "chronotopes", which highlights claims over "innovations" and the rights allocated to them, which change temporally and spatially (Valverde 2015). We highlight this methodology here as one of many that might be used in legal geography. While it is a quantitative methodology, our approach is to use the patent results to identify specific case studies, in order to do additional qualitative analysis about the use of particular plant species.

We have been undertaking patent landscaping as a research methodology because there is limited quantitative evidence about the scale of the problem of biopiracy, except for a limited number of case studies (see Dutfield 2005; Blakeney 2004; Robinson 2010) and reports from NGOs (see RAFI 1995) and governments (see Peruvian submissions to the WTO 2005a, 2005b & 2007). The results can have significant impacts, supporting Indigenous peoples' claims surrounding Indigenous knowledge and innovations, and influencing policy-making in a number of global forums. With the Nagoya Protocol entering into force in 2014 and being gradually implemented around the world, it is timely to monitor and evaluate the operation of patent systems, laws, policies and regulations as they relate to genetic resources and Indigenous knowledge. Focused and purposive patent analysis can be used to identify where there is commercial use of plant species known to have associated Indigenous knowledge. Tracking patent applications over biological resources provides empirical evidence that can be used to determine how ABS standards could be implemented in countries – in this case, in Vanuatu – and if the system governing intellectual property rights (IPR) is failing to prevent biopiracy under "business as usual" scenarios.

Patent landscape analysis is an established methodology used by researchers examining the utilisation of biological resources in innovations registered and/or protected by a patent (Oldham 2006; Bubela et al. 2013; Oldham, Hall & Forero 2013). The most comprehensive quantitative studies relating to patents and biodiversity have been conducted at the global level by Oldham (2006; Oldham, Hall & Forero 2013). As industries become more knowledge-intensive, and the "value-added" component of their production expands, it is increasingly likely that commercial enterprises will invest in patents and other IPR protections in agri-food, medicines, cosmetics and related fields that use biological resources. As we have argued elsewhere (Robinson & Raven 2017), patent landscaping analysis offers one of the primary methods to quantitatively or qualitatively understand the scope of this expansion. It is a method which allows us to look at the legal human-plant geographies of specific plant species, as they transition from being traditionally used species on a relatively small scale, to being commercially used in many countries and in global supply chains for foods, cosmetics, medicines and other products.

Our specific approach in this case study involved identifying a number of species that are reputedly native and endemic to the South Pacific country of Vanuatu. Forty-four species were identified from a recent publication by Bradacs, Helimann and Weckerle (2011) "Medicinal Plant Use in Vanuatu: A Comparative Ethnobotanical Study of Three Islands" in the *Journal of Ethnopharmacology*. This article was chosen as an indicative (not exhaustive) review of Vanuatu's medicinal plants – one of very few that has been recently published. Our inference from using this article is not that these researchers may have triggered biopiracy, but rather that more generally the disclosure of knowledge of the traditional uses of plants by other authors, researchers or even tourists can then lead to further research and the patenting of different uses of those plants, for a range of products or methods. Of relevance to the discussion in this paper, the researchers acknowledge that they were given permission to conduct the research by a number of departments and ministries in Vanuatu (Bradacs, Helimann & Weckerle 2011, p.447).

The scientific names (and some synonyms of those names) were then searched in a patent search tool: Patent Lens.[6] Patent Lens provides a meta-search tool which can identify keywords in the main national and global patent databases. We specifically searched these species names by "Title, Abstract and Claims" to narrow down cases where the species are specifically germane to the claims of the patent. Species were searched using the simple binomial species name in inverted commas to avoid, as far as possible, spurious "hits" of a particle of the species name. Without access to the high-end-computing and linguistics software packages utilised by Oldham, Hall and Forero (2013), manual searches were conducted. This involved laborious visual inspection of patent search results but allowed greater qualitative analysis and interpretation of the relevance of specific results and "hits" where a patent acknowledged use of Indigenous knowledge (as per Robinson & Raven 2017). By doing a "structured search" in Patent Lens, we limited the possibility of spurious mentions of the species in the patent documents or cases where it is not critical to the patent. While some of these patents may be on processes or methods of producing a product for different uses, some of them are explicitly on extracts derived from the plant biological material itself. The patents vary in terms of the field of use, the part of the plant used, the purpose of intended use, as well as many other variables. Where there are patents identified, we note that this does not explicitly indicate biopiracy, but it does provide an indication of commercial interest in a species. Then further detailed analysis on each patent and the claims therein is required to make further inferences about the researcher's activity. In many cases, it is impossible to identify where a researcher obtained the genetic resources and any associated knowledge. This highlights one of the gaps in the international and national regulatory regimes existing surrounding biological resources and intellectual property.

Because patents are often filed in multiple jurisdictions, they can be described in "families". The figures reported below in terms of "patents" need to be understood as meaning all unique patents identified in all jurisdictions, whereas the term "patent families" gives an indication of the number of discrete inventions filed. From each family there might be filings in multiple jurisdictions, for which the patent documents contain the same information. The patents and patent families might be considered "chronotopal" assemblages and expressions of "innovation" as they bundle socio-technical information with rights and must be assessed in national phases with different jurisdictional results, with changing results over time (Valverde 2015).

In previous research, the patent landscaping results have led us to case studies of particular species, whereby there may be patents "of concern" (see Robinson & Raven 2017). In some cases, there have been opportunities to challenge these using administrative provisions in the patent laws of the countries in question. As is discussed in Robinson, Raven & Hunter (2018), an administrative challenge was previously filed by some of the research team in relation to a patent application on uses of Kakadu plum as a cosmetic cream, by the Mary Kay company. This submission of evidence was successful in generating a negative report for novelty and obviousness from IP Australia and the subsequent withdrawal of the patent application (Robinson, Raven & Hunter 2018). This was possible in Australia where there are two administrative options in the *Patents Act 1990* (Cth):

1. Section 27(1) submissions (standard patents) (and Section 28 submissions apply also for an Innovation Patent) are allowed pre-grant:

   A person may, within the prescribed period after a complete specification filed in relation to an application for a standard patent becomes open to public inspection, notify the Commissioner, in accordance with the regulations, that the person asserts, for reasons stated in the notice, that the invention concerned is not a patentable invention.

2. Section 97(2) (standard patents) (and Section 101G submissions also apply for innovation patents) post-grant re-examination request:

   Subject to this section and the regulations, where a patent has been granted, the Commissioner may, and must if asked to do so by the patentee or any other person, re-examine the complete specification.

These provisions allow a simple challenge process for Indigenous knowledge-holders, those involved in local industries and other concerned stakeholders who want to challenge a patent or application, either pre-grant or post-grant. In other jurisdictions there may not be these same provisions, meaning that concerned third parties must challenge patents through the courts at considerable expense.

Results

Table 5.1 represents a new data set from our patent landscape sample described above, using species keywords from Bradacs, Heilimann & Weckerle (2011) and by searching native and endemic species only. We have sorted the data to present those at the top of the table with the most patent families, and we have excluded species from the table that have a zero-patent count.

We can see from the results that some of the most commonly patented species are those that have a wide bio-geographic distribution in the global tropics and parts of Asia, as well as Vanuatu (which is part of Melanesia) and the Pacific. We searched native species, but this does not preclude these species (e.g. *Cocos nucifera* – the common coconut palm) being found in other countries. We did, however, search for endemic species, and none of those species that appear to be endemic to Vanuatu appear to have been patented yet. This may be due in part to the limited biochemical research undertaken in the Pacific region to date, when compared to the extensive research being undertaken in parts of East and Southeast Asia on native species. As such, it is understandable that *Centella asiatica* – commonly known in Asia as *Gotu kola* – has been widely studied to determine its effectiveness, following its traditional use in these countries as a medicinal herb. Similarly,

TABLE 5.1 Patent Landscape Results from our Sample of Traditional Medicines Identified in Vanuatu

Species name	Local name (approx.)	Patents	Families	Known distribution (approx.)
Centella asiatica (L.) Urb. (n.c)	(Gotu kola)	2017	1305	Global tropics
Achyranthes aspera L.	Nabudschata, nokorin	398	379	Native to Asia, now global tropics
Morinda citrifolia L.	Yalatri, yelawud, nouras	613	316	Tropical Asia, Australia and Pacific and now global tropics
Cocos nucifera L.	Lihol, natora, kau(u)ra, navara, samsam, kokonas	494	233	Global tropics
Vitex trifolia ssp. Trifolia L.	Limadnobnob	64	59	Southeast Asia, Melanesia
Zingiber zerumbet (L.) Roscoe ex Sm.	Liwolängdob, billo	87	52	Southeast Asia, Melanesia, Polynesia and Tropical Australia
Crinum asiaticum L.	Lili, naha, wael, litainbop, mamwenlake	59	41	Tropical Asia and the Pacific
Cassytha filiformis L.	(Love vine)	44	40	Global tropics
Casuarina equisetifolia L.	Na(m)bangura, tamanu blong, solwota, nambakura, nepugre, inmobolhat	46	36	Global coastal tropics and temperate regions
Cordyline fruticosa (L.) A. Chev.	Nitschatimi, neggurrie, nara, nangaria	39	36	Eastern Asia, East Indies and South Pacific Islands to Hawaii, now global tropics
Macaranga tanarius (L.) Muell. Arg.	Navenu, livinu, leviunu tahor, nehivaing, nevingne	54	31	Tropical Asia to Northern Australia and Polynesia
Epipremnum pinnatum (L.) Engl.	Rop blong pik, nekamuro, nekaumro	14	12	Tropical Asia to Northern Australia and Polynesia
Saccharum robustum Brand. & Jesw. Ex	Tschib, sugaken, pitpit	14	7	Indonesia and Melanesia
Trema orientalis (L.) Bl.	Lirpilu	6	5	Global tropics
Barringtonia asiatica (L.) Kurz	Fis posentri, navele blong, solwota, nữt, neteng	5	5	Tropical Asia and the Pacific

(Continued)

TABLE 5.1 (Continued)

Species name	Local name (approx.)	Patents	Families	Known distribution (approx.)
Ficus septica var. cauliflora Burm. F.	Libäla, nälmaha	6	5	Tropical Asia and the Pacific
Syzygium malaccense (L.) Merr. & Perry	Nahabika, (na) kavika, negebige, hawei	5	5	Global tropics
Micromelum minutum (Forst. f) Wight & Arn.	Wael pima, nerrenärre	4	4	Tropical Asia, Australia and the Pacific
Tabernaemontana pandacaqui Lam.	Newawedäl, litschi, inmathethi	1	1	Southeast Asia, Tropical Australia and the Pacific islands
Garuga floribunda Decne	Namalaos, neradou, namalaus	1	1	Tropical Asia to Northern Australia and Polynesia
Terminalia catappa L.	Mataboa, natapoa, natalie	1	1	Tropical Asia to Northern Australia and Polynesia, now global tropics
Ficus wassa Roxb.	Newua	1	1	Melanesia and Indonesia
Drynaria rigidula (Sw.) Bedd.	Nässäi	1	1	Tropical Asia, Australia and Pacific

we have noted previously (Robinson & Raven 2017) that *Morinda citrifolia* – known commonly in the Pacific as *noni* or *nono*, and in Asia as Indian mulberry or cheese fruit – has been heavily researched in Asia for a range of medicinal and "health beverage" purposes particularly. Across parts of East and Southeast Asia and the Pacific islands, there seems to be similar traditional knowledge about the use of *noni*, and this could have occurred through trade in the region or through simultaneous experimentation by traditional medicines practitioners (see e.g. Whistler 1992, for a discussion of the range and variation of uses of *noni* in Polynesia).

The closest patent "hits" to endemic species that we searched are the near-endemic plants *Ficus wassa* and *Saccharum robustum*. We examine these in more detail below to provide examples of the scope of patent claims in relation to the species.

Near-endemic species patents

The *Ficus wassa* patent identified is a World Intellectual Property Organization (WIPO) patent with application number WO 2012/032494 A1 for a "Composition Comprising a Fig Plant Material Extract and Use Thereof in The Treatment of Benign Prostatic Hyperplasia". The applicants appear to be Swiss researchers from Geneva, and they have designated for protection in many countries through the 1970 WIPO *Patent Cooperation Treaty* (PCT). The patent claims "a composition comprising an extract of fig plant material, preferably extract of fig leaves, and the use thereof in a method of preventing and/or treating benign prostatic hyperplasia and/or symptoms of benign prostatic hyperplasia" and lists 221 *Ficus* species which could be used in the invention. This sort of broad-range listing of species names has become a strategy by patent attorneys to widen the potential scope of the patent and to give the inventor flexibility in the way they formulate their composition. It also highlights an absurdity of the patent system – that such a broadening of the inventor's interest may occur when, in all likelihood, the real interest would be in a small number of *Ficus* species. In any case, by listing *Ficus wassa* in such a long list the patentee has diluted any real monopoly claim over use of this species, which may be seen positively by those who utilise it currently for other purposes. In Vanuatu, *Ficus wassa* is reputedly used to stimulate fruiting of a watermelon plant (Bradacs, Helimann & Weckerle 2011, p.443), which is a completely different use. However, Bradacs, Helimann & Weckerle do note that other *Ficus* species more generally may include use of the inner-bark or leaf in a cold maceration taken internally for "postpartum abdominal pain", "taken internally for childhood diseases caused by spirits" and for other treatments. Despite some internal uses, there are no other commonalities between the patent and traditional uses, and so there is little likelihood of this patent impacting upon other users or other uses of *Ficus* species in Vanuatu.

A search for *Saccharum robustum* provides 14 hits from seven patent families for this near-endemic species found only in Melanesia and Indonesia, according to the Global Biodiversity Information Facility (GBIF). Bradacs, Helimann and Weckerle (2011, p.443) note that a variety of *Saccharum robustum* can be used by chewing the stalk as a remedy against Ciguatera, a food-borne illness caused by eating contaminated fish. None of the filed patents resemble this use, and the majority of patents are for very specific methods and genetic manipulations of a broad range of *Saccharum* species. For example, one targets the "Isolation and Targeted Suppression of Lignin Biosynthetic Genes from Sugarcane",[7] and another focuses on "Transgenic Plants for Nitrogen Fixation" in which the species is mentioned as one of many possible species to be used. It seems likely that *Saccharum robustum* – which is an isolated Melanesian species of sugarcane – has been "lumped together" with other *Saccharum* species. It is hard to know if these patents would be problematic for any producers in Vanuatu or Melanesia, but given the narrow specificity of the patents, they seem unlikely to be problematic in terms of limiting any "freedom to operate" in Melanesia.

We next analysed *Macaranga tanarius*. Some other species that are more widely distributed in Southeast Asia, Australia and the Pacific Islands have patents that show similarities with the uses described in Bradacs, Helimann and Weckerle (2011). For example, there are a number of patents filed by the Pokka Corporation in Japan which relate to "Periodontal Bacterial Growth Inhibitor, Oral Hygiene Product, and Food and Drink"[8] as well as other similar and alternative uses, all citing the use of *Macaranga tanarius*. The abstract of this above-titled patent indicates:

> A periodontal bacterial growth inhibitor contains as an active ingredient Macaranga tanarius extract extracted from Macaranga tanarius with an

extraction solvent including at least an organic solvent. Alternatively, the periodontal bacterial growth inhibitor contains as an active ingredient at least one selected from nymphaeol-A, nymphaeol-B, and nymphaeol-C. The periodontal bacterial growth inhibitor is used by being blended to, for example, an oral hygiene product or a food and drink.[9]

This patent is particularly interesting because the description from Bradacs, Helimann and Weckerle (2011, p.442) indicates that a subspecies of *Macaranga tanarius* has been used as a "mouthwash with decoction" for toothache in Aneityum, Vanuatu, as well as for wounds and other treatments. The traditional healers of Aneityum had knowledge of the oral healing qualities of the plant, which is similar to the more technical "periodontal bacterial growth inhibitor" for oral hygiene concept being used in the patent. Given that this species is found in tropical Asia, North-eastern Australia, Melanesia and Polynesia, it is likely that the lead for investigation of the plant for these qualities had come from traditional uses in parts of Asia, possibly from Japan where the researchers are based. The patent document notes that the plant has a wide distribution across this region and also notes that the plants are found in Okinawa – the most southerly and tropical part of Japan. In this case, we cannot draw any inferences except to assume that is quite possible that there might be similar traditional knowledge and uses of the plants across tropical Asia and the Pacific. If we apply "chronotopal" thinking here, these are some of the spatio-temporal dynamics that Valverde (2015) discusses that are hard to understand or predict without using a method like patent landscaping, plus case study. This then helps to inform our thinking about how international law and policy could be improved to better protect Indigenous knowledge and innovations.

This *Macaranga tanarius* example does highlight that it is not always possible to tell from a search of the patent document what the source or origin of the plant or genetic resource samples were for the research. This fact undermines ABS processes discussed earlier, and has been one of the reasons that many countries have been calling for patent reform. In the WIPO Intergovernmental Committee on Intellectual Property and Genetic Resources, Traditional Knowledge and Folklore (IGC), many countries have debated the possibility of using a "disclosure of origin" patent requirement for genetic resources and traditional knowledge, to help ensure that benefit-sharing occurs with the providers of the plant or genetic resource (see Robinson & Chiarolla 2017; Bagley 2017). However, these negotiations have been stalled for many years, with advanced economies unwilling to amend the global patent laws for fear of delays to patent filings and protections for their researchers and companies (Robinson & Chiarolla 2017).

Kava patents

It should be noted that this is not an exhaustive study of all medicinal species from Vanuatu. In another paper we chose to focus solely on the use of *Kava (Piper methysticum*) which has customary uses as a relaxant and has been used for a range of ceremonial purposes. By searching the species name, we identified 200 patents (including current applications) from 132 patent families. Kava is thought to be endemic to Melanesia and parts of Polynesia, but Vanuatu particularly is seen as a centre of diversity, with approximately 44 local "noble" varieties. There have been both economic and cultural concerns about the appropriation of *Kava* for decades, which have been raised by many stakeholders during our visits to countries in the Pacific Islands region. When we analysed the patents, many of the patented uses were for very different purposes and may be related to new plant cultivars, or new uses of *Kava*. For example, there are some patents that apply *Kava* to cosmetic and skin-care applications. Other patents are more concerning, as they relate more closely to the traditional uses or the drink as a relaxant (see Robinson et al. 2019 and Figure 5.1).

Through patent laws we see the transformation of *Kava* into the subject of a range of commodified inventions. Alternatively, traditional Vanuatu *Kava* origin myths and stories often speak to wider cultural notions in *Kastom* about proper relations between men and women, leaders and followers and between the living and the dead, in which Lindstrom (1997) refers to *Kava* as the "germinant corpse". *Kastom* sees *Kava* as embroiled in the linkage between death and life, fertility and growth (Turner 2012) and was traditionally used to enhance communication with ancestral spirits (Taylor 2010; Robinson et al. 2019). Thus, as the plants travel and are traded in different places, they take on different meanings and uses. But this transformation can also potentially be culturally offensive to the original users and custodians, according to their customary laws and traditions (*Kastom*). What seems clear is that there needs to be a broadening

FIGURE 5.1 Mature Kava Plants, Espiritu Santo, Vanuatu. Source: Daniel Robinson.

of legal imagination, such that state laws can accept or re-envision a place for Indigenous customary laws and worldviews (such as Indigenous-led "rights-to-nature" approaches).

Conclusions: between the incongruous legal human-plant geographies

While there are no concerning patent hits for endemic species to Vanuatu in this study, it is not an exhaustive list of species – in fact it is probably just a tiny sample. Indeed, other species such as *Kava* are being researched and patented, and species such as *noni* (*Morinda citrifolia* – found across the tropics) are being bought in Vanuatu by foreign companies, and potentially being used overseas for research. From an ABS perspective, both the transboundary nature of the plants and the difficulty in traceability pose significant challenges for enforcement. While the ABS process has a Clearing House for the checking of certificates of compliance from permits issued in Nagoya Protocol-compliant countries, this is still a globally evolving lawscape, with many countries still yet to implement the Nagoya Protocol thus leaving potential gaps in implementation into the future (e.g. the US is unlikely to ratify anytime in the near future – see Keating 2017, for a US perspective).

Our patent landscape methodology highlights the legal assemblages produced by the patent system and the "chronotopic" nature of patents (as per Valverde 2015). It draws attention to a socio-temporal dynamic in which injustices continue to occur (biopiracy) due to the privileging of technical scientific knowledge over Indigenous knowledges and innovations, as well as the divergent legal systems. These two sets of legal human-plant-geographies (the laws and connections allowing patentability of life forms, and the locally embedded customary law significance of these same life forms) are in many ways incongruous. This has seen many efforts at reform such as the "disclosure of origin" patent requirement noted above in the WIPO IGC. It has also led to the development and evolution of the ABS system, which at least creates a pathway for recognition of customary law relating to plants. But perhaps what is also needed is what Braverman (2016) describes as new "lively legalities" whereby "rights-to-nature" approaches are adopted, as we have seen in the recognition of legal personhood for the Whanganui River in Aotearoa New Zealand (Charpleix 2018). These sorts of approaches, which would cede political space to Indigenous worldviews of plants and nature, are certainly worth exploring to break down the conflicts that arise through the dominant nature–culture binaries.

Acknowledgements, funders and partners

The research with Vanuatu communities is part of the Australian Research Council (ARC) Discovery Project Indigenous Knowledge Futures (DP DP180100507). It is also part of the activities in the Pacific conducted by the multi-donor funded ABS Capacity Development Initiative (www.abs-initi-ative.info) which is managed by the Deutsche Gesellschaft für Internationale Zusammenarbeit (GIZ) GmbH. The research is funded and supported by: the ARC; University of New South Wales Australia; Macquarie University, Australia; and through the ABS Initiative by the 11th European Development Fund (through the ACP-EU Partnership Agreement). Some of our project partners are: Secretariat of the Pacific Regional Environment Programme (SPREP); United Nations Environment Programme; Department of Environmental Protection and Conservation, Vanuatu; Vanuatu Cultural Centre.

Notes

1 Notably, "biogeography" is its own sub-discipline studying the geographical distribution of biological diversity in physical geography, so here we have used "human-bio-geography" in a hyphenated form to emphasis the relationalities between humans and the "bio".
2 WTO Legal Texts. TRIPS Agreement, viewed 7 January 2019, <www.wto.org/english/docs_e/legal_e/27-trips_04c_e.htm#5>.
3 WTO TRIPS: Reviews, Article 27.3(b) and Related Issues, viewed 7 January 2019, <www.wto.org/english/tratop_e/trips_e/art27_3b_background_e.htm>.
4 Cambridge Online Dictionary, 'Patent', viewed 7 January 2019, <https://dictionary.cambri dge.org/dictionary/english/patent>.
5 Cambridge Online Dictionary, 'Landscape', viewed 7 January 2019, <https://diction ary.cambridge.org/dictionary/english/landscape>.
6 Patent Lens: <www.lens.org/lens/new-search>.
7 AU 2010/256356 A1, 'Isolation and Targeted Suppression of Lignin Biosynthetic Genes from Sugarcane'
published 2 February 2012, viewed 14 January 2019, <www.lens.org/lens/patent/072-533-907-528-190>.
8 US 2013/0115175 A1 'Periodontal Bacterial Growth Inhibitor, Oral Hygiene Product, and Food and Drink' Published 9 May 2013, viewed 14 January 2019, <www.lens.org/lens/patent/189-581-175-677-500>.
9 US 2013/0115175 A1 'Periodontal Bacterial Growth Inhibitor, Oral Hygiene Product, and Food and Drink' Published 9 May 2013, viewed 14 January 2019, <www.lens.org/lens/patent/189-581-175-677-500>.

References

Bagley, M. 2017, 'Of disclosure "Straws" and IP System "Camels": Patents, certainty, and the disclosure of origin requirement,' in *Protecting Traditional Knowledge: The Future of the WIPO Intergovernmental Committee on Intellectual Property & Genetic Resources, Traditional Knowledge and Folklore*, eds D.F. Robinson, P. Roffe & A. Abdel-Latif, Routledge, Oxon, pp.85–107.

Bartel, R., & Graham, N. 2016, 'Property and place attachment: A legal geographical analysis of biodiversity law reform in New South Wales', *Geographical Research*, vol. 54, no. 2, pp.267–284.

Bavikatte, K., & Robinson D. F. 2011, 'Towards a people's history of the law: Biocultural jurisprudence and the Nagoya Protocol on access and benefit sharing', *Law, Environment & Development Journal*, vol. 7, no. 1, pp.35–55.

Bennett, L., & Layard, A. 2015, 'Legal geography: Becoming spatial detectives', *Geography Compass*, vol. 9, no. 7, pp.406–422.

Blakeney, M. 2004, 'Bioprospecting and biopiracy', in *Intellectual Property and Biological Resources*, ed. B. Ong, Marshall-Cavendish, Singapore, pp.393–424.

Bradacs, G., Heilmann, J., & Weckerle, C. S. 2011, 'Medicinal plant use in Vanuatu: A comparative ethnobotanical study of three islands', *Journal of Ethnopharmacology*, vol. 137, no. 1, pp.434–448.

Braverman, I. 2016, 'Introduction: Lively legalities', in *Animals, Biopolitics, Law. Lively Legalities*, ed. I. Braverman, Routledge, Oxon, pp.3–18.

Bubela, T., Gold, E. R., Graff, G. D., Cahoy, D. R., Nicol, D., & Castle., D. 2013, 'Patent landscaping for life sciences innovation: Toward consistent and transparent practices', *Nature Biotechnology*, vol. 31, no. 3, pp.202–206.

Charpleix, L. 2018, 'The Whanganui River as Te Awa Tupua: Place-based law in a legally pluralistic society', *The Geographical Journal*, vol. 184, no. 1, pp.19–30.

Delaney, D. 2015a, 'Legal geography I: Constitutivities, complexities, and contingencies', *Progress in Human Geography*, vol. 39, no. 1, pp.96–102.

Delaney, D. 2015b, 'Legal geography II: Discerning injustice', *Progress in Human Geography*, vol. 40, no. 1, pp.267–274.

Doha Declaration on the TRIPS Agreement and Public Health, 2001.

Dominic, K. 2017, 'The WIPO IGC: A US perspective', in *Protecting Traditional Knowledge: The WIPO IGC*, eds D.F. Robinson, P. Roffe, P., A. Abdel-Latif, Routledge, Oxon, pp.265–276.

Dutfield, G. 2005, *Intellectual Property, Biogenetic Resources and Traditional Knowledge*, Earthscan, London.

Environment Protection and Biodiversity Conservation Act 1999 (Cth).

Graham, N. 2011, *Lawscape: Property, Environment, Law*, Routledge, London.

Head, L., & Atchison, J. 2009, 'Cultural ecology: emerging human-plant geographies', *Progress in Human Geography*, vol. 33, no. 2, pp.236–245.

Head, L., Atchison, J., & Gates, A. 2012, *Ingrained: A Human Bio-geography of Wheat*, Routledge, Oxon.

Head, L. 2016, *Hope and Grief in the Anthropocene: Re-conceptualising human–nature Relations*, Routledge, Oxon.

Lindstrom, L. 1997 'Anthropology: The cultural significance and social uses of Kava', in *Kava, The Pacific Elixir: The Definitive Guide to its Ethnobotany, History and Chemistry*, eds V. Lebot, M. Merlin & L. Lindstrom, Yale University Press/ Healing Arts Press, Vermont, pp.119–174.

Marrakesh Agreement Establishing the World Trade Organization, 1994, Annex 1C Agreement on Trade-Related Aspects of Intellectual Property Rights (TRIPS).

Nagoya Protocol on Access to Genetic Resources and the Fair and Equitable Sharing of Benefits Arising from the Utilization, 2010.

Oldham, P. 2006, 'Biodiversity and the patent system: Towards international indicators', *Global Status and Trends in Intellectual Property Claims*, UNEP/CBD/WG-ABS/5/INF/6, paper no. 3, pp.1–88.

Oldham, P., Hall S., & Forero, O. 2013, 'Biological diversity in the patent system', *PLoS ONE*, vol. 8, no. 11, e78737.

Patents Act 1990 (Cth).

Peru, 2005a, *Analysis of Potential Cases of Biopiracy*, WTO Document: IP/C/W/458, 7 November.

Peru, 2005b, *The Patent System and the Fight Against Biopiracy – The Peruvian Experience*, WTO Document: WIPO/GRTKF/IC/8/12, 30 May.

Peru, 2007, *Combating Biopiracy – The Peruvian Experience: Communication from Peru*, WTO Document: IP/C/W/493, 19 September.

Rural Advancement Foundation International, 1995, 'Biopiracy Update: A Global Pandemic', *RAFI Communique*, September–October.

Robinson, D. F. 2010, *Confronting Biopiracy: Challenges, Cases and International Debates*, Earthscan, London.

Robinson, D. F. 2014, *Biodiversity, Access and Benefit-sharing: Global Case Studies*, Routledge, Oxon.

Robinson, D. F., & Forsyth, M. 2016, 'People, plants, place, and rules: The Nagoya Protocol in pacific island countries', *Geographical Research*, vol. 54, no. 3, pp.324–335.

Robinson, D. F., & Chiarolla, C. 2017, 'The role of databases, contracts and codes of conduct', in *Protecting Traditional Knowledge: The WIPO Intergovernmental Committee on Intellectual Property and Genetic Resources, Traditional Knowledge and Folklore*, eds D. F. Robinson, P. Roffe & A. Abdel-Latif, Routledge Research in International Environmental Law, Oxon, pp.108–121.

Robinson, D. F., & Raven, M. 2017, 'Identifying and preventing biopiracy in Australia: Patent landscapes and legal geographies for plants with Indigenous Australian uses', *Australian Geographer*, vol. 48, no. 2, pp.311–331.

Robinson, D. F., Raven, M., & Hunter, J. 2018, 'The limits of ABS laws: Why gumbi gumbi and other bush foods and medicines need specific indigenous knowledge protections', in *Biodiversity, Genetic Resources and Intellectual Property: Developments in Access and Benefit Sharing*, eds C. Lawson & K. Adhikari, Edward Elgar, London, pp.185–207.

Robinson, D. F., & Graham, N. 2018, 'Legal pluralisms, justice and spatial conflicts: New directions in legal geography', *Geographical Journal*, vol. 184, no. 1, pp.3–7.

Robinson, D. F., Raven, M., Tari, T., Hickey, F. & Kalfatak, D. 2019, 'Kava, Kastom and indigenous knowledge: Next steps under the Nagoya protocol', *Journal of South Pacific Law*, accepted and forthcoming.

Taylor, J. P. 2010 'Janus and the siren's call: Kava and the articulation of gender and modernity in Vanuatu', *Journal of the Royal Anthropological Institute*, vol. 16, no. 2, pp.279–296.

Turner, J. W. 2012, 'Listening to the ancestors: Kava and the Lapita peoples', *Ethnology*, vol. 51, no. 1/2 (Winter/Spring 2012), pp.31–53.

United Nations Convention on Biological Diversity, 1992.

Valverde, M. 2015, *Chronotopes of Law: Jurisdiction, Scale and Governance*, Routledge, Oxford.

Whistler, A. 1992, *Polynesian Herbal Medicine*, National Tropical Botanical Garden, Hawaii.

6

CONSULTING THE CONSULTATORS

A Kaupapa Māori-informed approach
to uncovering Indigenous jurisdiction
and shifting the research gaze

Maria Bargh and Estair van Wagner

Introduction

This chapter examines the methodological approach applied in phase one of our
project examining Māori consultation in New Zealand's minerals and mining
regime. The project was borne out of a friendship between the authors, and
our bicultural Kaupapa Māori-informed research methodology is shaped by that
relationship. Here we explore how legal geography research can be informed and
enriched by Kaupapa Māori approaches to research related to Indigenous peoples
and their lands and resources. Our project uniquely demonstrates how bringing legal geography and Kaupapa Māori approaches into conversation can shift
necessary attention to exposing the day-to-day administration of settler-state
law while also upholding the role of Indigenous legal orders and jurisdiction in
environmental and natural resource governance.

Minerals and mining in Aotearoa New Zealand are regulated primarily
through national law, specifically the *Crown Minerals Act 1991* (NZ) (*Crown
Minerals Act*). However, as with other areas of law, minerals and mining directly
engage Te Tiriti o Waitangi (the Treaty of Waitangi), signed by the British and
Māori in 1840, and therefore Māori legal orders. The project described below
takes Te Tiriti and Māori legal orders as the starting point for Crown–Māori
relations about minerals and their extraction and use. It explores the structure
of settler-state "consultation" to uncover how law and legal processes are used
to shape and constrain Māori people–place relations. We outline what we have
learned by bringing relational and place-based approaches from Kaupapa Māori
and legal geography together. Reflecting on our bicultural examination of
Indigenous consultation, we offer insights for researchers who aim to expose
the operation of settler law on Indigenous lands while supporting Indigenous
assertions of jurisdiction and self-determination. We specifically consider how

researchers can use these methodological approaches to critically examine consultation processes without adding to the burden Indigenous communities face to engage in such processes.

Who we are: research relations

The research project described in this chapter was born out of a friendship between the authors – two academics then at Victoria University of Wellington with common interests and overlapping concerns about Indigenous environmental governance and self-determination. The authors are Māori (Te Arawa (Ngāti Kea/Ngāti Tuarā), Ngāti Awa) and non-Māori (Canadian, with Dutch, Scottish, Irish and Spanish ancestry), one from Aotearoa New Zealand, the other from Canada but living in New Zealand at the time of the research. We are also from different academic disciplines – one an Indigenous studies scholar trained in political science and international relations, the other a legal scholar with a background in environmental studies. These different positionalities and experiences mean we bring different skills, methodological approaches, knowledge and understanding to the project. At the same time, we are both engaged with critical theory and motivated by contributing to projects that further Indigenous environmental justice and self-determination, though in different ways as Māori and settler persons. Our bicultural methodological approach draws on both our academic and personal backgrounds to contribute to supporting Māori assertions of jurisdiction and people–place relations. It is also grounded in our friendship and mutual trust about doing this work in a good way. In this context, it remains an on-going process of relating through research "to shape a socially and culturally just present and future" (Amundsen 2018, p.149).

The 2013 Epithermal Gold Block Offer: a case study

In 2013, the New Zealand government opened up a "block offer", a national minerals tendering process, for epithermal gold over a large area the Central North Island of Aotearoa New Zealand.[1] The proposed area encompassed the *rohe*, or traditional geographical territory, of several *iwi* (tribe) (Te Arawa, Ngāti Awa, Ngati Rangi, Ngāti Ranginui, Tuhoe, Ngāti Tuwharetoa, Ngāti Maniapoto, Tainui). This includes the rohe of one of the co-authors (Bargh). It also included areas governed by several local authorities, including a large area of the Bay of Plenty Regional Council's jurisdiction. Approximately 31.5% of the Bay of Plenty area is land held by Māori owners, either in Māori Land title or general title.[2] While the Crown selected the Block Offer area "to take account of geology and prospectivity" in a "prime region for epithermal gold and silver deposits" (Darby 2013, p.6), the people–place relations of the proposed area are much more complex.

Given the strong and overlapping Māori political, economic and spiritual relationships with the area outlined above, the region has been the subject of a number of legal and political inquiries and agreements. As one local council noted in their submissions on the Block Offer, "the Treaty settlement landscape is a significant feature of our region" (Bay of Plenty Regional Council 2013). The Waitangi Tribunal has held 16 historical inquiries into breaches of the Treaty in the area, including the five-volume *He Maunga Rongo: Report on Central North Island Claims*, in which they concluded the Crown had failed to protect the *tino rangatiratanga* (self-determination/chiefly authority) of Central North Island iwi and hapū. This failure, the tribunal concluded, included the Crown having actively undermined Māori legal and governance systems and facilitating the loss of Māori control over natural resources (Waitangi Tribunal 2008, p.1674).[3] In particular, the tribunal noted the loss of control over geothermal resources, which have been subject to "the exercise of [Māori] authority ... that has remained unbroken for hundreds of years" (Waitangi Tribunal 2008, p.1543). The area has also been the subject of eight major Treaty Settlements, with approximately another ten being negotiated.[4] Existing settlements include landmark Crown–Māori agreements, including the 2008 Central North Island Iwi Collective Crown Forest Settlement and the Te Arawa Lakes Settlement in 2006. The latter included the first iwi-Council natural resource co-governance model in New Zealand.[5]

In addition to the complex social, cultural and legal context, the Central North Island is geologically unique, with significant geothermal resources and features caused by the region's volcanic nature. Indeed the presence of shallow epithermal gold deposits is directly linked to the unique geothermal volcanic landscape (Darby 2013; Ralph 2017). As noted by the Tribunal, the Central North Island's unique geology is intimately linked to Māori cultural and spiritual relations, and therefore local *tikanga* (Māori law). For Central North Island *iwi* and *hapū*, these geothermal features are *taonga* (treasured resources) (Waitangi Tribunal 2008, p.1543). As explained in the following section introducing key elements of the Māori worldview, such taonga are at the heart of kinship relationships and responsibilities set out in *tikanga*, and therefore inextricably linked to Māori jurisdiction and governance of land and resources (Waitangi Tribunal 2008, p.1543).

The Block Offer consultation process

Notification of iwi and hapū took place in March 2013, followed by approximately two months of consultation. Several in-person meetings were held between the Te Arawa Coalition[6] and New Zealand Petroleum and Minerals, the Minister of Energy and Resources and Minister of Māori Development. In addition, Te Arawa hapū held separate *hui* (meeting) in their rohe. At the Te Arawa hui-a-iwi a resolution was passed rejecting "epithermal gold mining in the Te Arawa rohe" (Te Arawa Hui-a Iwi 2013).

During the consultation period 15 iwi and hapū made written submissions. Seven local councils also made submissions. As we have discussed elsewhere, the

15 Māori submitters overwhelmingly rejected the proposed offer area (Bargh & van Wagner 2019). All submitters, including local councils, expressed significant concerns about the Block Offer area proposed by the Crown through both written submissions and at face-to-face meetings (Darby 2013, p.2). As permitted by the *Crown Minerals Act* process detailed below, many iwi and hapū requested the Crown exclude either their entire *rohe* or specific significant areas from the Block Offer.[7] Notably local councils also requested exclusions despite not having a statutory right to do so. The combined requested exclusions cover the total area of the Block Offer (Bargh & van Wagner 2019).[8]

Results of the consultation process

New Zealand Minerals and Petroleum completed the required Results of Consultation Report to the Minister in 2013. On the recommendation of the report, the Block Offer proceeded (Darby 2013, p.2). No iwi and hapū exclusion requests were granted from the final offer area (Darby 2013, pp.19–20). The only exclusions were small areas requested by local councils, both of which where considered as having low or no prospectivity by the report. Silver City Ltd was granted a five-year permit in 2014 for a 33-kmsq portion of the final Block Offer area (Bridges 2014). Silver City relinquished their permit in 2016, and no exploration has taken place to date ("Going for Kawerau Gold" 2015).

Kaupapa Māori-informed legal geography: law, space and Indigenous jurisdiction

Our project uses this case study to examine the *Crown Minerals Act* consultation process in order to understand whether and how Māori can meaningfully influence decisions about mineral exploration and mining. The proliferation of Indigenous consultation and partnership opportunities with respect to mining in a number of jurisdictions has not generally resulted in meaningful opportunities to consent to or even influence decision-making. According to Huntington et al., for a consultation process to be meaningful, "the time and effort that local people put into participation must be met with clear opportunities to influence outcomes, and with a transparent path to final decisions so that the role of local input can be understood and appreciated" (Huntington et al. 2012, p.46). As they conclude after examining Arctic oil and gas consultation processes in Canada, the United States and Greenland, "[t]he line between meaningful and burdensome consultations depends in part on whether those consultations have positive outcomes from the local point of view" (Huntington et al. 2012, p.46). In the context of mining in Aotearoa New Zealand, Ruwhiu and Carter et al. argue consultation with Indigenous communities must be "closely connected to and driven by the appropriate worldview of each respective Indigenous community" (Ruwhiu & Carter 2016, p.651). The 2013 Epithermal Gold Block Offer was selected as a case study because it exposed a mismatch between the levels of Māori engagement in the process and the level of influence on the outcome. For us this indicated the need for a better understanding of the nature of the consultation process under the Act and the way the consultation process impacts final decisions about mineral exploration. Our findings from this case study demonstrate "that while Māori do contribute in terms of time and resources and are active participants in the process of Block Offers, their views are ultimately routinely and systematically excluded from key decisions" (Bargh & van Wagner 2019, p.138).

In the following sections we briefly examine Kaupapa Māori methodologies and legal geography separately before detailing how our own approach brings them together to uncover the people–place relations of minerals and mining in Aotearoa New Zealand. Through the development of this hybrid methodological approach we explicitly aim to use our scholarship to highlight and prioritise Māori aspirations and assertions of jurisdiction and *tino rangatiratanga* (self-determination/chiefly authority).

Kaupapa Maori research

Kaupapa Māori methodologies emerged as part of a broader cultural and political shift in Crown–Māori relations in Aotearoa New Zealand in the late 1970s and 1980s. Māori scholars were responding to generations of activism and resistance from Māori since colonisation began in the 19th century. As Māori scholar Mason Durie notes, Kaupapa Māori research is situated within this context of "Māori rejuvenation" and reconsideration of the Treaty relationship based on the understanding that "achieving best outcomes for Māori across a range of endeavours needs to take account of a Māori worldview" (Durie 2017, p.2). Linda Tuhiwai Smith is a Māori education scholar who has written extensively about Indigenous research methodologies. She notes Kaupapa Māori research is based on "the assumption that research that involves Māori peoples, as individuals or communities, should set out to make a positive difference for the researched" and acknowledge that Māori peoples have not historically benefited from research about them (Smith 2017, p.20). For Smith, it is part of Māori reclaiming space and reimaging the world: "Imagining a different world, or reimagining the world, is a way into theorizing the reasons why the world we experience is unjust, and posing alternatives to such a world from within our own world views" (Smith 2017, p.204). In addition to being situated within this political context, Indigenous methodologies are also rooted in place-based Indigenous worldviews, grounded in specific relationships with the lands and resources of a particular place (Kovach 2010).

Kaupapa Māori research is applied and developed in a range of contexts and locations and has evolved in diverse ways as demonstrated through the work of different scholars. The collective scholarship of Linda Tuhiwai Smith, Fiona Cram and Graham Smith, who have written extensively about Kaupapa Māori methodologies, identify core principles of Kaupapa Māori research: (a) the importance of relationships with Māori; (b) grounding the research in Māori worldviews and principles; (c) ensuring respectful and culturally appropriate behaviours; and (d) acknowledging and reflecting on the political nature of research and both its constraints and its transformative potential (Smith 1999; Cram 1997; Smith 2009; Smith 2012; Smith 2007). In the context of our research, we deliberately chose to focus our fieldwork on interviewing a small group of key government officials rather than Māori community members as Māori perspectives on the minerals process were readily available through submissions to the *Crown Minerals Act* consultation process. Therefore, our methodological emphasis was on principles (b) and (d) to ensure that our work was grounded in Māori worldviews and would directly contribute to Māori political aspirations and assertions of jurisdiction. Our project also engages concerns about the place of non-Māori in Kaupapa Māori research.

Māori worldviews and concepts

A central element of Kaupapa Māori research is its grounding within Māori worldviews. As Russell Bishop points out, it situates research outside the dominant paradigm and therefore can generate alternative solutions and aspirations (Bishop 1994; Smith 2017). *Tikanga* (Māori law and protocols) in relation to people–place relations and natural resource governance are of particular importance (Jones 2016). Māori are the *tangata whenua* (people of the land) of Aotearoa New Zealand, and their systems of law are the "first law" of the land – developed with and in response to the land itself (Williams 2013, p.2). We note there are regional variations in terms of *tikanga*, as appropriate to particular places and communities. However, shared concepts and principles inform this diversity on the ground (Tomas 2011). While a detailed review of Māori worldviews and *tikanga* is beyond the scope of this chapter, we describe several key concepts that underpin the perspectives and claims expressed by Māori submitters in the case study we examined: *whanaungatanga* (relationships); *manaakitanga* and *kaitiakitanga* (generosity and guardianship); *mana* (authority); *tapu* (spiritual quality); and *utu* (balance or reciprocity) (Durie 1999; Mead 2003).

Maintaining balanced relations between humans, and between humans and other physical and metaphysical entities, is central to *tikanga* (Jones 2016, pp.65–86; Mead 2003, pp.32–3). Relationships are the source of rights and obligations of kinship, "the glue that held, and still holds, the system together", and *whanaungatanga* can be understood as "the fundamental law of the maintenance of properly tended relationships" (Williams 2013, pp.4–5). Caring, kind and generous relations, as well as guardianship obligations, encourage and reinforce sustainability within both human and environmental relations (Jones 2016, pp.71–2). Māori legal scholar Carwyn Jones notes how *whānau* and *hapu* have specific *kaitiaki* or guardianship responsibilities that must be upheld to avoid harm and the loss of *mana* (authority) (Jones 2016, p.73; Tomas 2011, p.228). Another Māori legal scholar, Nin Tomas, explains how humility and acknowledgment of dependence on *Paptuanuku* (Earth mother) guide the exercise of rights to use natural resources and the fulfilment of corresponding obligations (Tomas 2011, p.226). *Mana* is understood as the source of the rights and obligations of both spiritual and democratic leadership in Māori society, including the authority that particular people and groups have with respect to resources. This *mana whenua* is "sourced in the land itself" and must be maintained through connection to the land and upholding obligations (Jones 2016, p.69). The concept of *tapu* or the spiritual element of all things is also central to Māori society providing for the regulation of the sacred or spiritual, as is *noa*, the ability to remove restrictions and restore "a normal, everyday state" (Jones 2016, pp.73–4). Finally, *utu* the principle of reciprocity or restoring balance requires an action (*utu*) to restore balance if rules are breached, or a loss or change occurs. The appropriate action, Jones notes, will differ depending on the context and the relationships involved (Jones 2016, pp.75–6).

Bicultural Kaupapa Māori research

The grounding of Kaupapa Māori research in Māori identity and worldviews has raised questions about whether non-Māori involvement is appropriate (Smith 2017, p.12), including the collaboration between us for this research. As described above, our research partnership grew out of a friendship and common concern about the need for critical scholarly engagement with the mining regime in Aotearoa New Zealand. This grounding in relationship and shared political goals is consistent with Kaupapa Māori approaches and Indigenous methodologies more broadly, which emphasise relationships, trust and reciprocity in research (Kovach 2010; Smith 1999). Nonetheless, even in the context of personal relationships bicultural research requires relational accountability and "reflection on self-location, purpose, and sources of knowledge" (Latulippe 2015, p.6).

Smith concludes that while "being Māori, identifying as Māori and as a Māori researcher is a critical element of Kaupapa Māori research", *Pākehā* [white New Zealander settlers] researchers can be appropriately involved in Kaupapa Māori research. Of particular relevance to this project is Bishop's model of Kaupapa Māori research, which he frames through the Treaty of Waitangi. In this model *Pākehā*, as Treaty partners, can participate in Kaupapa Māori research as "useful allies and colleagues in research" (Smith 2011, p.227). Maui Hudson and Khyla Russell, advocate for Treaty-informed engagement between Māori and non-Māori researchers to "share skills and technologies that will address research questions generated by Māori" (Hudson & Russell 2009, p.64). They reinterpret the Treaty relationship from a Kaupapa Māori perspective to emphasise reciprocity, equity and benefit sharing in research relationships. In the Canadian context, Latulippe has also argued for Treaty relations as a basis for bicultural research as it "encourages relational accountability, acknowledges difference, and bridges interpretive communities" (Latulippe 2015, p.7). By explicitly centring Treaty relations in settler-colonial contexts, researchers can foreground the mutuality of ongoing obligations between settlers and Indigenous people – both as researchers and as members of overlapping political communities. Further, as discussed below, a Treaty-based approach encourages maintenance through ongoing reflexivity about the nature of bicultural research as well as care for relationships as they evolve and change over time.

We centre the Treaty as a foundational legal document shaping natural resource governance in Aotearoa New Zealand, and consultation with Māori in particular. As outlined below, a Treaty-informed bicultural approach to research is particularly appropriate to our goal of examining settler-state legal processes while also recognising and supporting assertions of Māori jurisdiction. As Latulippe concludes, "a treaty perspective affords the space from which to decolonize research practice, make connections, and develop mutually beneficial outcomes through the respectful and reciprocal interplay of distinct knowledge systems" (Latulippe 2015, p.11). In the final section of the paper both authors reflect on what this means for this project and us as researchers.

Legal geography

Irus Braverman et al. describe the focus of legal geography as "interconnections between law and spatiality, and especially their reciprocal construction" (Braverman et al. 2014, p.1). Rooted in the work of critical scholars who pointed to the socially constructed nature of space and the mutually constitutive relationship between the spatial, the social and the temporal, legal geography explores the ways in which space and law are produced through social relations (Martin et al. 2010, p.234). What is critical for legal geography is that law and space are not simply brought together; they are understood as "enmeshed" and mutually constitutive (Delaney 2015; Martin et al. 2010, p.117). Law not only produces and is produced by the social, it is spatialised and as such can reinforce power relations through processes of exclusion, coding or locating (Blomley 2005, p.283). This grounding in critical theory and attention to power compliments the explicitly political orientation of Kaupapa Māori approaches as described above. Indeed, making visible the ways in which state law is spatialised can help to denaturalise and unsettle state claims to exclusive jurisdiction. This unsettling can create space for pluralism and Indigenous legal orders. Recent engagements by legal geographers with legal pluralism point to its potential as a framework for the development of "new strategies to challenge anthropocentric, Euro-American, neoliberal, binary and partial approaches to people and place within dominant legal regimes" (Robinson & Graham 2018, p.6). In the context of Aotearoa New Zealand, Liz Charpleix points to the enactment of Te Awa Tupua, recognising the Whanganui River as a legal person, as "an interstitial legal structure" arising "between, and separate from, the dominant legal system and tikanga, without either being absorbed into it" (Charpleix 2018, p.26).

In our work we have particularly built on the attention to "place" and relationality in legal geography. Deborah Martin summarises place as, "a setting for and situated in the operation of social and economic processes, and it also provides a 'grounding' for everyday life and experience" (Martin 2003, p.732). This conception of space builds on the work of Doreen Massey who understands the politics of place as relational, precisely the challenge of continuing to negotiate how we will live together. Places, she argues, "implicate us, perforce, in the lives of human others, and in our relations with nonhumans they ask how we shall respond to our temporary meeting-up with these particular rocks and stones and trees" (Massey 2005, p.141, 2004). Disputes about natural resource development in settler states are disputes about the negotiation of place and how we will live together, both as human and more-than-human entities and as entanglements between. Yet legal scholarship has rarely attended to the complexity of people–place relations (Graham 2011). Indeed the "dephysicalised" conception of property relations in dominant settler legal frameworks intentionally severs non-instrumental relations between people and places to produce land and resources as alienable commodities (Graham 2011). By bringing space and place back into the frame, legal geographers expose property as spatially contingent. As Sarah Keenan concludes, property can thus be reshaped to produce "alternative spaces of belonging" that can unsettle and reshape the world around them (Keenan 2010, pp.93, 96). Thus assertions of Indigenous jurisdiction, such as those identified in our case study, cannot be understood purely as a response to settler-state law, in that they may also unsettle and reshape. They are part of the reimagining of relations with minerals and of the places in which they are situated as part of broader geological and ecological systems (Tomas 2011; Watson 2007; Watson 2002).

Martin et al. argue that legal geographers have focused on the framing of geographical concepts in the outcomes of legal process to the neglect of the process through which individuals and groups make legal claims and use the law. This framing, they argue, "implies the raising and contesting of claims, but it does not trace this process through in detail so as to understand the shaping influence of various actors" (Martin et al. 2010, p.179). Land use disputes, they argue, expose the "discontinuity between place identity and legal regulation of place" (Martin et al. 2010, p.182). Settler legal frameworks do not account for the range of concerns and attachments expressed by parties to such disputes. Thus while parties seek ways to use these laws and legal processes to advance their claims and express their concerns these interests do not necessarily translate into existing legal narratives (Martin et al. 2010). In the context of Indigenous people–place relations this has particular significance as relational rights and obligations under place-based Indigenous legal orders are subjugated to placeless colonial legal frameworks within which their meaning is obscured and even denied. In this way opportunities for formal consultation in settler-state legal processes not only fail to provide for meaningful participation in environmental decision-making, they actively undermine Indigenous jurisdiction and the place-based relations on which Indigenous legal orders are founded (Bargh & van Wagner 2019).

For legal geographer Nicholas Blomley, this work to organise disputes and the consequences through law operates by bracketing specific types of relations and placing them "outside the frame" (Blomley 2014, p.136). Bracketing does not simplify, he argues, rather it draws on specific and often complex relations while attempting to sever others and therefore to succeed in successfully producing law in particular ways (Blomley 2014, p.145). This, he notes, is always a political process, as the power to frame the boundaries of a particular dispute is not evenly distributed (Blomley 2014, p.139). In the context of our case study, the concept of bracketing assists in understanding how conceptions of jurisdiction are deployed by both the Crown and Māori to frame relations with mineral resources and expose when and how particular assertions are successful. Canadian geographer Shiri Pasternak argues jurisdiction claims are central to settler–colonial relations as Indigenous assertions of ownership and authority are transformed to "reflect state frames of recognition" (Pasternak 2014, p.17). Examining the specific work that is done through law to make, remake and sustain particular jurisdictional claims and shape people–place relations can help us disentangle the "success" of a particular representation and its "truth" (Blomley 2013, p.42). In particular, this approach compliments how Kaupapa Māori research aims to disrupt dominant colonial paradigms and universal truth claims and reimagine the world (Mahuika 2008, p.6).

Bringing methodologies into conversation

How can legal geography and Kaupapa Māori approaches work together in a project like ours? In our view, these are distinct but complementary approaches to uncovering the complexity of the people–place relations at stake in decisions about minerals and mining. While our focus is examining and exposing the structure of state mining law and not a detailed examination of *tikanga* (Māori system of law) in relation to minerals and land use, our research intentionally foregrounds Māori perspectives and jurisdictional claims. This is an explicit decision to support Māori aspirations and assertions of jurisdiction and *tino rangatiratanga* (self-determination/chiefly authority) in relation to minerals and other natural resources. However, this is also a means to denaturalise and critique the administration of settler-state mining law and to imagine alternative models of natural resource governance. Legal geography's attention to the place and spatialisation of power through law complement the relational orientation of Kaupapa Māori approaches and the need for pluralistic approaches to legal research. Reflecting on our learning through this project we have drawn out three areas of learning: (1) the importance of centring Māori perspectives without burdening Māori communities; (2) decentring state claims to jurisdiction by prioritising *tikanga*; and (3) friendship as a foundation for bicultural research.

Centring Māori perspectives without burdening Māori communities

While Kaupapa Māori research is generally understood to emphasise community-based research and face-to-face relations with Māori, our project deliberately shifts the focus to non-Māori spaces within the settler state. These spaces are implicated in Māori life by virtue of their power to designate, allocate and ultimately transform the lands and resources of Māori communities. Yet they are rarely the subject of scholarly inquiry and are often characterised by a lack of transparency about how decisions are made and implemented. As Belinda Borell argues, the effect is to both hide "key determinants of outcomes for Māori" and the potential to limit the legitimate subjects of Māori research to projects involving Indigenous peoples (Borell 2017, pp.33–4). She notes the unrealised potential for Kaupapa Māori research to be employed to "understand, deconstruct and critique wider environmental structures and norms that frame the long term interests" of Indigenous people in reimagining society to reflect their experiences and values (Borell 2017, p.45).

This shift does not mean that Māori perspectives are not centred in our project. Indeed, consistent with Kaupapa Māori we started our project with a case study in the rohe of one of the authors for the purpose of centring iwi and hapū perspectives and grounding our inquiry in relationship with a particular place and community. However, the shift in focus did have important methodological implications in terms of our sources and methods. We explicitly chose to rely on existing documentation of Māori perspectives on minerals and mining rather than undertaking new interviews with Māori participants. In the context of researching consultation practices this seemed appropriate for two reasons. First, because consultation is already a significant burden on iwi and hapū, who are constantly being asked to participate and comment on often highly complex and technical proposals in their rohe (Ruckstuhl et al. 2014). As Ruwhiu and Carter note, capacity issues have a significant impact on the ability of Indigenous communities to effectively engage with consultation process even where they are available (Ruwhiu & Carter 2016, p.647). In our view, and based on Bargh's direct experience as a community member, iwi and hapū had already expended considerable time participating in *Crown Minerals Act* processes and have clearly expressed their views on both specific proposals and minerals and mining regulation more generally. In this context it was both appropriate and respectful to iwi and hapū to rely on their existing work rather than add another burden, no matter how well-intentioned.

Second, because our concern was uncovering how the settler-state administration interprets and responds to these Māori perspectives and assertions of jurisdiction, it was appropriate for our interview-based research to focus on talking to government decision makers. In contrast to Māori participants in the *Crown Minerals Act* process there is little documentation of government decision-making, and what little exists had to be extracted through *Official Information Act* requests. The work of some legal geographers such as Blomley and Mariana Valverde has similarly been directed towards exposing specific "legal knowledges" and the work of particular legal actors (Nicholas Blomley 2011; Valverde 2011; Valverde 2005). As David Delaney notes, "[l]egal geographers take us into the workshops where space, law and (in)justice are the means of the co-production of each other" (Delaney 2016, p.268). In this way legal geography supports the goals of Kaupapa Māori in an unexpected way – by demonstrating the potential of reversing the research gaze towards the actors within the everyday operation of the settler state. Rather than exacerbate the deep historical harms research "inextricably linked to European imperialism and colonialism" has caused to Indigenous peoples (Smith 2012, p.1), we aimed to make visible the specific ways settler law is used to place Māori legal relations and perspectives outside the frame of legal relevance. By documenting how Māori legal relations are placed outside the frame we can demonstrate the presumptions of settler-state sovereignty and a fundamental disregard for Māori law and jurisdiction, and therefore the Treaty partnership. In doing so, we can support the work of Māori communities to disentangle their lands and resources from settler law and assert their place as Treaty partners in natural resource decision-making.

Prioritising *tikanga*: framing minerals jurisdiction

Our intention in this project was to focus on a critical examination of the *Crown Minerals Act* processes and the actors who implement them. However, our examination of the submission documents revealed iwi and hapū rely on a range of sources of law and jurisdiction beyond the *Crown Minerals Act* and other state legislation. The legal pluralism in the iwi and hapū submissions included assertions of *tikanga* and inherent jurisdictions such as *mana whenua* and *kaitiakitanga* as the primary source of authority, as well as reliance on legal relations flowing from the Treaty relationship. References to Treaty Settlements and statutory acknowledgments, Agreements with the Crown about particular places, Accords with the Crown, Waitangi Tribunal reports; and instruments under the *Crown Minerals Act* and the Resource Management Act, such as iwi management plans, or official plans, all demonstrate the importance of the Treaty partnership as the starting point for natural resource governance and as recognition of the *mana* of iwi and hapū.[9] Jurisdictional claims were linked to assertions of the ownership of minerals in some instances.[10] However, even submitters who did not reference ownership asserted jurisdiction over resources in their *rohe*.[11]

In contrast, the Crown relied solely on the *Crown Minerals Act* and its instruments as the source of legal authority. In the final Report to the Minister, no reference was made to *tikanga* or to relevant Tribunal reports. State law, and specifically mining law, is the dominant if not only relevant legal framework, and Māori are submitters who must fit into its frame or risk being left outside. Even as Māori engage with the Crown-defined and determined process, the people–place relations they assert are "balanced" against a variety of other Crown-determined values (Bargh & van Wagner 2019). As we have argued elsewhere, this balancing act is expressly structured by the *Crown Minerals Act* to prioritise the extraction of minerals. The purpose of the Act is the promotion of "prospecting for, exploration for, and mining of Crown owned minerals for the benefit of New Zealand" (s. 1A). According to former New Zealand Minerals and Petroleum National Manager Sefton Darby, who led the 2013 Block Offer process, the role of the Ministry can be characterised as "cheerleading instead of regulating" (Darby 2017, p.23). Ministry officials described themselves as having "their hands tied because of our legislation" (Interview, Ministry of Business, Innovation and Employment Officials 2017). In this context, we have concluded consultation with Māori about mineral exploration is structured to bracket not only assertions of ownership and jurisdiction that unsettle state sovereignty over natural resources, but also the ecological and metaphysical relations that underpin Māori legal and social relations (Bargh & van Wagner 2019).

While our focus has been on exposing the legal structure of relations created and enforced through the *Crown Minerals Act*, our research also reveals the ways in which iwi and hapū reject the *Crown Minerals Act* framing. There are a range of differing perspectives on mining within and between Māori communities, and in response to particular proposals; however, there is a common commitment to sustainable use of mineral resources for the benefit of both Māori and the more-than-human world (Ruckstuhl et al. 2013). Māori relations with mineral resources are embedded in *te ao Māori* (the Māori worldview) and the use of and benefit from natural resources results in a "reciprocal obligation to care for those resources, their environment, and even enhance their 'energy' (*mauri*)" (Ruckstuhl et al. 2013, p.310). The responsibility to maintain and protect these sustainable relations with the places in an iwi or hapū's rohe remain central to claims to authority and jurisdiction – both as *kaitiaki* and as Treaty partners.

Trust, friendship, humility: bicultural research collaboration

Kaupapa Māori research has been described as research by Māori, for Māori (Jones 2012, p.81; Walker et al. 2006). As Jones notes, some interpret this as exclusionary: non-Māori "do not have the authority conferred by whakapapa to contribute to what might constitute Māori thought and practice" (Jones 2012). Yet, she notes this framing can also be understood as inclusive rather than exclusive if one is able to decentre the dominant group as the audience for Kaupapa Māori: "Māori researchers primarily and deliberately address Māori in Kaupapa Māori discourse; they do not usually seek to address Pakeha. That is their point". The classic text by Smith on Indigenous research methodologies captures this orientation: "The book is written primarily to help ourselves" (Smith 2012).

What then of the kind of bicultural collaboration described here? Can this be a Kaupapa Māori project? Smith herself notes the promise of emerging "bicultural research, partnership research and multi-disciplinary research" in New Zealand (Smith 2012, p.17). Nonetheless, as Jones notes in reflecting on her own involvement in Kaupapa Māori research as a Pākehā researcher, such collaboration "requires justification and care" (Jones 2012, p.183). It is, as Graham Smith notes, risky, but it is about people, it is about relationships (Smith et al. 2012). While fundamentally informed by Kaupapa Māori, legal researchers Rachel Parr and Paul Meredith situate their Māori–Pākehā collaborations more broadly as *"collaborative cultural research"*, which they describe as a *"bridge-building* exercise" that "seeks to inform debates, particularly where there is obvious ignorance or limited information to make good judgments" (Parr & Meredith 2001, p.2). As with our collaboration, they build on Bishop's Treaty-informed research framework in which all parties have responsibilities as partners.

Our collaboration is both scholarly and relational: it is based on a friendship developed through common interests and political and academic goals. It began with a conversation about common concerns and possibilities for sharing resources and skills, though necessarily from different experiences and understandings.

For the author Bargh, I recall our friendship arising initially out of sharing coffee, herbal tea and food, as well as political strategies. When we have travelled to the Bay of Plenty, van Wagner has accompanied me on visits to my marae, church and cemetery near the base of our mountain, to my grandparent's house where she met members of my *whānau* (extended family) and to sites of tribal significance in the Rotorua area. We both have two young children around a similar age and have had our children and families together for food and play.

I mention these events in which food appears to play a central part, because building the reciprocal and trusting research relationship has involved creating deeper and personal connections, which include nurturing *whānau*. Hirini Moko Mead has explained that the principle of *manaakitanga* (hospitality) commonly involves sharing food to help nurture and facilitate relationships (Mead 2003, pp.28–9). Relationships are central to the ways that *tikanga* Māori operates. The *whanaungatanga* (relationship) principle creates obligations, in particular that individuals will be supported by their kin but can also extend to obligations for those who are not strictly speaking kin (Mead 2003, p.28). By van Wagner becoming part of relationships with my whānau and sharing the dilemmas of the extended whānau, *hapū* and *iwi*, she has in some ways now become entangled in the task of helping to support the resolution of these challenges.

For the author van Wagner, the friendship was born out of the generosity and hospitality of the co-author Bargh who welcomed me as guest and ally to Aotearoa New Zealand shortly after my arrival from Canada, despite my limited understanding of Māori language and worldviews. She provided valuable mentorship in relation to engaging with Māori perspectives and aspirations in both teaching and research. As a guest and a settler the invitation to collaborate provided a unique opportunity to engage in community-linked research. It was however a humbling proposal requiring self-reflection about what I could bring to the table to ensure some measure of reciprocity.

As we move to stage two of the project, I, van Wagner, remain uncertain about whether this reciprocity has been achieved and thoughtful about how I might do better going forward. I believe, and hope, that my ability to engage with settler-state law and legal systems from a critical, though Western, perspective, is strategically useful in supporting efforts to disentangle that law from Māori places and people–place relations. Indeed it is this work of disentangling the legal frameworks that continue to facilitate dispossession of Indigenous lands and resources that I see as one of my responsibilities as a Treaty person, both as a guest in Aotearoa New Zealand hoping to maintain ties and solidarity, and now back in Canada where my ancestors have been bound by Treaties with Indigenous peoples for generations.

Our collaboration is based on deliberate engagement between our different knowledges and experience, and not on a desire to collapse or erase it. Because of this, trust and humility remain central to the research relationship in a number of practical ways. For van Wagner, humility is crucial to awareness of the limits of my role in the project as part of the broader work of *iwi* and *hapū* environmental governance. For example, reflecting on the limits of what can and should be shared with me as a non-Māori requires a reorientation from a traditional researcher role towards one characterised by allyship, not Western academic presumptions about "the ultimate *knowability* of things, and our *entitlement* to know" (Jones 2012, p.189). This is part of a broader and ongoing process of self-reflection about where, when and how I, van Wagner, fit into the research.

In seeking to resolve some of the political challenges of the whānau, hapū and iwi, our collaboration has enabled us access and made us privy to a range of conversations that might otherwise have been closed off or concealed from us as separate individuals. In different interviews or when requesting information from particular sources we have at different times foregrounded one or other of our identities, van Wagner as a "legal" scholar and Bargh as a "Māori Studies" scholar. As we surmised early in our project planning, our different identities receive different receptions, as with Chapters 4, 7 and 8 of this edited collection. In this way, we understand our deliberate and conscious engagement *with* our differences as productive both personally and for the research we undertake together.

Conclusions

Bringing legal geography and Kaupapa Māori approaches together has been both challenging and fruitful for this project and for us as researchers. While it may have been possible to undertake this work from either of these methodological approaches in isolation, we view the conversation between them to be a core outcome of the project. Here we have drawn out three lessons from this hybrid methodological engagement. First, in setting out to research community consultation, our research has aimed to centre Māori perspectives without burdening Māori communities. Legal geography approaches have enhanced the community-based orientation of Kaupapa Māori by demonstrating how to shift the research gaze to non-Māori spaces, such as state consultation processes, which nonetheless have profound impacts on Māori life. The seemingly mundane day-to-day operation of such spaces is often characterised by a lack of transparency and is thus often difficult to effectively critique or ultimately transform. Second, by retaining a Kaupapa Māori-informed approach we were able to make this shift in focus while still decentring state claims to jurisdiction. Prioritising *tikanga* and the pluralistic Treaty-based perspectives of iwi and hapū highlights the way communities seek to reframe and reimagine relations with and about minerals in their territories. Understanding the specific ways in which the Crown responds to iwi and hapū assertions of people–place relations and jurisdiction is essential to this unsettling work and to successfully reshaping Treaty relations from a pluralistic perspective. Finally, we see the role of friendship, trust and humility as a foundation for bicultural research as a core finding of this project. Engagement with our differences and cross-cultural entanglement in the dilemmas of whānau, hapū and iwi require ongoing reflexivity and justification. This work reflects Kaupapa Māori principles and upholds *tikanga* by nurturing relationships characterised by reciprocity, humility and hospitality. As we move to phase two of this project, it is this relational work that will continue to guide us as we expand our inquiry, deepen our friendship and continue to support Māori aspirations for *tino rangatiratanga* (self-determination/chiefly authority).

Notes

1 While New Zealand Petroleum and Minerals does post information about Block Offers online, information about the 2013 Epithermal Gold Minerals Block Offer is no longer available on their website. Indeed much of the information collated for this article required numerous *Official Information Act 1982* (NZ) requests between 2016 and 2017 to obtain.
2 Māori Land is a specific title of land governed under the Māori *Land Act/Te Ture Whenua Māori 1993* (NZ). Te Puni Kōkiri, *Report on the Māori Asset Base in the Waiariki Economy* (2010) at p.4, online: <www.tpk.govt.nz/en/a-matou-mohiotanga/business-and-economics/te-ripoata-ohanga-maori-mo-te-waiariki>.
3 All five volumes of the He Maunga Rongo: Report on Central North Island Claims, Stage 1are available online: <www.waitangitribunal.govt.nz/publications-and-reso urces/waitangi-tribunal-reports/>.
4 Bay of Plenty Regional Council, *Treaty of Waitangi Toolkit* (2015), online: <www.b oprc.govt.nz/about-council/kaupapa-maori/treaty-of-waitangi-toolbox/>.
5 *Central North Island Forests Land Collective Settlement Act 2008* (NZ) 2008/99; *Te Arawa Lakes Settlement Act 2006* (NZ) 2006/43.
6 The Coalition is comprised of representatives from Te Arawa Federation of Māori Authorities, Te Arawa Lakes Trust, Te Pūmautanga o Te Arawa, Te Arawa Primary Sector Inc, Te Arawa River Iwi Trust and Tapuika Iwi Authority.
7 The following iwi groups requested their entire tribal areas to be excluded: Te Arawa, Raukawa Charitable Trust, Tapuika Iwi Authority, Te Maru o Rereahu, Maniapoto Māori Trust Board, Ngāti Koroki Kahukura and Ngāti Tūwharetoa.
8 Ngāti Rangiwewehi, Te Maru o Rereahu, Te Arawa River Iwi Trust, Tūwharetoa Māori Trust Board, Bay of Plenty Regional Council, Matamata-Piako District Council, Waipa District Council and the Waikato Regional Council.
9 The majority of submissions pointed to existing recognitions by the Crown of their mana, through Deeds of Settlement or protocols or statutory acknowledgments. Of the 15 submissions, eight have Deeds of Settlement or Protocols or Statutory Acknowledgments or Iwi Management Plans that set out protections or acknowledgments of the mana of those groups over their particular areas. Included amongst these are: Ngāti Rangiwewehi Te Tāhuhu o Tawakeheimoa Trust – Crown Minerals Protocol and Conservation Protocol, Ngāti Rangiwewehi Deed of Settlement; Ngāti

Tūwharetoa Statutory acknowledgements for Tarawera and Rangitaiki Rivers and Kawerau Geothermal System. Deed of Settlement 2005; Ngāti Tuwharetoa Hapū Forum Terms of Negotiation 2013; Raukawa Deed of Settlement 2012 (includes statutory acknowledgement areas formally registering Raukawa's connection to these places), Raukawa Energy Accord with Minister of Energy and Resources and Ministry of Business, Inovation and Employment; Tapuika Iwi Authority Trust, Deed of Settlement (before parliament 2013 and includes Crown Minerals Protocol); Te Arawa Lakes Settlement 2006; Ngāti Tuwharetoa, Raukawa, *Te Arawa River Iwi Waikato River Act 2010* (NZ); Te Arawa River Iwi Energy and Resources Accord – with Minister of Energy and Resources and MBIE 2012; Te Mana o Ngāti Rangitihi Trust, Iwi Environmental Management Plan (lodged with Whakatane District Council and Bay of Plenty Regional Council 2012); Tūwharetoa Environmental Iwi Management Plan. Many of these high-level agreements or Settlements have taken years to negotiate and some involve legislative backing (e.g. *Waikato-Tainui Raupatu Claims (Waikato River) Settlement Act 2010* (NZ)). Several of these Accords commit the Crown and iwi to "giving effect to the principles of Te Tiriti o Waitangi/Treaty of Waitangi".

10 Ngāti Tuwharetoa Settlement Trust, Response to Crown Consultation on the "Proposed Competitive Tender Offer for Metallic Minerals Exploration Permits – Part of Bay of Plenty and Waikato" (19 July 2013); Tūwharetoa Māori Trust Board, Submission on Proposed Tender, Epithermal Gold 2013 (19 July 2013).

11 Ngāti Kea Ngati Tuara, Submission on Sichuan Tianboa Minerals Exploration Permit Application (20 April 2015); Te Arawa River Iwi Trust, Submission to the Ministry of Economic Development Regarding the Epithermal Gold 2013 Proposed Tender (2013); Te Mana o Ngāti Rangithi Trust, Submission to the New Zealand Epithermal Gold 2013 – Proposed Competitive Tender Allocation of Exploration Permits in the Central North Island (May 2013).

References

Amundsen, D. 2018, 'Decolonisation through reconciliation: The role of Pakeha Identify', *MAI Journal*, vol. 7, pp.139–154.

Bargh, M., & van Wagner, E. 2019, 'Participation as exclusion: Māori engagement with the Crown Minerals Act 1991 Block Offer process', *Journal of Human Rights and the Environment*, vol. 10, pp.118–139.

Bay of Plenty Reginal Council, 2013, *Bay of Plenty Regional Council, Submission to Proposed Competitive Tender – Epithermal Gold 2013*, Released under the Official Information Act.

Bishop, R. 1994, 'Initiating empowering research?', *New Zealand Journal of Educational Studies*, vol. 29, no. 1, pp.175–188.

Blomley, N. 2005, 'Flowers in the Bathtub: Boundary crossings at the public–private divide', *Geoforum*, vol. 36, pp.281–296, doi: 10.1016/j.geoforum.2004.08.005.

Blomley, N. 2011, *Rights of Passage: Sidewalks and the Regulation of Public Flow, Social Justice*, Routledge, New York.

Blomley, N. 2013, 'Performing property, making the world', *Canadian Journal of Law & Jurisprudence*, vol. 26, no. 1, pp.23–48.

Blomley, N. 2014, 'Disentangling law: The practice of bracketing', *Annual Review of Law and Social Science*, vol. 10, pp.133–148, doi: 10.1146/annurev-lawsocsci-110413-030719.

Borell, B. 2017, *The Nature of the Gaze: A Conceptual Discussion of Societal Privilege from an Indigenous Perspective* (Unpublished PhD Thesis). PhD Thesis, Massey University, Wellington, New Zealand.

Braverman, I., Blomley, N., Delaney, D., & Kedar, A. 2014, 'Introduction: Expanding the spaces of law', in *The Expanding Spaces of Law. A Timely Legal Geography.* eds I.

Braverman, N. Blomley, D. Delaney & A. Kedar, Stanford University Press, Stanford, pp.1–29.

Bridges, S., 'Epithermal Gold Exploration Permit Awarded'. Media statement, 9 May. http://www.scoop.co.nz/stories/PA1405/S00173/epithermal-gold-exploration-permit-awarded.htm.

Charpleix, L. 2018, 'The Whanganui River as Te Awa Tupua: Place-based law in a legally pluralistic society', *The Geographical Journal*, vol. 184, pp.19–30, doi: 10.1111/geoj.12238.

Cram, F. 1997, 'Developing partnerships in research: Pākehā researchers and Māori research', *Sites*, vol. 35, pp.44–63.

Darby, S. 2017, *The Ground Between: Navigating the Oil and Mining Debate in New Zealand*, Bridget Williams Books, Wellington, New Zealand.

Darby, S. 2013, *New Zealand Epithermal Gold 2013 Minerals Competititve Tender: Results of Consultation (Released under the Official Information Act)*, New Zealand Minerals and Petroleum, Wellington, New Zealand.

Delaney, D. 2016, 'Legal geography II: Discerning injustice', *Progress in Human Geography*, vol. 40, pp.267–274.

Delaney, D. 2015, 'Legal geography I: Constitutivities, complexities, and contingencies', *Progress in Human Geography*, vol. 39, pp.96–102.

Durie, M. 1999, *Te Mana, Te Kawanatanga: The Politics of Māori Self Determination*, Oxford University Press, Auckland.

Durie, M. 2017, 'Kaupapa Māori: Indigenising New Zealand', in *Critical Conversations in Kaupapa Maori*, ed. A. Jones, Huia Publishers, New Zealand, pp.1–10.

'Going for Kawerau Gold', 2015. *Rotorua Daily Post*, 6 May, viewed 11 June 2019 <https://www.nzherald.co.nz/rotorua-daily-post/news/article.cfm?c_id=1503438&objectid=11443969>.

Graham, N. 2011, *Lawscape: Property, Environment, Law*, Routledge, London.

Hudson, M. L., & Russell, K. 2009, 'The treaty of waitangi and research ethics in Aotearoa', *Journal of Bioethical Inquiry*, vol. 6, pp.61–68.

Huntington, H. P., Lynge, A., Stotts, J., & Hartsig, A. 2012, 'Less ice, more talk: The benefits and burdens for arctic communities of consultations concerning development activities thematic focus: Climate change, arctic change: Law and policy', *CCLR*, vol. 33, pp 33–46.

Jones, A. 2012, 'Dangerous liaisons: Pākehā, kaupapa Māori, and educational research', *New Zealand Journal of Education Studies*, vol. 47, no. 2, pp.100–112.

Jones, C. 2016, *New Treaty, New Tradition*: Reconciling New Zealand and Maori Law, UBC Press, Canada.

Keenan, S. 2010, 'Subversive property: Reshaping malleable spaces of belonging', *Social & Legal Studies*, vol. 19, pp.423–439.

Kovach, M. E. 2010, *Indigenous Methodologies: Characteristics, Conversations, and Contexts*, University of Toronto Press, Canada.

Latulippe, N. 2015, 'Bridging parallel rows: Epistemic difference and relational accountability in cross-cultural research', *The International Indigenous Policy Journal*, vol. 6, no. 2, art. 7, pp.1–17.

Mahuika, R. 2008, 'Kaupapa Māori theory is critical and anti-colonial', *MAI Revie*, vol. 3, pp.1–16.

Martin, D., Scherr, A., & City, C. 2010, 'Making law, making place: Lawyers and the production of space', *Progress in Human Geography*, vol. 34, pp.175–192.

Martin, D. G. 2003, '"Place-framing" as place-making: Constituting a neighborhood for organizing and activism', *Annals of the Association of American Geographers*, vol. 93, pp.730–750.

Massey, D. 2005, *For Space*, Sage, London.

Massey, D. 2004, 'The political challenge of relational space: Introduction to the Vega Symposium', *Geografiska Annaler: Series B, Human Geography*, vol. 86, no. 1, pp.3–3.

Mead, H. M. 2003, *Tikanga Maori: Living by Maori Values*, Huia Publishers, New Zealand.

Parr, R., & Meredith, P. 2001, *Collaborative Cultural Research for Laws and Institutions for Aotearoa/New Zealand: A Summary Paper*, Te Matahauariki Institute Occasional Paper Series, University of Waikato, Te Matahauariki Institute. Retrieved from http://www.lianz.waikato.ac.nz/.

Pasternak, S. 2014, 'Jurisdiction and settler colonialism: Where do laws meet?', *Canadian Journal of Law and Society/Revue Canadienne Droit et Société*, vol. 29, pp.145–161.

Ralph, C. 2017, 'Epithermal Gold and Silver Deposits', *ICMJ Prospecting and Mining Journal*, viewed 11 June 2019 <https://www.icmj.com/magazine/article/epithermal-gold-and-silver-deposits-3618/>.

Robinson, D. F., & Graham, N. 2018, 'Legal pluralisms, justice and spatial conflicts: New directions in legal geography', *The Geographical Journal*, vol. 184, pp.3–7, doi: 10.1111/geoj.12247.

Ruckstuhl, K., Carter, L., Easterbrook, L., Gorman, A. R., Rae, H., Ruru, J., Stephenson, J., Ruwhiu, Diane, Suszko, A., Thompson-Fawcett, M., & Turner, R. 2013, *Maori and Mining*, University of Otago, Dunedin, New Zealand.

Ruckstuhl, K., Thompson-Fawcett, M., Carter, L., Ruwhiu, D., & Stephenson, J. 2014, 'Where go the indigenous in the mining nation?', in *International Indigenous Development Research Conference 2014* Proceedings, presented at the International Indigenous Development Conference, University of Auckland, pp.79–86.

Ruwhiu, D., & Carter, L. 2016, 'Negotiating "meaningful participation" for indigenous peoples in the context of mining', *Corporate Governance: The International Journal of Business in Society*, vol. 16, pp.641–654, doi: 10.1108/CG-10-2015-0138.

Smith, G., Hoskins, T. K., & Jones, A. 2012, 'Interview: Kaupapa Maori: The dangers of domestication', *New Zealand Journal of Educational Studies*, vol. 47, no. 2, pp.10–20.

Smith, G. H. 2009, 'Mai i te maramatanga, ki te putanga mai o te tahuritanga: From conscientization to transformation', in *Social Justice, Peace, and Environmental Education: Transformative Standards*, eds J. Andrzejewski, M. Baltodano & L. Symcox, Routledge, London, pp.31–40.

Smith, L. T. 2007, 'On tricky ground', *The Landscape of Qualitative Research*, vol. 1, pp.85–113.

Smith, L. T. 1999, *Decolonizing Methodologies: Research and Indigenous Peoples*, Zed books, London and New York.

Smith, L. T. T. R. 2011, 'Kaupapa maori research', in *Reclaiming Indigenous Voice and Vision*, ed. M Battiste, UBC Press, Canada, pp.225–247.

Smith, L. T. 2012, *Decolonizing Methodologies: Research and Indigenous Peoples* (2nd edition), Zed Books, London and New York.

Smith, L. T. 2017, 'Towards developing indigenous methodologies: Kaupapa Māori research', in *Critical Conversations in Kaupapa Maori*, eds T. K. Hoskins & A. Jones, Huia Publishers, New Zealand, ch. 2.

Te Arawa Hui-a Iwi, 2013, *NZ Epithermal Gold Hui-a-Iwi o Te Arawa Resolution*, Unpublished Resolution of Meeting held in Rotorua, May 2013.

Tomas, N. 2011, 'Maori concepts of rangatiratanga, kaitiakitanga, the environment, and property rights', in *Property Rights and Sustainability: The Evolution of Property Rights to Meet Ecological Challenges*, eds D. Grinlinton & P. Taylor, Brill Nijhoff, Netherlands, pp.219–248.

Valverde, M. 2005, 'Taking land use seriously: Toward an ontology of municipal law', *Law Text Culture*, vol. 9, no. 1, pp.34–59.

Valverde, M. 2011, 'Seeing like a city: The dialectic of modern and premodern ways of seeing in urban governance', *Law & Society Review*, vol. 45, pp.277–312.

Waitangi Tribunal, 2008, *He Maunga Rongo: Report on Central North Island Claims, Stage 1 (No. Wai 1200)*, Waitangi Tribunal, Wellington, New Zealand.

Walker, S., Eketone, A., & Gibbs, A. 2006, 'An exploration of kaupapa Maori research, its principles, processes and applications', *International Journal of Social Research Methodology*, vol. 9, pp.331–344.

Watson, I. 2007, 'Aboriginal women's laws and lives: How might we keep growing the law?', *Australian Feminist Law Journal*, vol. 26, pp.95–107, doi: 10.1080/13200968.2007.10854380.

Watson, I. 2002, 'Buried alive', *Law and Critique*, vol. 13, pp.253–269, doi: 10.1023/A:1021248403613.

Williams, J. 2013, 'The harkness henry lecture lex aotearoa: An heroic attempt to map the Maori dimension in modern New Zealand law', *Waikato Law Review: Taumauri*, vol. 21, pp.1–34.

PART 3

Investigating the legal geographies of regulation

Resource, risk and resilience

7

INSIDE-OUTSIDE

An interrogation of coastal climate change adaptation through the gaze of 'the lawyer'

Tayanah O'Donnell

Introduction

The lens of legal geography offers a new, insightful approach to scholarly and applied understandings of climate change adaptation for two reasons. First, law, however defined, plays a central role in influencing social, cultural and political barriers and drivers in relation to climate change adaptation. Second, legal geography uniquely illuminates law's interaction with overlapping and complementary spatial, relational and temporal frames as played out in a variety of empirically rich and societally important research contexts. This chapter explores these themes through a personal lens to encourage legal geographers (and others) to continually engage in self-reflection as we undertake our research. As is powerfully and persuasively argued by Osbourne (2019), existential crises including climate change demand that we make our research personal, for our work is nothing if not personal. So, let us then query and reflect on how the "personal", and our positionality, may influence our scholarly trajectories.

While writing this chapter during the warmer-than-usual Australian summer of 2019 – as the land, the animals and the people here survived yet another severe heatwave with temperatures up to 49.5 degrees Celsius in some parts of the country – my phone laughed at me with a Twitter notification. It prompted a momentary pause. I opened the social media platform to the top tweet in my newsfeed. This re-tweet – a share from Professor Lesley Head, author of *Grief and Hope in the Anthropocene* (2016) – contained a thread by American environmental anthropologist Dr Melissa R. Poe. Poe's tweet eloquently unravelled the all-consuming guilt associated with being an environmental or climatic scientist. On opening the full thread, Poe had tweeted: "immersive study takes a toll. We are insider-outsider" (Poe 2018). I pondered that there is indeed minimal doubt that immersing oneself in research about climate change impacts takes a toll.

Previews into an increasingly likely bleak future world is emotionally exhausting (Osbourne 2019). What caused me to pause and ponder this tweet longer than usual, however, was the "insider-outsider" description. It is an apt description of my own experience in researching coastal management and climate change adaptation. Governments and communities are bracing for coastal climate change impacts, such as sea level rise and increased coastal erosion. Alongside these physical changes are past, present and future legalities: in particular, an anticipated swathe of litigation as land use planning and ideas of property rights collide (O'Donnell 2017; O'Donnell 2019a). In this complex social and regulatory environment, many localities are working hard on ways to implement climate change adaptation to navigate this complexity.

I am insider-outsider. This is in many ways due to my own professional status as a lawyer. I practised law between 2007 and 2010. Back then, climate change considerations were beginning to make inroads into legal cases; for some, this brought a hope that the law would force societal change via legal precedent. Since then, progress in this regard has been slow. Even as reminders of the already-present and imminent losses due to climate change become more frequent and more urgent (described throughout Head's 2016 book; see also Anderson & Smith 2001, for a discussion of hope, grief and loss due to climate change), the role of law in shifting regulatory responses to climate change is still increasingly looked to as a force for change. As scientists we are desperate for the deep transformational change required to stave off the worst impacts of environmental change.

Harm-Benson (2014) implicitly explores the insider-outsider concept, with an exploration of the types of knowledge held by those involved in legal arenas, such as judges and lawyers, and argues that such knowledge can be categorised as specialist knowledge. Such specialists are said to provide additional insights on research questions precisely because of their attainment of specialist knowledge. In addition, specialists also have a unique entry or pathway into research contexts where they are simultaneously specialist and researcher. And while reflexivity and positionality are not new concepts to the social sciences (Rose 1997), nor to social research in law (Watkins & Burton 2013), the multiple social meanings given to law, legal and lawyer warrant specific consideration (Blomey, 2011).

These multiple ways of describing and perceiving law are ultimately about power. For the purposes of this chapter, in talking about power I mean both power relations between individuals and within and across institutions, as well as how we, at individual and societal levels, come to attach meanings to particular terms, and the institutions we construct them to represent. Harm-Benson's (2014) depiction of litigation as a window into the operation of law and law's key actors explores in depth the role these actors play in orchestrating, interpreting and creating law in powerful legal arenas (in her example, courtrooms). Delaney's (2010) detailed analysis of who does what, where and when characterises these actors as "nomospheric agents or technicians" who, as specialists, are holders of power. These agents or technicians move between the legal and spatial (and other geographies) even as they co-create their meaning(s). Valverde (2015) explores legal networks as "chronotopes", for which power is manifest in temporal, spatial and jurisdictional scales, the foregrounding of each dependent in part on power relations as they manifest in legal and social networks.

An "inside-outside" binary describes the effects on lived experiences before and after the arrival of climate science to localities, linked to time (Fincher et al. 2014), social values (Graham et al. 2014; Kreller & Graham 2018) and land use planning (Hurlimann et al. 2014). For climate change adaptation, the multiple applications of an inside-outside framing seek, if not demand, reflection on those deemed to hold specialist knowledge – what does this specialist status mean for the construction of research questions, the subsequent approach(es) to answering these, and for how the researched subject and context responds?

Scholars working in or around the edges of legal geography recognise that legal geography is "a dynamic 'field' of analysis that overlaps, bridges, under-mines, divides, disconnects, and reconnects with other symbolic domains and imagined visions of nature ... landscape, and legal practice" with the field being a confluence of "braided lines of inquiry" (Braverman et al. 2014, p.1) that draws on multiple disciplines (as shown by the breadth and depth of this edited collection alone). Braverman et al. (2014), in their introductory chapter of *The Expanding Spaces of Law: A Timely Legal Geography*, note that the rethinking of law requires more than a manifestation of social context. Drawing attention to the importance of "the particular ways in which legal actors (however defined) think and act", Braverman (2014, p.15) invites atten-tion to and reflection of the "how" as well as the "why". Attention to enact-ments of law (including by legal actors) to understand more deeply questions of the "how" shines an oft-forgotten light on broader remits of the role of law in our social and political world.

The New South Wales coast and challenges for adaptation

The Australian coast is a culturally and ecologically significant geographical context, for which climate change impacts will be extreme. These impacts are further problematised because the majority of the Australian population reside in close proximity to the coastline (Gurran et al. 2011). As Australian weather is already highly variable, the impacts of climate change to those weather patterns will see increased precipitation in some areas and a decline in others, along with increased bushfire occurrence and intensity, increased drought and then signifi-cant flood events; for the coast, this includes increased frequency and severity of coastal storm events and the resulting erosion and coastal flood impacts (see gen-erally, Palutikof et al. 2015). In addition, coastal locations directly face the tem-poral threat of sea level rise. The economic impacts of climate change to coastal locations are expected to be in the hundreds of millions of dollars (Australian Government 2009). The cultural impacts, although harder to quantify, are also well-documented (Graham et al. 2014).

Empirical exploration of the role of law for residential climate change adaptation has involved several years of research in vulnerable coastal localities. The Australian context requires recognition of added regulatory complications brought about by our federated system of government, meaning local governments play a critical role in implementing climate adaptation policy (Measham et al. 2011; O'Donnell 2019a; O'Donnell, 2019b, O'Donnell 2017). Furthermore, a focus on residential coastal property and climate change adaptation requires consideration of land use management and property rights, each as manifestations of law and regulation. I have explored these themes empirically in numerous localities on the east coast of Australia, with close attention to the state jurisdiction of New South Wales.

The focus on land use planning and property rights in the context of sea level rise, and other climate change impacts, serves to illustrate the direct connection between law and spatiality. Though not an exhaustive analysis (instead see O'Donnell & Gates 2013; O'Donnell 2016; O'Donnell 2017; O'Donnell 2019a), this can be summarised as follows. First, colonial property law in Australia assigns rights to "private property", including rights of alienation, enjoyment and exclusion. Land use planning and coastal management laws and regulations can, and in some cases do, restrict these rights. Additional complications arise when material environment – in this case, the coastline and surrounds – change. Coasts are already dynamic places; climate change impacts including erosion, flood and sea level rise further exacerbate this. The threat of litigious waterfront property owners suing local governments, for land use planning decisions they say have negatively impacted their property, is not foreign in Australia (Gibbs 2016; O'Donnell 2016; O'Donnell & Gates 2013). The anticipation of litigation against government authorities as the relevant decision-makers required to consider climate change under environmental or administrative law is also well-documented (McDonald 2007; McDonald 2010; McDonald 2011; McDonald 2014; Bell 2014; O'Donnell 2019a, Peel and Osofsky 2015).

This imagined nomosphere – that is, the potential for litigation arising as a result of land use planning decisions that have failed to take proper account of climate change impacts in coastal locations – has had a cascade effect on climate adaptation policy in Australia, particularly for coastal policy. Legal liability concerns have been raised in independent government inquiries and ongoing iterations of law reform (Australian Government 2009; Productivity Commission 2012; O'Donnell 2019a). In undertaking fieldwork broadly concentrated on the role of law for coastal climate change adaptation in vulnerable coastal localities since 2010, it is apparent that the relative novelty (or not) of legal liability concerns remains. This is notwithstanding the considerable research and thought that has been afforded to this area by legal scholars working in the Australian context (McDonald 2011; Macintosh et al. 2013; O'Donnell & Gates 2013; Bell 2014; Peel & Osofsky 2015; O'Donnell 2016; O'Donnell 2017; O'Donnell 2019a). A legal geography lens shows that this is because law operates in specific ways, in specific places and at specific points in time.

Linking law and climate change adaptation

A legal geography lens brings to climate change adaptation the necessary linkage between law as a mechanism of the state and the relationship between law, time and space. Legal geography can also offer insights into "systemic spatiopolitical dynamics" (Pierce & Martin 2017, p.456), recognising that "adaptation is as much about actions in time as in space" (Barnett et al. 2014, p.1104). Moreover, it can bring a recognition of the critical policy, governance and legal elements needed to enable climate change adaptation practices. In introducing their cornerstone text for climate change adaptation scholarship, *Adapting to Climate Change: Thresholds, Values, Governance*, Adger et al. (2009, p.1) define adaptation to climate change:

> Look out the window and assess the weather. If it is hot, change into a lighter shirt. If it is raining, take an umbrella. This is adaptation to changing weather. Adaptation to a changing climate is a different matter. The climate may change either slowly or rapidly, and the changes may be irreversible and impossible to predict with any accuracy. The simple principles of adapting to changing weather begin to break down when the climate changes. In the context of climate change the options for adaptation may involve relocating homes, moving cities, changing the foods we grow and consume, seeking compensation for economic damages, and mourning the loss of our favourite place or iconic species. The difference between adapting to changing weather and adapting to a changing climate lies both in the timeframe and in the significance of the changes required. Moreover, the consequences of *not* adapting to climate change may be far more serious than not adapting to changing weather. (emphasis in original)

Climate change adaptation is defined by the Intergovernmental Panel on Climate Change (IPCC) as an "adjustment in natural or human systems in response to actual or expected climatic stimuli or their effects, which moderates harm or exploits beneficial opportunities" (IPCC 2007; see also Palutikof et al. 2015). As argued by Harman et al. (2013, p.797), coastal climate change adaptation faces dual scalar problems: the need for localised adaptation and the need for political leadership and robust law-making from higher governance scales – mainly state and federal governments (Peel and Osofsky 2015). Adaptation to a changing climate will not occur in the same way in every geographical location, nor will it occur evenly across both temporal and spatial scales (Adger et al. 2009). Exploring the role of law in coastal climate change adaptation illustrates the versatility and ubiquitous role that legal geography enables in light of arguably the most complex challenge of our times (i.e. climate change and the adaptation imperative: O'Donnell 2016; O'Donnell 2017; O'Donnell 2019a, O'Donnell 2019b).

Moreover, law functions in specific ways. Law can be thought of as an institution trying to maintain objectivity as this institutional system imparts rulings and, perhaps for some, justice. For legal geographers, law is a discourse that derives its power from the perception of objectivity; that is, law is premised on the idea of objectivity. Geographer Nicholas Blomley argues that this is a process of "bracketing" (Blomley 2014). Bracketing, in Blomley's definition, is a normative process in which law abstracts a set of legally relevant relations, linked to social interactions that conflate perceptions of law's objectivity. Bracketing aims to position law as objective, or somehow removed from societal influence. Several scholars, including Blomley in his bracketing thesis, have argued the impossibility of removing law from its social and material surrounds (Graham 2011; Bennett & Layard 2015; Pue 1990). This is notwithstanding that law is heavily implicated in social ordering (Latour 2017; Forsyth 2017; Delaney 2010; Delaney 2017); it is a system that is continually formed, informed and co-constituted by the social world.

In the context of climate change, law has long been viewed as an important facilitator of social change necessary to enable both mitigation and adaptation (Adger et al. 2005; Moser & Ekstrom 2010; Hinkel et al. 2015; Waters et al. 2014; Peel & Osofsky 2015; Peel & Godden 2009; McDonald 2007; McDonald 2010; McDonald 2011). However, law as an institution, and its relationship with climate change, can be categorised as one that creates societal, economic, cultural and political drivers and barriers in responding to climate change impacts and responses (see also Barnett et al. 2014; Wise et al. 2014). Therefore, the linking of law to spatiality is of continued importance in the contemplation of one of the greatest challenges for humanity. This is particularly so for how we as legal geographers consider law in a normative sense, as well as law's agency in the so-called nomosphere.

Methodology

My ongoing research into coastal climate change adaptation has drawn on multiple social research methods, including interviews, surveys, document and legal analysis and ethnography (see Table 7.1). Multiple social research methods recognise the importance and value of social scientific research methods informing legal research, and the value that legal training can bring to social research methods (Landry 2016; Boer et al. 2016; Darian-Smith 2013; Tamanaha 1997). Adaptation scholars, such as Moser (2010), have argued that inter- and multi-disciplinary approaches are critical to addressing climate change adaptation scholarship and practice, resulting in the blurring of methodological lines of enquiry (Castree 2014; Geels et al. 2016).

Geels et al. argue that inter-disciplinary approaches to research provide a "more differentiated set of analytical approaches [and] enables a more differentiated approach to climate policy making" (2016, p.576). In exploring the "inside-outside" theme for this chapter, I am focussing primarily on the

TABLE 7.1 Methods and Rationale

Method	Rationale
Ethnography	The ethnographic approach grounds the research in real world experiences, and is a critical component of the reflexive approach to the research. Captured via a field diary, the ethnographic approach provides researcher-led perspectives of characteristics of the localities and reports on participant observation(s). The field diary also records post-interview observations, my observations of local government day-to-day workings and conversations with residents (Atkinson and Hammersley 1994).
Semi-structured, in-depth, one-on-one recorded and transcribed interviews with a range of anonymous key actors assigned pseudonyms (residents, local council staff, local council elected officials, state government and insurance sector representatives – total 38 across three localities)	The interviews expand upon publicly available knowledge on coastal climate change adaptation and land use planning law by going beyond the "public face" of an organisation and obtaining perspectives on how participants really think and feel, aided by their de-identification (Silverman 2006). The interviews can be transcribed and then analysed either manually, whereby key themes were identified and grouped, or using coding software. In my research, interview data was analysed and coded thematically and manually. Transcripts were printed and read several times in order to immerse myself in the data in identifying key themes, as per Mansvelt and Berg (2005). All interviewees were contacted personally, provided with the ethics approved information sheet and consent forms and made aware that they would be de-identified and could withdraw from the project at any time.
Postal survey: residents	The mixed methods survey distributed to residents was designed to obtain broad "resident perceptions" on climate change and on private property, to explore their understanding of key pieces of law and policy and to enable residents to self-select for interviews. The survey also received ethics approval. Design of the survey yielded quantitative and qualitative data. The quantitative data were coded in Microsoft Excel. The qualitative data were reviewed for key themes and narratives, similar to the interviews (Silverman 2006).
Document analysis: thematic analysis of policies, guidelines, land use planning instruments and selected submissions to the Productivity Commission	Textual analysis of these documents provides insights into the language used in written materials and into the intent and purpose such documents can serve. This approach is concerned with how the analysis of discourse and language can enrich both the analysis of text itself (as language), and of social theoretical issues (Fairclough 2003).

(*Continued*)

TABLE 7.1 (Continued)

Method	Rationale
Legal analysis	Descriptive and doctrinal legal analysis provides important information about which legal frameworks are applicable in specific contexts, at specific points in time.
Case studies	The use of case studies as a research method enables in-depth engagement with a particular context (in this case, coastal locations). Multiple case studies can enable comparisons to be made between multiple research contexts (Yin 2009).

ethnographic element insofar as it highlights the positionality of a "lawyer" in exploring law and climate change adaptation. As a research method, ethnography is a continuous learning process, requiring ongoing observation and attentiveness to social context, situations and actions: a "contextual, dynamic, reflexive [process] ... open to all sorts of stimuli" (Flood 2005, p.40; Darian-Smith 1999). It is also alert to the element of participant observation and the processes of recording this (Atkinson & Hammersley 1994). As ethnography is concerned with "interpretation and not causal analysis", it can be messy (Flood 2005, p.47). An ethnographic approach also provides "insights into the richness of social life", which is an important viewpoint for research involving law (Flood 2005, p.33). Ethnography accounts for perspectives of, and stories by, people at particular times and in particular places, and there are numerous examples of ethnography informing knowledge that contributes to policy (Robinson 2013; Griffiths 2005). Both the auto-biographical origins of my research and my insider-outsider status each provide an important foundation for my research.

Into the field...

In 2009, I worked as a tipstaff[1] to Justice Sheahan AO in the Land and Environment Court of New South Wales. Prior to this, I had been employed as a practising solicitor in New South Wales, Australia. The segue of employment at the court seemed a viable pathway to pursuing a career at the New South Wales Bar in becoming a barrister. However, in October 2009, the case of *Vaughan v Byron Shire Council* [2009] NSWLEC 88 was listed for hearing before my judge. This litigation required judicial interpretation of conditions in a development consent that a New South Wales coastal council had issued to itself in 2001. The development consent had authorised the building of a sandbag wall to protect private property on a beach known for coastal erosion events. A storm in May 2009 led to the sandbag wall collapsing and to the subsequent erosion of the public beach, a public access road and several metres of the Vaughans' private land. My observations of this litigation sparked my interest in coastal governance and legal geography (see O'Donnell 2016, for analysis of the 2009 Vaughan litigation) and instead led me to pursue a PhD.

In 2010, as part of my PhD candidature, I attended the annual *New South Wales Coastal Conference*, held that year in Batemans Bay on the New South Wales south coast. Three keynote addresses were given on sustainable planning in the context of coastal growth, adaptation strategies and insurance. Later, solicitors from the Environmental Defenders' Office spoke on New South Wales coastal law reform and the new sea level rise policy. When it came time for questions, several hands shot up from the audience. Questions kept returning to the same theme – legal liability – and the role of local government as the key decision-maker in the context of land use planning, coastal management and climate change risk.

My attendance at this conference afforded two important insights for this research. First, I was introduced to the person who would become my key contact at Lake Macquarie City Council, through whom I gained the opportunity to observe the inner workings of a local council. Second, my access to Port Stephens Council was aided by a recommendation of a participant at Lake Macquarie City Council to a council officer at Port Stephens, to "speak with this lawyer doing research on planning and sea level rise". Prior to this, when contacting the Port Stephens Council, I had received no response. Third, my attendance at the conference labelled me as a "lawyer" due to introductions made by professional

FIGURE 7.1   Coastal protection works on Belongil Beach, Byron Bay, 2011. Source: author provided.

colleagues from my legal practitioner days. From that point on, the label of "lawyer" has influenced many of my research interactions, including detailed fieldwork in Eurobodalla Shire Council, in other local councils across Australia, and more recently with other Australian state jurisdictions. Participants seek me out and engage with me because of my legal training, my interdisciplinary coastal research and, my now perpetual insider-outsider status.

"You're a lawyer": positionality and the nomosphere

Undertaking research in the climate adaptation field with the inside-outside status of both researcher and "lawyer" has allowed participants to engage with me in ways they might not have if I had not held this status. This is complemented by Turton's analysis in Chapter 13, and Schenk's observations of bureaucratic and administrative knowledges in Chapter 4 of this book. Harm Benson (2014) unpacks this in her discussion of nomospheric agents as those who can move between the discipline of law with all its quirks and specificities of language, and other disciplines (in my case, a PhD in human geography).

In considering what these insights might mean for a particular empirical context, we can build on Harm-Benson's observations regarding the production of space. Here, the key question is: what happens when your participants assign to you nomospheric agent status? Such a question opens the door for detailed reflection. In pausing for reflection here, I again draw attention to positionality in legal geography research. Those legal geographers who are legally trained hold a privileged vantage point (Harm Benson 2014, p.228).

While my empirical research, as outlined above, has been conducted across numerous coastal locations, I focus here on one aspect of my work in Port Stephens, as it illustrates well the "inside-outside" positionality and how specialist agency influences the trajectories of semi-structured interviews. The following interviews took place in 2011–2012. It is firstly important to provide a brief history of Port Stephens Councils' experience with litigation.

In 2005, Port Stephens Council lost a judicial review case. In *Port Stephens Shire Council v Booth and Gibson* [2005] NSWCA 323 (*Booth*), Port Stephens Council was found to have breached the law by permitting the development of a holiday resort to operate in the vicinity of the Air Force Base in Williamstown, Port Stephens. A development consent had authorised the subdivision of land, the construction of individual cabins and then the sale of these individual lots. The close proximity to the airfield meant there were noise complaints to the council, and lot owners and the resort's managers eventually sued the council based on two causes of action: first, that the council erred in granting development consent of this type so close to an airfield and second, that the council failed to fulfil their obligation to warn of the risk of associated noise on the s.149 certificates (which accompany land title information searches and disclose known risks to property, as per the *Conveyancing Act 1919* (NSW)). It was alleged that these actions each amounted to breaches of Port Stephens Council's duty of care under the *Local Government Act 1993* (NSW). The court found against Port Stephens Council and included an order that the council pay the legal costs of the proponent. Leave to appeal to the High Court of Australia was denied.

This particular litigation came up several times during my interviews with Port Stephens Council staff and elected officials, prompted by participants. During my interview with Bob,[2] a Port Stephens elected official, we discussed *Booth*. Bob spoke loudly, arms flailing as he described his observations and his perspective on the practice of elected officials reviewing council staff decisions on development applications (known as "calling up" an application):

> I mean, because every time you raise it [climate change and sea level rise] [REDACTED] leads his band and says oh, it'll never happen, it'll never happen. It's like approving the houses right under the flight path. One of his mates right under it. Oh, he's a bloody farmer; he's always out in the paddock. He doesn't worry about the planes, but as soon as the next bloke buys it, different story. Different story. As soon as he dies, and the next bloke buys it, he can—you know, we can be—get our pants sued off. We've already lost one of those cases where it cost us multi-million dollars for one of their mates.

Booth also came up during my interview with a Port Stephens Council officer. Kelly[3] mentioned *Booth* and asked if I knew the case, given that I was a trained lawyer (she said, "you're a lawyer, do you know it?"). I confirmed that I did know the case. Kelly became animated: "Oh, let me tell you about it!" she said, and continued on:

> Council was deemed to have a report – constructive knowledge of this report that had been done for a place down the road which was in a different zone and a whole host of things. But the planners had not been given that report. It was sitting in another arm of council. But it was deemed to be constructive knowledge and; therefore, they should have taken that into account when providing information on s.149 Certificates and in relation to the development application. So, it's kind of you're damned if you do and you're damned if you don't for councils. So, in terms of general liability questions, yeah, it's a real fine line. Whenever the major issue sort of comes up when it comes to the crux we will get [legal] advice. As you know, each thing turns on its own facts.

Whether the same conversations would have been had with a different, non-legally trained researcher is hard to know. However, it is telling that Kelly and I engaged in a detailed interaction about case law, prompted by both my insider-outsider status and by her assignment of me as "specialist". It is also telling that participants wanted to discuss the *Booth* case with me when the discussion of case law, let alone such a specific piece of litigation, was not even in my interview schedule – my interview prompts were related mainly to the role of law generally. It is also telling that there was talk of being sued. It is possible, if not probable, that these discussions came about solely because of my legal experience.

Setting aside the content of these empirical explorations, there are important reflections to be had on the process behind these interactions. Comments like "you're a lawyer" demonstrate the interview took this path because of my positionality. Additional observations can also be made. In presenting this data in my dissertation and now here in this chapter, I am careful to meet ethical requirements in not identifying these participants (Silverman 2006). As an admitted solicitor I am acutely aware of my professional obligations and the importance of confidentiality. I have therefore redacted identifying information in Bob's quote, not only to deidentify the data but to also protect myself and the participant due to its content. These actions are, as a trained lawyer and academic researcher, almost unconscious acts as they are required professionally.

Legal geographers must not only reflect seriously on our engagement with the spatial, temporal, political, cultural and more, but consider carefully how we engage with research participants on topics related to law. In doing so, we must attend closely to reflection on the process by which we engage empirically. This is especially so for those undertaking empirical research ethnographically, which of itself presents a significant opportunity for the legal geography field. Reflecting on method in legal geography cannot ignore the powerful impetus that "law" holds, in whatever guise she wears – this includes prefacing geography with "legal".

Conclusion

Engaging with law and the legal brings with it a necessary reflection on the power that is attached to law. This power is attached not only to the letter of law, in terms of interpreting "the law", but also to the very idea of law – that law as an institution is somehow objective, and to discourses that seek to maintain this objectivity. Both law as a system and discourses of law hold a strong societal position in defining the perimeters of social conduct, in line with Blomley's bracketing (2014). As legal geographers, we are actively engaged with this as we define ourselves with law, through the "legal" in legal geography. Doing so not only affords us a unique and important lens with which to better understand our social world, but also assigns to us some of the power that is attached to law. Overt recognition of this remains important for legal geography scholarship.

The implications of my dual status as "insider-outsider" as reflected in this chapter encompasses numerous important layers related to the above. First, there is significance for legal geography scholarship insofar as the blurring of the line between disciplines, expertise and participatory social research approaches is concerned. Second, the assignment of "lawyer" or "legal" to an academic researcher brings with it significant power, and so a concomitant responsibility, especially in terms of enabling access to sites, data and participants, in how we treat and report on research outcomes. Third, social research pertaining to law in its myriad of forms is also pushing scholarly boundaries – especially disciplinary boundaries – as, increasingly, research demands inter-, multi- and trans- disciplinary epistemologies.

Does a scholar need legal training to do legal geography? This is a question students and scholars alike have often asked of me, to which the answer is a definitive *no*. However, there is no denying that legal training affords scholars specialist skills. On the other hand, there are many accomplished legal geographers who do not hold formalised legal training. Each legal geographer will hold different world views; a legally trained legal geographer will likely hold some world views consistent with that training. The same can be said for all of us: our individual positionality can and will influence the research trajectories we embark upon, in specific places, and at specific points in time. And so in pondering this, it is important to not lose sight of the power of our positionality. If you do hold formalised legal training, your view of the world is necessarily and irreversibly different to those who do not. But this need not be a dichotomous position. Rather, acknowledging and reflecting on this is important for legal geography scholarship, and for all empirical and theoretical endeavours, as we catapult toward an uncertain future.

Notes

1 "Tipstaff" is the name given to an Australian judge's primary legal researcher. Such positions are highly competitive and as such are usually held by each Tipstaff for one year.
2 This is a pseudonym.
3 This is a pseudonym.

References

Adger, W. N., Arnell, N. W., & Tompkins, E. L. 2005, 'Successful adaptation to climate change across scales', *Global Environmental Change*, vol. 15, pp.77–86.

Adger, W. N., Lorenzoni, I., & O'Brien, K. 2009, 'Adaptation now', in *Adapting to Climate Change: Thresholds, Values, Governance*, eds W. N. Adger, I. Lorenzoni, & K. O'Brien, Cambridge University Press, Cambridge, UK.

Anderson, K., & Smith, S. 2001, 'Editorial: Emotional geographies', *Transactions of the Institute of British Geographers*, vol. 26, no. 1, pp.7–10.

Atkinson, P., & Hammersley, M. 1994, 'Ethnography and participant observation', in *Handbook of Qualitative Research*, eds N. K. Denzin & Y. S. Lincoln, London, UK, pp.248–261.

Australian Government, House of Representatives Standing Committee on Climate Change, Environment and the Arts, 2009, *Managing Our Coastal Zone in a Changing Climate: The Time to Act is Now*, Canberra, Australia.

Australian Government Productivity Commission, 2012, *Inquiry into Barriers to Effective Climate Change Adaptation.*

Barnett, J., Graham, S., Mortreux, C., Fincher, R., Waters, E., & Hurlimann, A. 2014, 'A local coastal adaptation pathway', *Nature Climate Change*, vol. 4, no. 12, pp.1103–1108.

Bennett, L., & Layard, A. 2015, 'Legal geography: Becoming spatial detectives', *Geography Compass*, vol. 9, pp.406–422.

Bell, J. 2014, *Climate Change and Coastal Development Law in Australia*, The Federation Press, Leichardt.

Blomley, N. 2011, *Rights of Passage: Sidewalks and the Regulation of Public Flow*, Routledge, Abingdon.

Blomley, N. 2014, 'Disentangling law: The practice of bracketing', *Annual Review of Law and Social Science*, vol. 10, no. 1, pp.133–148.

Boer, B., Hirsch, P., Johns, F., Saul, B., & Scurrah, N. 2016. *The Mekong: A Socio-Legal Approach to River Basin Development*, Routledge, London.

Braverman, I. 2014, 'Who's afraid of methodology? Advocating a methodological turn in legal geography', in *The Expanding Spaces of Law: A Timely Legal Geography*, eds I. Braverman, N. Blomley, D. Delaney, & A. Kedar, Stanford Law Books, Standford, CA.

Braverman, I., Blomley, N., Delaney, D., & Kedar, A. (eds) 2014, *The Expanding Spaces of Law: A Timely Legal Geography.* Stanford Law Books, Standford, CA.

Castree, N. 2014, 'Changing the intellectual climate', *Nature Climate Change*, vol. 4, pp.763–768.

Conveyancing Act 1919 (NSW).

Darian-Smith, E. 1999, *Bridging Divides: The Channel Tunnel and English Legal Identity in the New Europe*, University of California Press, Berkeley, CA.

Darian-Smith, E. 2013, *Law and Societies in Global Contexts: Contemporary Approaches*, Cambridge University Press, Cambridge, UK.

Delaney, D. 2010, *The Spatial, the Legal and the Pragmatics of World-Making: Nomospheric Investigations*, Routledge, Abingdon.

Delaney, D. 2017, 'Legal geography III: New worlds, new convergences', *Progress in Human Geography*, vol. 41, no. 5, pp.667–675.

Fairclough, N. 2003, *Analysing Discourse: Textual Analysis for Social Research*, Routledge, Abingdon.

Fincher, R., Barnett, J., Graham, S., & Hurlimann, A. 2014, 'Time stories: Making sense of futures in anticipation of sea-level rise', *Geoforum*, vol. 56, pp.201–210.

Flood, J. 2005, 'Socio-legal ethnography', in *Theory and Method in Socio-Legal Research*, eds R. Banakar & M. Travers, Hart Publishing, Oxford.

Forsyth, M. 2017, 'Legal pluralism: The regulation of traditional medicine in the cook islands', in *Regulatory Theory: Foundations and Applications*, ed. P. Dragos, ANU Press, London, pp.233–246.

Geels, F. W., Berkhout, F., & van Vuuren, D. P. 2016, 'Bridging analytical approaches for low-carbon transition', *Nature Climate Change*, vol. 6, pp.576–583.

Gibbs, M. 2016, 'Why is coastal retreat so hard to implement? Understanding the political risk of coastal adaptation pathways', *Ocean and Coastal Management*, vol. 130, pp.107–114.

Graham, N. 2011, *Lawscape: Property, Environment, Law*, Routledge, Abingdon.

Graham, S., Barnett, J., Fincher, R., Hurlimann, A., & Mortreaux, C. 2014, 'Local values for fairer adaptation to sea level rise: A typology of residents and their lived values in lakes entrance', *Global Environmental Change*, vol. 29, pp.41–52.

Griffiths, A. 2005, 'Using ethnography as a legal tool in legal research', in *Theory and Method in Socio-Legal Research*, eds R. Banakar & M. Travers, Hart Publishing, pp.113–131.

Gurran, N., Norman, B., Gilbert, C., & Hamin, E. 2011, *Planning for Climate Change Adaptation in Coastal Australia: State of Practice*, Report No 4 for the National Sea Change Taskforce, Faculty of Architecture, Design and Planning, University of Sydney.

Harman, B. P., Heyenga, S., Taylor, B. P., & Fletcher, C. S. 2013, 'Global lessons for adapting coastal communities to protect against storm surge inundation', *Journal of Coastal Research*, vol. 31, no. 4, pp.790–801.

Harm-Benson, M. 2014, 'Rules of engagement: The spatiality of judicial review', in *The Expanding Spaces of Law: A Timely Legal Geography*, eds I. Braverman, N. Blomley, D. Delaney, & A. Kedar, Stanford Law Books, Stanford, CA.

Head, L. 2016, *Grief and Hope in the Anthropocene*, Routledge, Abingdon.

Hinkel, J., Jaeger, C., Nicholls, R. J., Lowe, J., Renn, O., & Peijun, S. 2015, 'Sea-level rise scenarios and coastal risk management', *Nature Climate Change*, vol. 5, pp.188–190.

Hurlimann, A., Barnett, J., Fincher, R., Osbaldiston, N., Mortreux, C., & Graham, S. 2014, 'Urban planning and sustainable adaptation to sea-level rise', *Landscape and Urban Planning*, vol. 126, pp.84–93.

Intergovernmental Panel on Climate Change, 2007, WRII Chapter 11, Contribution of Working Group II to the Fourth Assessment Report of the Intergovernmental Panel on Climate Change. Cambridge University Press.

Kreller, A. M., & Graham, S. 2018, 'Fair for whom? How residents and municipalities evaluate sea level rise policies in Botany Bay, Australia', in *Routledge Handbook of Climate Justice*, Routledge, Abingdon, ed. T. Jafrey, pp.337–353.

Latour, B. 2017, *Facing Gaia: Eight Lectures on the New Climatic Regime*, Polity Press, Cambridge.

Landry, R. J. III. 2016, 'Empirical scientific research and legal studies research — A missing link', *Journal of Legal Studies Education*, vol. 33, no. 1, pp.165–170.

Local Government Act 1993 (NSW).

Macintosh, A., Foerster, A., & McDonald, J. 2013, *Limp, Leap or Learn? Developing Legal Frameworks for Climate Change Adaptation Planning in Australia*, Final Report for the National Climate Change Adaptation Research Facility, Gold Coast.

Mansvelt, J., & Berg, L. D. 2005, 'Writing qualitative geographies, constructing geographical knowledge', in *Qualitative Research Methods in Human Geography*, Oxford University Press, Oxford, ed. I. Hay, 2nd ed, pp.252–273.

McDonald, J. 2007, 'A risky climate for decision-making: The liability for development authorities for climate change impacts', *Environment and Planning Law Journal*, vol. 24, no. 6, pp.405–416.

McDonald, J. 2010, 'Mapping the legal landscape of climate change adaptation', in *Adaptation to Climate Change: Law and Policy*, eds T. Bonyhady, A. Macintosh, & J. McDonald, Federation Press, Alexandria.

McDonald, J. 2011, 'The role of law in adapting to climate change', *Wiley Interdisciplinary Reviews: Climate Change*, vol. 2, no. 2, pp.283–295.

McDonald, J. 2014, 'A short history of climate adaptation law in Australia', *Climate Law*, vol. 4, pp.150–167.

Measham, T. G., Preston, B. L., Smith, T. F., Brooke, C., Gorddard, R., Withycombe, G., & Morrison, C. 2011, 'Adapting to climate change through local municipal planning:

Barriers and challenges', *Mitigation and Adaptation Strategies for Global Change*, vol. 16, no. 8, pp.889–909.

Moser, S. C. 2010, 'Now more than ever: The need for more societally relevant research on vulnerability and adaptation to climate change', *Applied Geography*, vol. 30, no. 4, pp.464–474.

Moser, S. C., & Ekstrom, J. A. 2010, 'A framework to diagnose barriers to climate change adaptation', *Proceedings of the National Academy of Sciences*, vol. 107, no. 51, pp.22–26.

O'Donnell, T., & Gates, L. 2013, 'Getting the balance right: A renewed need for the public interest test in addressing coastal climate change and sea level rise', *Environment and Planning Law Journal*, vol. 30, no. 3, pp.320–335.

O'Donnell, T. 2016, 'Legal geography and coastal climate change adaptation: The vaughan litigation', *Geographical Research*, vol. 53, no. 1, pp.201–213.

O'Donnell, T. 2017, 'Adaptation on the Australian East Coast', in *Contested Property Claims: What Disagreement Tells Us about Ownership*, eds M. H. Bruun, P. J. L. Cockburn, B. S. Risager, & M. Thorup, Routledge, pp.151–165.

O'Donnell, T. 2019a, 'Coastal management and the political-legal geographies of climate change adaptation in Australia', *Ocean and Coastal Management*, vol. 175, pp.127–135.

O'Donnell, T. 2019b, 'Contrasting land use policies for climate change adaptation: A case study of political and geo-legal realities for Australian coastal locations', *Land Use Policy*, vol. 88, pp.104–145.

Osbourne, N. 2019, 'For still possible cities: A politics of failure for the politically depressed', *Australian Geographer*, vol. 50, no. 2, pp.145–154.

Palutikof, J. P., Barnett, J., Boulter, S. L., & Rissik, D. 2015, 'Adaptation as a field of research and practice: Notes from the frontiers of adaptation', in *Applied Studies in Climate Adaptation*, eds J. P. Palutikof et al., Wiley Blackwell, Chichester, pp.6–20.

Participant interview, Kelly, Port Stephens Council Officer (3 August 2011).

Participant interview, Bob, Port Stephens Councillor (21 May 2012).

Peel, J., & Godden, L. 2009, 'Planning for adaptation to climate change: Landmark cases from Australia', *Sustainable Development Law and Policy*, vol. 9, no. 2, pp.37–41.

Peel, J., & Osofsky, H. M. 2015, *Climate Change Litigation: Regulatory Pathways to Cleaner Energy*, Cambridge University Press, Cambridge.

Pierce, J., & Martin, D. 2017, 'The law is not enough: Seeking the theoretical "frontier of urban justice" via legal tools', *Urban Studies*, vol. 54, no. 2, pp.456–465.

Poe, M. R. (2018). *Immersive Study Takes a Toll. We are Insider-outsider*, 29 December [Twitter]. https://twitter.com/mpoetree/status/1078671204783779840.

Port Stephens Shire Council v Booth and Gibson [2005] NSWCA 323.

Pue, W. 1990, 'Wrestling with law: (Geographical) specificity vs. (Legal) abstraction', *Urban Geography*, vol. 11, no. 6, pp.566–585.

Robinson, D. F. 2013, 'Legal geographies of intellectual property, "traditional" knowledge and biodiversity: Experiencing conventions, laws, customary law and karma in Thailand', *Geographical Research*, vol. 51, pp.375–386.

Rose, G. 1997, 'Situating knowledges: Positionality, reflexivities and other tactics', *Progress in Human Geography*, vol. 21, no. 3, pp.305–320.

Silverman, D. 2006, *Interpreting Qualitative Data*, Sage, London.

Tamanaha, B. Z. 1997, *Realistic Socio-Legal Theory: Pragmatism and a Social Theory of Law*, Clarendon Press, Oxford.

Waters, E., Barnett, J., & Puleston, A. 2014, 'Contrasting perspectives on barriers to adaptation in Australian climate change policy', *Climatic Change*, vol. 124, pp.691–702.

Watkins, D., & Burton, M. (eds) 2013, *Research Methods in Law*, Routledge, Abingdon.

Wise, R. M., Fazey, I., Stafford Smith, M., Park, S. E., Eakin, H. C., Archer Van Garderen, E. R. M., & Campbell, B. 2014, 'Reconceptualising adaptation to climate change as part of pathways of change and response', *Global Environmental Change*, vol. 28, pp.325–336.

Valverde, M. 2015, *Chronotopes of Law: Jurisdiction, Scale and Governance*, Routledge, Abingdon.

Vaughan v Byron Shire Council [2009] NSWLEC 88.

Yin, R. K. 2009, *Case Study Research: Design and Methods* (4th ed), Sage, London.

8

LEGAL GEOGRAPHY – PLACE, TIME, LAW AND METHOD

The spatial and the archival in "Connection to Country"

Lee Godden

Introduction

"Connection to Country" is a fundamental, constitutive element of establishing native title in Australia under the *Native Title Act* 1993 (Cth) (NTA). It is used both instrumentally and rhetorically in Australian settler law as a signifier to reference the Aboriginal law and custom that connects Aboriginal peoples[1] to their traditional places. It is NOT to be confused with that law and custom. Rather, the construct provides a splice of space–time, in which that law and custom must flow from the past to the present, to gain recognition under settler law as native title. By contrast, connection for Aboriginal peoples is more firmly embedded in sacred obligations to place. As Aboriginal leader and anthropologist Marcia Langton explains, Aboriginal proprietary interests are "a spiritual bequest and subservient to the social relationships with the sacred ancestral past from which the legacy is derived". The bequest bestows not just rights but "a social world and for particular people arduous duties and responsibilities that are transmissible across generations" (Langton 2010, p.76).

Yet, as connection to country remains central to settler recognition of native title (Strelein 2001), this chapter seeks to unravel settler/Aboriginal law interconnections through the lens of legal geography method. Legal geography typically seeks to make apparent how "time", "space" and "law" crystallise around a conception of place (see Braverman et al. 2014, Blomley et al. 2001). This chapter brings together perspectives on methodology that reference the transdisciplinary fields of geography, law and history – denoted methodologically as referencing the spatial, the interpretive and the archival (Keenan 2015; Luker 2016), although law as much as history can be understood in its methodology as archival (Birrell 2010).

Analysing place as connection to Country canvasses a wide range of attendant epistemological and disciplinary knowledge practices that includes Aboriginal peoples' narratives (Black 2011), historical documents and anthropological "family trees" to define apical ancestors to determine the claimant group, as well as voluminous legal materials ranging from legislation to tenure histories. These sources of knowledge must be referable to the spatial dimensions of a native title claim for determination or compensation. The native title claims process and agreement-making are heavily reliant upon geographical methods of geospatial mapping and representation (Bowen 2010). Databases of registrations of claims and determinations, complete with online visualisation, mapping and search tool, are the techno-spatial underside to the judicial, litigation and administrative processes that constitute the conventional legal methods used to determine connection to Country. Further, all sources must be legitimated, and attachments to place must be ordered and presented according to the strictures of evidential and substantive native title law.

Native title, then, is a claim of Indigenous peoples' belonging to a specific place where the contextual and affective practices of connection must be demonstrated to exist from time immemorial to the present (ALRC 2015, Chapter 4). Povinelli (2002, p.49) argues that the sustained performance of these practices enlivens a "discursive passage into being" predicated not on the actual Aboriginal claimants, but rather upon the artefact of the Indigenous subject constructed by such traditional practices. Notwithstanding, such practices must be sufficiently robust to have endured the imposition of a colonial jurisdiction, and the ongoing impacts of an imposed settlement.

Analysing connection to Country as place

Native title typically has been analysed by critical legal and legal geography scholarship with reference to competing claims of sovereignty and the displacement of the legal categorical of *terra nullius* (see for example, Wolfe 1994). *Terra nullius* has become more than a legal fiction anchored in common law cases applicable to the colonisation of Australia (Attwood 2004, pp.7–8). The construct has become a template for a universalised methodology that has become a generic form of postcolonial deconstruction to reveal an underlying Indigenous law and material attachment to place (Kedar et al. 2018).

The methodology of "unsettling" colonial law is apparent in various postcolonial settings – typically relying on the prior idea that the settler law was emplaced via the assertion of colonial jurisdiction (Dorsett & McVeigh 2012). This iteration between colonial jurisdiction and sovereignty implies a certain jurisdictional method; such a method is replete with embedded assumptions about space–time relationships that enabled Britain as a colonising power to institute law and order from imperial centres to colonial peripheries (see Benton & Ford 2016, pp.1–6). That settling of the dominant law also can be envisaged by another movement analogy. Barr (2017, p.222) takes the methodology of law's movement a "step" further to argue that the common law was transferred by the bodies of common law subjects – principally their feet! "More than just questions of place, through

Blackstone's formulation, common law walking – our legal footprints – become a place-*making* and arguably a place-*taking* activity" (Barr 2017, p.230). There may be a debate over whether the law attached to the feet of colonists or to their status as British subjects. Nonetheless, the merging of movement, legal status and embodied carriage of law allows us insight into a different methodological position on space, time and law. Similarly, unpacking the layering of settler law (see for law and tenure in south-east Asia, Gillespie 2016, pp.258–60) envisages a method where the slices of space through time are exhumed by a forensic judicial analysis to reveal Indigenous relationships to land (Godden 2003), to reveal land as a narrative source of law (Black 2011) and to reveal networks of dispossession (Lester & Laidlaw 2015). Such methods largely still retain the primacy of law as an ethical conduct in structuring the space–time dimensions of place.

If we examine how place figures in the interstices of settler law, then defining Country becomes vital. It is always for Aboriginal peoples and Torres Strait Islanders to speak for their traditional places or to maintain its silences. As Kevin O'Brien notes:

> Country is an aboriginal Idea. It is an Idea that binds groupings of aboriginal people to the place of their ancestors, past, current, and future. It understands that every moment of the land, sea and sky, its particles, its prospects and its prompts, enables life. (cited in Barr 2017, p.232)

Country is "located in a material place as always being somewhere, Country is not only temporal, but spatial too ... the land itself can be thought of ... as a material form of Aboriginal law" (Barr 2017, p.232). Traditional Country may be a material expression of Aboriginal law, nonetheless, an intrusion of the native title process is the requirement for Aboriginal peoples to make manifest to the settler law the physical and spiritual meaning pertaining to Country. The NTA was initially drafted to reflect the common law in *Mabo v Queensland (No. 2)* (1992) 175 CLR 1 (Mabo [No 2]). Subsequent judicial interpretation of native title in Section 223 of the NTA has significantly altered that formulation. A series of High Court decisions narrowed both the scope and propriety character of native title, conceiving it as a "bundle of rights", rather than a place or even a title (Strelein 2009, pp.38–40). In association with legislative amendments in 1998 that delivered extensive grounds for extinguishment of native title, the relationship between Aboriginal peoples and Country, realised through native title law, has become increasingly fractured and partial, where native title coexists with, but yields to, other interests, and where the threads of belonging are attenuated.

While ostensibly the law determines the connection to Country for any group of native title claimants, the constitutive dimension inferred from this term is undercut by a legal test that emphasises first the need for "continuity" of law and custom since prior to the assertion of colonial sovereignty (*Members of the Yorta Yorta Aboriginal Community v Victoria* (2003) 194 ALR 538 (Yorta-Yorta)). It simultaneously emphasises the additional, onerous requirement that the connection has been "substantially uninterrupted" since prior to the assertion of sovereignty (*Bennell v State of Western Australia* (2006) 230 ALR 603, 609). Thus, an interruption to connection acts as a potential rupture – the coming into law of a non-place.

Understanding how substantial interruptions break connection to Country is the specific methodological focus for the remainder of this chapter. Thus, attention is turned from continuity through time, toward ruptures and discontinuities – to a point in time where there is no longer an Aboriginal law and custom extant in a place that can be made recognisable to settler law. The native title narrative of law, space and time reveals the requirement for "a legally prescribed 'traditional' indigeneity, which requires the performance of an unachievable embodiment of indigeneity" (Birrell 2010, p.81). The demand for an authentic cultural practice across time is made of many Indigenous communities that claim land and resources. In this manner, place is understood as coextensive with identity: tying peoples to specific sites and localities (McNeil 1989). Extensive commentary has illuminated the limitations inherent in the tests for continuity and connection (Strelein 2009, pp.74–80). The conundrum at the heart of native title law is revealed in that those Aboriginal communities most impacted by colonisation are least able to make a claim to place.

The intention here is not to simply rehearse these compelling issues of inequity and dispossession although they reveal much about "the relation between spatiality and subjectivity and the role of law in producing that relation" (Keenan 2015, p.63). Rather, the intention is to probe by what method(s) are ruptures in the continuity of connection achieved? How is the space–time of Aboriginal relationships to place rendered discontinuous? How does law simultaneously perpetuate discontinuities while ensuring narrative continuity with respect to place and property?

Places of belonging and method

To comprehend disconnection requires investigation of the methods that define space–time in its relationship with law, as indicative of the spatial turn in the social sciences and humanities (Keenan 10, p.17). Valverde (2015, pp.31–2) notes the need to jointly consider space–time, and to be aware that spatialisation affects temporalisation, and vice versa. She also urges not to reify space per se. By contrast, the consideration of "space" by law and the turn to materiality (Davies 2017b) was predicated partly on the idea that law as a discipline, and as matter of professional practice, is spatially abstract (Keenan 2015, p.10). Blomley (1994, p.76), stated that "[t]he essence of law is always understood as disembodied rather than formed in material and historical conditions". At the very least, law is seldom distinguished as a rhetorical methodology that references the affective, social and material spheres (Manderson 2015, pp.1301–2).

The space–time dimension in legal scholarship is now widely explored (see generally Davies 2017, pp.436–50). As Davies (2017, p.437) notes, "in thinking about the materiality of law ... recent theory has moved beyond human interactions to conceptualising the connections between law and the physical world". This inquiry asks how law "shapes" physical or material space, and in turn how law manifests in material form (Davies 2017, p.437). Legal geography scholars have long examined how space is entangled with the social, material and political sphere in which law inheres (Keenan 2015, pp.36–7; Bartel 2016). Drawing on a long tradition of geographical method, they have mapped key aspects of the interdisciplinary research field (Bartel et al. 2013) and provided a theoretical framing (Holder & Harrison, 2003; Graham & Bartel 2016). Legal geography posits space as integral to disciplinary epistemologies, and as linked ontologically with existence; or conceived as unbounded, open and malleable.

Geography – space and place

Indeed, the valorisation of place and community has been a means to subtend the dualism of Cartesian space and time constructs in geography (see for example Lefebvre's three forms of space, Lefebvre 1991). It has formed a bridge across the methodological divides of physical and human geography. Doreen Massey (2005) has been highly influential in encouraging research on space as a series of connections, patterns or nodal points. This conception owes as much methodologically to the quantitative "revolution" in 20th-century geography, as to the humanistic, social space of feminist geographers, to actor network theories or even to Foucauldian knowledge structures, with embedded socio-spatial power relationships.

Harvey's approach to spatial justice drew from Marxist theory and the emergent interest in quantification across "space", utilising computer technologies to undertake research with distributive justice implications (Harvey 1981). For Harvey (1996, p.207), such socio-spatial constructs operated as a "reference system by which we locate ourselves with respect to that world". Materiality, in this sense, still implicitly references Western scientific thought even if no longer as rudimentary as Cartesian space constructs. The culmination of these methodological trends has been to encourage research on space, place and time that transcends static scientific understandings. There have been other efforts to develop comprehensive positions on law and geography, such as the nomosphere (Delaney 2010). Experimentation continues with articulating a coherent methodology to comprehend the diverse approaches (Keenan 2015, Philippopoulos-Mihalopoulos & Brooks 2017). In tandem with legal geography's concern with new materialisms is increasing attention to alternative conceptions of time and space offered by Spinoza's philosophy, along with a resurgence of interest in the work of Deleuze and Guattari (1987) with their signature concepts, such as rhizomes, capturing a sense of movement, entanglement and affect.

"Affect" has become a prominent methodological position in legal geography with scholars exploring a concern with how landscapes are experienced emotionally and how an embodied organism, such as a human, is affected by the smells, sounds, touch – and by its interactions with other organisms and things (Philippopoulos-Mihalopoulos 2013). Such themes in legal geography and law have merged with long-standing scholarship, largely from a feminist standpoint, on the body. Indeed Olson (2016) argues that the affective dimension and its understanding of materiality and body has challenged the classic reliance on the subject, rationality and human agency as an explanatory mode in law and society. She asks, "is this a more radically de-historicized notion of affect?" (Olson 2016, p.350).

The "placeness" of connection to Country

Given this convergence theoretically and methodologically between space, materiality, affect and situated "bodies", where does this leave Aboriginal peoples' disconnection to Country? We begin with the view that place is subsidiary to space (Keenan 2015, p.40). The reversal of the space–place continuum reflects Massey's theoretical position that we should not seek bounded places but rather their articulation "as moments in networks of social relation and understandings" (Keenan 2015, p.41). Such a move from positionality to spatiality, and studying space per se, contrasts with conceptions of place as somewhere definite and tangible, and productive of memory in everyday life (Keenan 2015, p.41). While this conception of place reflects both affective and embodiment tropes, places are dependent on "space" as the reference system. Thus, "[p]lace is a particular point in space, whether that point is understood as an intersection, a node, a switching point, a location, a feeling or some other referent" (Keenan 2015, p.12). Native title law has drawn extensively upon spatial representational methods such as geographic information systems (GIS) mapping and registration systems within institutional settings, such as the National Native Title Tribunal, to give tangible form to nodes, locations and intersectional points between settler and Aboriginal law – again privileging the Western methods and knowledge practices, while rhetorically situating native title in traditional "space".

In this methodological structuring of the relationship between space and place, we can discern native title law's conception of connection and continuity. The referent for connection is Aboriginal traditional law and custom (Young 2008), but it must be experienced as a process in continual, iterative flow to engage "[t]he common association of place with exclusive belonging" (Keenan 2015, p.41). If place is held to be dependent on iteration of law and custom, and associated affect that situates such places materially and discursively vis-à-vis a wider settler jurisdictional space, then it also becomes feasible to conceive of breaks or interruptions to that process, feeling or flow of affect and attachment. Significantly, the substantially uninterrupted test is remarkably reticent in directly identifying the specific acts of physical violence, and the regulatory controls over Aboriginal peoples' lives that constituted those breaks to the process or flow. In short, it becomes possible to instigate a test for continuity that contains the potential to precipitate substantial interruptions to connection when that process is conceived as an embodied and material articulation of the practices of belonging.

Much analysis of place by writers aligned to legal geography seeks to contrast places of "belonging" (Cooper 1998) with marginalisation (i.e. as a dichotomous positionality of the subject). Native title law has been strongly critiqued as a form of marginalisation of Aboriginal subjects and places. While that critique may be valid, it does not deal fully with the prior question of how the law both realises places and non-places for Aboriginal peoples due to loss of connection. Having examined space and place and the implications for connection to Country there is an equally urgent imperative to think through how method, "time", "history" and "materialism" are implicated in postcolonial scenarios of connection and dispossession.

The historical turn

The relationship between space, history and law has been deeply theorised, and full explication is beyond the scope here. If we are to unravel some of the complex threads of the interrelationships between space, time and law, then an obvious starting point is Marxist theory. Marxism has been highly influential in historical discourse and method and, even if somewhat obliquely, in legal geography and the spatial turn in law. It is axiomatic to Marxist positions that there is a dialectical relationship between history, society and materialism. Marx's early writings realigned the dialectical method but rejected the Enlightenment assumptions about "reason" as the driving impetus for history and law. As Douzinas and Geary (2005, p.203) suggest, Marxist theory claims "that the inevitable triumph of socialism would confirm that Marxism correctly explains history and society. The link between theory and reality is provided by political practice". As the predicted revolution has not occurred, it challenges the extent to which Marxism accurately reflects social reality. Thus, Marxist analysis later reified law as integral to the ideological superstructure to the economic base of society. Efforts to suggest a less instrumental function favour a "reinvented Marxist legal theory that can articulate the complex relations and differences between law and economy – providing a place for the subject and politics" (Douzinas & Geary 2005, p.204). The subject is interpellated into a jurisdictional "space" in answering the call of law (Douzinas & Geary 2005, p.204). There are clear resonances of a material understanding of the situated subject, with a gesture also to the constitutive performative of law.

From a methodological standpoint, classic Marxist methodology sits firmly within 19th-century positivism. Positivism emphasises the empirical analysis of causation and thereby the purported explanation of social phenomena (Douzinas & Geary 2005, p.203). While sharing positivist method (empiricism), mainstream historical positions embed a specific modernist conception of time and history, based on periodisation and narrative explanatory form. This implicit model was typically accompanied by detailed examination of the places and contextual factors that shaped that historical change. The "empirical" scale within the historical narrative is disparate - embracing Big History, Deep History, Global History

and Local History, as well as histories "located" by reference to thematic structures of gender, race and class. These scales typically are conjoined by a narrative
method that seeks to elucidate the political, social and economic characteristics
of that period by reference to historical data.

Legal history's conventional methodology largely reflected the positivist legacy
of the two convergent disciplines. The method constituted various phenomena as
historical in character by locating them in temporally discrete empirical contexts or
periods. In turn, "[i]t explains the reality of law, by assessing change over time in law
relative to the contextualizing domain (society, polity, economy) from which it is held
relationally distinct" (Tomlins 2017, p.3). The resurgence of critical legal histories in
late 20th-century scholarship produced a more relationally complex sense of history
in law and society scholarship (Tomlins 2016). History (or perhaps historicism) was
integral to this revival. The view of law as contingent, as being made and unmade,
as well as being embedded in the specificities of context (Tomlins 2017, p.58) made
the present state of law appear more open to change (Gordon 2017, p. 285). Tomlins
(2017, p.67) suggests that "[h]istoricist law is plural, contested, socially constructed,
vernacular. Its meanings are produced from, created by, circumstances (contexts) that
it simultaneously moulds, to such an extent that if context changes then meanings
change with it". Historicism allows a critique of legal method that can destabilise
law, where the present law is perceived as reliant upon an authoritative past (Gordon
2017). Critical legal theorists identified how lawyers sought to use history through
precedent as a justificatory rationale for the legal status quo. Indeed, Gordon argued
that lawyers used history "to endow currently dominant claims of entitlement and
distributions of legal advantage and modes of legitimating property and power with
the authority of the past" (Gordon 2017, p. 288; see also Gordon 2012).

Applying such insights to native title and settler history, legal case method
– tied tightly to Privy Council jurisprudence in the late 19th century – was
instrumental in justifying the colonial distribution of entitlement and power
across the British Empire, and in confirming the dispossession of Indigenous
peoples. While the general proposition of British colonial territorial acquisition
was largely undisturbed in *Mabo [No 2]* (Ritter 1996 p. 6), the authority of the
past was sufficiently destabilised to allow judicial interpretation of *terra nullius*
to be rendered historically contingent (Attwood 2004, p.4), made referable to a
colonial context and no longer perceived to be in accordance with contemporary
moral and national values (Attwood 2004, p.13). Such an appeal to a judicial historical method is not simply a displacement of existing law as spatially and temporally dependent (Attwood 2004, p.3). As Tomlins (2017) notes, "[m]ateriality
is not context" – it is not reducible to the backdrop against which subjects initiate
actions with legal effect. Yet for all its limitations, juridical history has given the
potential to realise a more diverse materiality of Aboriginal peoples' presence in
the land and the law (see e.g. the allegorical materiality in Wright 2013).

Indeed, if matter has its own "ongoing historicity" (Olson 2016, p.821),
then conventional legal-geographical-historical analysis becomes problematic.
Rather than identifying a standpoint of subjectivity and historicising from that

positionality, we need different methodological practices (Olson 2016, p.822). Attention to new materialisms, where legal geography seeks to uncover the intersectionality of the spheres of the physical (material), social and ethical-legal across a space–time continuum, opens the possibility of considering the materialisation of places, but equally the potential for discontinuities. In short, we might examine method in native title law as an archive of documentary sources, discursive and professional practices and performances from which lawyers, but most obviously judges, select from that archive to rhetorically fashion the contours of the connection to Country and the substantially uninterrupted "test".

The progress model of history and decolonising method

Before considering native title law through the rubric of the archive, we need to disengage the progress model of history. As a testimony to the power of progress in the 19th century, Darwin's evolutionary theory was aligned to the social objective of proclaiming that a select group of humans (those from civilised or developed countries) were innately superior to nature – and therefore to other peoples (Tuhiwai Smith 2013, p.94). From an Indigenous perspective, Williams' (2012, see especially pp.206–10) seminal research challenges the assumptions of that historical explanation to reveal how the historical trajectories of Western civilisation are founded upon an anxiety about savagery. Classic Western histories of the world seek to distance themselves from the barbarian – including the Indigenous "other" – in the polarities of the natural and the civilised. Thus, historical progress encompasses culturally specific conceptions of time, space, movement and subjectivity. Moreover, as Tuhiwai Smith (2013) argues, history in contemporary society is a contentious issue for many Indigenous communities:

> It is not only the story of domination; it is also a story which assumes that there was a 'point in time' which was 'prehistoric'. The point at which society moves from prehistoric to historic also is the point at which tradition breaks with modernism. (Tuhiwai Smith 2013, p.113)

Where tradition breaks with modernism is also problematic within native title law for Aboriginal peoples who were strongly impacted by colonisation and settler development. It is the point where the continuity of connection can become "substantially interrupted" by an intrusion of modernism (see generally, Fitzpatrick 2001).

Critiques of the historical progress model are well-established. Legal geography scholarship has sought to pre-empt the founding assumptions of the progress model (Massey 2005). Postcolonial scholarship in law has critiqued Modernity's narrative of progress (e.g. Kapur 2006, pp.668–9) and the purported historical trajectories used to subjugate colonised peoples (Darian-Smith 2013, p.252). Recent attention is directed to the methodological implications, as Indigenous scholars call for a decolonisation of research:

> From an indigenous perspective, Western research is more than just research that is located in a positivist tradition. It is research which brings to bear, on any study of indigenous peoples, a cultural orientation, a set of values, a different conceptualization of such things as time, space and subjectivity, different and competing theories of knowledge, highly specialized forms of language, and structures of power. (Tuhiwai Smith 2013, p.92)

The collection of documents and assemblage of practices that constitute knowledge of Indigenous peoples have embedded "rules of practice" (Tuhiwai Smith 2013, p.93). Following Foucault's invocation of the episteme, Western civilisation (and arguably its researchers) is not directly cognisant of "taken for granted rules" of classification and representation. Such assemblages comprise perceptions of human nature, human morality and conceptions of space and time, gender and race (Tuhiwai Smith 2013, p.96). The rules of practice allow knowledge systems to stabilise the modes of classifications and systems of representation. Western settler law for example, relies upon a complex mixture of positivist science intermingled with rhetorical systems of language and affective description for realising the materiality of space and time.

Moreover, the stability of such systems of classification and representation are the constitutive fabric against which different cultural traditions are selectively archived. The archive is significant in Western research, including legal geography, which seeks to assemble text, images and artefacts pertinent to Indigenous cultures (see e.g. Geertz 1979). Fragments and artefacts of other cultures can be "retrieved and reformulated in different contexts as discourses, and then [can] be played out in systems of power and domination" (Tuhiwai Smith 2013, p.93). An archive can be reformed around new administrative and socio-political requirements, but typically as a material method for describing and ordering the world, the archive embeds culturally specific values. As Birrell (2010, p.81) notes, "[w]ithin the Australian legal archive, in native title jurisprudence and beyond, the Indigenous subject is paradoxically positioned as, simultaneously, the subject of European imperial conquest and classification, and as a unique and transgressive alterity".

Growing awareness of the negative consequences of archives for Indigenous peoples has precipitated demands for control of archives, the return of records about Indigenous peoples, as well as Indigenous-controlled archives (Buchanan 2007). Postcolonial scholarship has challenged the neutrality of archives as repositories of information and has constituted them as historical forces in themselves – "less as stories for a colonial history than as active, generative substances with histories, as documents with itineraries of their own" (Stoler 2009, p.1). Drawing on feminist approaches to historiography, Luker (2017, p.109) argues that "attention to archival records have provoked affective responses that reveal the way archives have the capacity to 'motivate, inspire, anger and traumatize'". The politics of reconciliation has driven Indigenous peoples' demands for decolonisation of settler archives and sought to alter archival theory and practice. This opens up a further question of whether methodologies based in archival theories may offer a more nuanced accounting of time and space, and connection to Country in native title law.

Law and archive

Yet, still, "[i]n settler colonial contexts such as Australia and Canada, moves towards reconciliation between settlers and Indigenous peoples are betrayed in the archive" (Luker 2017, p.108). Many settler governments are reluctant to allow full transparency of, and access to archives in relation to events, such as forced Indigenous family separations, the movement of Indigenous peoples into reserves and the violence of colonial dislocation and dispossession more generally. The impacts of these historical events on Indigenous peoples are made material to settler law as the legitimate, i.e. authoritative, recording of their occurrence in archival form. Yet those actions, laws and regulation are exactly the type of events likely to give rise to the euphemised "interruptions" that sever the desired continuity in a claim for determination of native title. The ambivalent function of archives in Indigenous peoples' dispossession highlights the importance of using such a methodological lens to examine time, law, space, in relation to connection and the substantially uninterrupted test.

Law and the archive are integrally linked in postmodern legal theory in the scholarship of Derrida (Derrida 1992). Derrida's conception of the archive and law (Derrida 1995) is derived not so much discursively as ontologically. The archive comprehends a theory on the origins of law – how law becomes institutionalised or inaugurated as law (Birrell 2010, p.93). The archive traditionally delineates the site from which the law is drawn and manifests the metaphorical space of law's authority (Motha & van Rijswijk 2016, p.11). The archival moment is one where law simultaneously inscribes itself as law and establishes the right or authority of that law, while admitting a technical repository of official documents and records (Birrell 2010, pp.93–4). In relation to colonial settler law, the assertion of sovereignty by the colonial power simultaneously creates the legal archive (British colonial law) and sets its limits that are articulated, authored and authorised by the unlimitable sovereign (Birrell 2010, p.95). In this sense, the archival sits at the foundation of law and the constitution of the state.

Moreover "[t]here is no political power without control of the archive, if not of memory" (Birrell 2010, p.95). The archive at law is understood not just as a repository of memory, but as a site of simultaneous remembrance and forgetting. The archive's apparent completeness and authority act to constitute the past, necessitate a forgetting of the instigation of law, such that it becomes a mere "technology" of authority. The function of the archive in constituting the law, and the public memory of a people and nation through the authority to act to constitute the past, illuminates why the concerns of Indigenous scholars with archives are so pressing for persons marginalised from and by that authority. Addressing similar concerns, Motha and van Rijswick (2016, p.12) in a collection adopting a counter-archival stance, "interrogate the teleological narratives of progress that law constitutes after violent events" to examine potential points and means of resistance to the archival power of the state.

Native title law as archive

At a more immediate level for native title law, the relationship between law and archive is evident in the growing attention to the role of courts as repositories of a vast amount of information calibrated according to the exigencies of legal categories, classifications and evidential practices that accumulate as a documentary recording of places, claimants, agreements and determinations. The "Common law archives itself ceaselessly. Indeed, it exists in order to create an Archive" (Luker 2016, p.88). Thus, we can bring such insights around the overarching impulse in law to institute and archive to bear in terms of how it applies in respect of connection to Country. Native title law operates upon similar historically repressive archival rules to those critiqued by Tuhiwai Smith (2013). In the choices of archival sources, however, it is possible to stabilise or destabilise the teleological narrative of progress that has followed the violence of colonisation in Australia.

Aboriginal claimants' connection to Country under the NTA is formed from the materials and processes collated as evidence brought to the court for a determination of native title or presented to governments and other parties. A native title claim can be litigated, and the determination reached by the court on the evidence led, or a determination can be by consent of the parties (NTA ss. 61, 223, 225). Biber and Luker (2014, p.6) have examined how legal evidence can be deployed as archive. Similarly, we find an extensive archive of evidence in native title law that comprises traditional knowledge of law and custom (typically offered in oral narrative form by Indigenous witnesses), expert evidence from anthropologists, archaeologists, historians and other nominated experts, documentation – both historical and contemporary, and the vast accumulation of the registration processes pertaining to land law and tenurial systems. This archive creates a detailed but still selective accounting of the connection to Country for Aboriginal peoples in Australia. Further, the archive sources must be tested against a high evidential threshold where the failure to produce a continuous record – an assemblage of artefacts that produce a space–time continuum – is fatal to the claim. Scholars have argued the imperative of acknowledging law's counter-archive, through a reflection on the function of archival sources in government actions and in legal proceedings (Motha & van Rijswick 2016, p.13). Rather than regarding archival sources as simply documents, they are better understood as artefacts of knowledge, events, ideas and practices. Such methods bring with them "a distinctive epistemology, ethics, and aesthetics, demanding that scholars reflect on knowledge-making practices" (Biber & Luker 2014, p.8).

The Federal Court of Australia ("the Court"), with its specialist jurisdiction in native title, has been central to the jurisprudence on connection and continuity, and thereby in articulating where the discontinuities of time may interrupt traditional law and custom. While the Court has developed extensive jurisprudence on native title, simultaneously as a Court of Record it holds a large archive of evidence submitted for native title determinations and litigation on related matters. Debates continue over the ownership of and control over such materials between the Court and claimants. While much of the archive exists in documentary form, increasingly native title has an online presence that makes it not only tangible but also a physically searchable archive. In this way, law is a process of movement – "weaving", "shaping" and "formatting" – of material objects, including files, texts, litigants and decision-makers (Luker 2016, p.94). Aboriginal law and custom that establishes connection to Country therefore must be realised in the terms of Western knowledge and the spatial/historical archival method of the law.

Moreover, the practices of archival selection may actively work with the controls over the settler spaces that came into being through colonial territorial acquisition. Law's inauguration, as noted, legitimates the sovereign appropriations of space and time. The result across many societies has been the dispossession of Indigenous peoples from their traditional places, followed by an ad hoc gesture of "gifting back" small amounts of these lands and waters to Indigenous people (Tuhiwai Smith 2013, p.9; see also e.g. Deeds of Grant in Trust in Queensland). In native title, the process is not a "gifting back" per se as native title is not derived from the settler, tenurial land system. Rather, it is a recognition of the pre-existing (read perhaps prehistoric) rights. Thus, for space to become traditional place, it must be historicised as continuity within the legal archive. As Motha (2018, p.1) states, "[l]aw disavows violence in order to be free of an abhorrent past, and yet preserves it as such events are also the origin of present distributions of property, territory, and membership in community".

Courts as archives: the performance of connection in native title law

Bodney v Bennell [2008] FCAFC 63, known as the "single Noongar claim", marked the high water of a stringent substantially uninterrupted test for native title determinations, together with a high evidential threshold for establishing continuity in connection. Since that time there has been an engagement by the courts, and at times by governments (see e.g. the Victorian Land Justice process), with examining how evidential processes for the native title archive, including historical materials, can better represent Indigenous cultural values within the rigours of that "test". The Federal Court has since adopted a range of evidentiary and procedural changes in native title practice that, while still a far step from "decolonising the archive", can accommodate a more culturally relevant account of traditional law and custom in native title law (ALRC 2015, Chapter 7). While legal decisions may still performatively at one level reproduce the originary archive of sovereign violence of dispossession, there are measures in place to begin to "reorient the law in the wake of histories of violent sovereign impositions" (Motha & Rijswick 2016, p.11). These changes better capture the ongoing material and metaphysical presence of law and custom as embedded Aboriginal relationships to place (Clark et al. 2019, p.788). There is no longer the same prioritisation of documentary historical narratives over Aboriginal oral evidence that proved so devastating in *Yorta-Yorta* (Strelein 2009, p.78), where the claimants were held to have abandoned traditional law and custom for the benefits of modern agricultural society.

Expert evidence, particularly anthropological evidence, has expanded enormously in volume and scope in demonstrating and articulating Aboriginal claims to connection. The expansion of knowledge of connection has occurred alongside a growing awareness of the markers of traditional law and custom in the material landscape, such as evidence of Aboriginal trade networks, agriculture and aquaculture practices (Pascoe 2018) and extensive Aboriginal economies. Where once the native title "archive" typically included static archaeological and heritage sites caught in the warp of pre-historic time, punctuated by gaps and the voids filled by the march of Modernity, there are now vital songlines on country that performatively and materially illuminate the material and affective dimension of Aboriginal relationships to rivers, trees, mountains and other species (Clark et al. 2019, p.843). The materiality of Aboriginal connection to Country may have existed since time immemorial but Western culture and law remained largely oblivious and evidentially blind until "law" learnt to listen, touch and feel Country. The "unsettling" of settler law must occur not just at the originary moment of sovereign acquisition where the archive of violence is imbricated, but as an ongoing, open and grounded performative of reconciliation (McMillan & Rigney 2018).

Of particular significance in native title law and practice is the move by the Court to hear evidence on traditional Country. The performativity of judges walking on Country is not just an imposition of sovereign authority over Indigenous places. As Barr (2017) notes, walking carries the law and opens up the affective dimensions of the places in which it travels. The ceremonial act of a hearing on Country is an opportunity for a meeting of cultures and peoples in which law through the persona of the judge may understand Country and begin to realise transitional justice. "At its core, transitional justice works discursively to establish a break between the violent past and a peaceful, democratic future, and is based upon compelling frameworks of resolution, rupture and transition" (Kent 2016, p.1). The further step in that pathway is when the Federal Court convenes on the native title claimants' traditional Country to perform the emplacement of connection to Country associated with a determination of native title. To do so the Court, through its rhetorical exercise in judgment (Manderson 2015, p.1302), must sublimate the violence inherent to the substantially uninterrupted test in order to overcome the potential discontinuities of time and space; to find continuity and, then ritually in place, to confirm its "determination" of native title. The juridical history (Attwood 2004, p.2), that collapses the time from when Aboriginal peoples waved to Captain Cook to the present reality of law's normative demands, is shaped also by an implicit spatial method and ontology of places in confirming continuity (Attwood 2004, pp.6–7). Legal method may still elide the violence that is implicit to the substantially uninterrupted test, but from the abstractions of space, the court materialises a "place" of belonging; although sovereign attachment for First Nations is still elusive. A native title determination on Country constitutes a space in which law is performed, in which the performative subsumes both the narrative historical form of the evidential archive and the ceremonial of affect and sensation of being in place on Country.

Conclusion

The lens of legal geography has allowed insights into how the law and judicial practice governing connection to Country might be unsettled. From the vantage of legal geography method, working in combination with critical legal and historical approaches, connection can be understood at one level as a recognition space that represents a critical node or intersection of legal, economic and political systems. But that point of connection should not be understood as an abstract locator that traces a dominant settler thread in the land. Instead, the insights offered by the spatial and material turn in the social sciences and humanities suggest that such a space must reference an embodied and emplaced group of Aboriginal claimants with agency and resistance, not simply the abstracted subject of native title jurisdiction flattened by imperial law (van Rijswijk, H. 2014, p. 118). Moreover, rather than being identified with a static Indigenous identity over time, the scholarship of Massey and other postcolonial geographers is acutely relevant in challenging such research assumptions to begin decolonising methods and archives. Such intuitions also build on a large body of earlier legal geography scholarship that examined the spatial justice implications of law. Those methodological insights that highlighted how law is implicated in ongoing, structural disadvantage, with direct, material consequences, remain pertinent for Aboriginal communities in the native title space. A finding of substantial interruption to connection has significant financial and cultural ramifications. Engagement with more dynamic dimensions of connection to Country may also draw on the trend in legal geography that has extended its longstanding methodological concern with place to investigate the affective and grounded materiality of knowing that place. This chapter has not canvassed in any depth how Aboriginal peoples know place – that is always the prerogative of Aboriginal peoples to speak to their own methodology and traditional knowledge. A self-reflective engagement with method by researchers is important however to provide the "space" for those voices to be heard.

Note

1 In this chapter, the term Aboriginal peoples (which is intended to include Torres Strait Islanders where relevant) is used as the more locally appropriate language. Indigenous is used where the reference is intended to be of a more generic nature.

References

Attwood, B. 2004, 'The law of the land or the law of the land?: History, law and narrative in a settler society', *History Compass*, vol. 2, no. AU 082, pp.1–30.

Australian Law Reform Commission ('ALRC'), *Connection to Country: Review of the Native Title Act 1993 (Cth)*, ALRC Report 126, 2015, Sydney <https://www.alrc.gov.au/publications/alrc126>.

Barr, O. 2017, 'Legal footprints', *Law Text Culture*, vol. 21, pp.214–251.

Bartel, R., Graham, N., Jackson, S., Prior, J., Robinson, D., Sherval, M., & Williams, S. 2013. 'Legal geography: An Australian perspective', *Geographical Research*, vol. 51, pp. 339–353.

Bartel, R. 2016, 'Legal geography, geography, and the research-policy nexus', *Geographical Research*, vol. 54, no. 3, pp.233–244. 12p.

Bennell v State of Western Australia (2006) 230 ALR 603.

Benton, L., & Ford, L. 2016, *Rage for Order: The British Empire and the Origins of International Law, 1800–1850*, Harvard University Press, Cambridge, MA.

Biber, K., & Luker, T. 2014, 'Evidence and the archive: Ethics, aesthetics and emotion', *Australian Feminist Law Journal*, vol. 41, no. 1, pp.1–14.

Birrell, K. 2010, 'An essential ghost: Indigeneity within the legal archive', *Australian Feminist Law Journal*, vol. 33, pp.81–99.

Black, C. 2011, *The Land is the Source of the Law: A Dialogic Encounter with Indigenous Jurisprudence*, Routledge, London.

Blomley, N. 1994, *Law Space, and the Geographies of Power*, Guildford, New York and London.

Blomley, N., Delaney D., & Ford R. (eds), 2001, *The Legal Geographies Reader*, Wiley Blackwell, Oxford.

Bodney v Bennell [2008] FCAFC 63.

Bowen, P. 2010, 'Spatial technologies, mapping and the native title process', in *Comparative Perspectives on Communal Lands and Individual Ownership Sustainable Futures*, eds. L. Godden & M. Tehan, Routledge, Abingdon UK, pp.309–322.

Braverman, I., Blomley, N., & Delaney, D., & Kedar, A. 2014, *The Expanding Spaces of Law: A Timely Legal Geography*, Stanford University Press, Stanford.

Buchanan, R. 2007, 'Decolonizing the archives: The work of New Zealand's Waitangi Tribunal', *Public History Review*, vol. 14, pp.44–63.

Clark, C., Emmanouil, N., Page, J., & Pelizzon, A. 2019, 'Can you hear the rivers sing? Legal personhood, ontology, and the nitty-gritty of governance', *Ecology Law Quarterly*, vol. 45, no. 4, pp.787–844, doi: 10.15779/Z388S4JP7M.

Cooper, D. 1998, *Governing Out of Order: Space, Law and the Politics of Belonging*, Rivers Oram Press, London.

Darian-Smith, E. 2013, 'Postcolonial theories of law', in *An Introduction to Law and Social Theory*, eds. R. Banaker & M. Travers, Hart Publishing, Bloomsbury, UK, pp.1247–1264.

Davies, M. 2017a, *Asking the Law Question* (4th edition), Lawbook Co, Sydney.

Davies, M. 2017b, *Law Unlimited: Materialism, Pluralism, and Legal Theory*, Routledge Abingdon, UK.

Delaney, D. 2010, *The Spatial, the Legal and the Pragmatics of World-Making Nomospheric Investigations*, Routledge, Abingdon, UK.

Deleuze, G., & Guattari, F. 1987, *A Thousand Plateaus: Capitalism and Schizophrenia* University of Minnesota Press, Minneapolis.

Derrida, J. 1992, 'Force of law: The "mystical foundation of authority', trans. M. Quaintance, in *Deconstruction and the Possibility of Justice*, eds D. Cornell, M. Rosenfeld & D. Carlson, Routledge, New York, pp.3–35.

Derrida, J. 1995, *Archive Fever*, trans. E. Prenowitz, University of Chicago Press, Chicago.

Dorsett, S., & McVeigh, S. 2012, 'Conduct of laws: Native title, responsibility, and some limits of jurisdictional thinking', *Melbourne University Law Review*, vol. 36, no. 2, pp.470–493.

Douzinass, C., & Geary, A. 2005, 'Marxism, justice and the social order', in *Critical Jurisprudence: The Political Philosophy of Justice*, Hart, Oxford, pp.203–226.

Fitzpatrick, P. 2001, *Modernism and the Grounds of Law*, Cambridge University Press Cambridge.

Geertz, C. 1979, 'From the Native's Point of View: On the Nature of Anthropological Understanding', in *Interpretive Social Science*, eds R. Paul and W. M. A. Sullivan, University of California Press, Berkeley, pp.225–241.

Gillespie, J., 2016, 'A legal geography of property, tenure, exclusion, and rights in Cambodia: Exposing an incongruous property narrative for non-Western settings', *Geographical Research*, vol. 54, no. 3, pp.256–266.

Godden, L. 2003, 'Grounding law as cultural memory: A proper account of property and native title in Australian law and land', *Australian Feminist Law Journal*, vol. 18, pp.61–80.

Gordon, R. 2012, 'Critical legal histories revisited: A response', *Law & Social Inquiry*, vol. 37, pp.200–215.

Gordon, R. 2017, *Taming the Past: Essays on Law in History and History in Law* , Cambridge University Press, Cambridge pp.282–316.

Graham, N., & Bartel, R. 2016, 'Legal geography/ies', *Geographical Research*, vol. 54, no. 3, pp. 231–232.

Harvey, D. 1981, 'The spatial fix: Hegel, Von Thunen and Marx', *Antipode*, vol. 13, no. 3, pp.1–12.

Harvey, D. 1996, *Justice, Nature and the Geography of Difference*, Wood, Oxford.

Holder, J., & Harrison, C. (eds), 2003, *Law and Geography*, Oxford University Press, Oxford.

Kapur, R. 2006, 'Human rights in the 21st Century: Take a walk on the dark side', *Sydney Law Review*, vol. 28, pp.665–687.

Kedar, A., Amara, A., & Yiftachel, O. 2018, *Emptied Lands: A Legal Geography of Bedouin Rights in the Negev*, Stanford University Press, Stanford.

Keenan, S. 2015, *Subversive Property: Law and the Production of Spaces of Belonging*, Routledge, Abingdon, UK.

Kent, L. 2016, 'Introduction -Transitional justice in law, history and anthropology', *Australian Feminist Law Journal*, vol. 42, no. 1, pp.1–11.

Langton, M. 2010, 'The estate as duration: 'Being in place' and aboriginal property relations in areas of Cape York Peninsula in North Australia', in *Comparative Perspectives on Communal Lands and Individual Ownership Sustainable Futures*, eds L. Godden & M. Tehan, Routledge, London, pp.75–98.

Lefebvre, H. 1991, *The Production of Space*, trans. D. Nicholson Smith, Blackwell, Oxford.

Lester, A., & Laidlaw, Z. (eds), 2015, *Indigenous Communities and Settler Colonialism: Land Holding, Loss and Survival in an Interconnected World*, Palgrave Macmillan, Basingstoke, UK.

Luker, T. 2016, 'Animating the archive: Artefacts of law', in *Law, Memory, Violence: Uncovering the Counter-Archive*, eds S. Motha & H. van Rijswijk, Routledge, Abingdon, UK, pp.70–96.

Luker, T. 2017, 'Decolonising archives: Indigenous challenges to record keeping in 'reconciling' settler colonial states', *Australian Feminist Studies*, vol. 32, pp.108–125.

Mabo v Queensland (No. 2) (1992) 175 CLR 1.

Manderson, D. 2015, 'Literature in law – Judicial method, epistemology, strategy, and doctrine', *UNSW Law Journal*, vol. 38, no. 4, pp.1300–1315.

Massey, D., 2005, *For Space*, Sage, London.

McMillan, M., & Rigney, S. 2018, 'Race, reconciliation, and justice in Australia: From denial to acknowledgment', *Ethnic and Racial Studies*, vol. 41, no. 4, pp.59–777, doi: 10.1080/01419870.2017.1340653.

McNeil, K. 1989, *Common Law Aboriginal Title*, Clarendon Press, Oxford.

Members of the Yorta Yorta Aboriginal Community v Victoria (2003) 194 ALR 538.

Motha, S. 2018, *Archiving Violence: Law History Violence*, University of Michigan Press, Ann Arbor.

Motha, S., & van Rijswijk, H. (eds) 2016, *Law, Memory, Violence: Uncovering the Counter-Archive*, Routledge, Abingdon, UK.

Native Title Act 1993 (Cth).

Olson, G. 2016, 'The turn to passion: Has law and literature become law and affect?' *Law & Literature*, vol. 28, no. 3, pp.335–353.

Pascoe, B. 2018, *Dark Emu Aboriginal Australia and the Birth of Agriculture*, Scribe Publications UK Ltd, London, UK; Brunswick, Victoria, Australia.

Philippopoulos-Mihalopoulos, A. 2013, 'Atmospheres of law: Senses, affects, lawscapes', *Emotion and Society*, vol. 7, pp.35–44.

Philippopoulos-Mihalopoulos, A., & Brooks, V. (eds) 2017, *Research Methods in Environmental Law: A Handbook*, Edward Elgar Publishing, Northampton, MA.

Povinelli, E. 2002, *The Cunning of Recognition: Indigenous Alterities and the Making of Australian Multiculturalism*, Duke University Press, Durham, London.

Ritter, D. 1996, 'The rejection of terra nullius in *Mabo*: A critical analysis', *Sydney Law Review*, vol. 18, pp.5–33.

Stoler, A. L. 2009, *Along the Archival Grain: Epistemic Anxieties and Colonial Common Sense*, Princeton University Press, Princeton, NJ.

Strelein, L. 2001, 'Conceptualising native title', *Sydney Law Review*, vol. 23, pp.95–124.

Strelein, L. 2009, *Compromised Jurisprudence: Native Title Cases Since Mabo* (2nd edition), Aboriginal Studies Press, Canberra.

Tomlins, C. 2016, 'Historicism and materiality in legal theory', in *Law in Theory and History: New Essays on a Neglected Dialogue*, eds M. Del Mar & M. Lobban, Hart Publishing, Oxford, pp.58–83.

Tomlins, C. 2017, 'Law as ... IV: Minor jurisprudence in historical key. An introduction law text culture', *Law Text Culture*, vol. 21, pp.1–29.

Tuhiwai Smith, L. 2013, *Decolonizing Methodologies Research and Indigenous Peoples* (2nd edition), Zed Books, London & New York.

Valverde, M. 2015, *Chronotopes of Law: Jurisdiction, Scale, and Governance*, Routledge, London.

van Rijswijk, H. 2014, 'Archiving the northern territory intervention in law and in the literary counter-imaginary', *Australian Feminist Law Journal*, vol. 40, no. 1, pp.117–134.

Williams, R. 2012, *Savage Anxieties; The Invention of Western Civilization*, MacMillan, New York.

Wolfe, P. 1994, 'Nation and miscegenation: Discursive continuity in the post-mabo era', *Social Analysis*, vol. 34, pp.93–152.

Wright, A. 2013, *The Swan Book*, Giramondo Publishing, Sydney NSW.

Young, S. 2008, *The Trouble with Tradition: Native Title and Cultural Change*, The Federation Press, Leichardt NSW.

9

COMPARATIVE LEGAL GEOGRAPHY

Context and place in "legal transplants"

Liesel Spencer

> Indeed, geography is fate. Fate not only for a country but also for its culture and its law ... the geographic environment colours the law and enables or hinders the transfer of legal institutions.
>
> (Grossfeld 1984, pp.1512–3)

Introduction

Legal transplants, or law reform via the borrowing of legal ideas from another jurisdiction, are an established method in comparative law. In response to a problem identified in the home jurisdiction, lawmakers may seek possible law reform solutions by looking at what other jurisdictions do in response to that problem, with a view to importing the foreign solution (Danneman 2008, p.400). This chapter considers what legal geography has to offer in mitigating a problem with legal transplants as a method of law reform. The borrowed laws of the "transplantor" jurisdiction may be insufficiently adapted to the human and environmental contexts of the "transplantee" jurisdiction – there may be a poor "'fit' between the transferred law and the local context" (Graziadei 2008, p.472). This chapter proposes incorporating legal geography into comparative law methodology, as a means to bring a critical perspective to the ways in which social and environmental context complicates law reform proposals based on legal transplants.

Bringing legal geography into comparative law puts "place" – and contextually specific adaptation of laws to place – at the centre of legal transplant proposals. A legal geographical approach to legal transplants interrogates not just the laws being transplanted from one jurisdiction to another but also the diversity of environments to which they are being transplanted, the diversity of people to whom the transplanted laws might apply and the "interrelationships among the environment, people and social institutions" (Bartel et al. 2013, p.340).

Place-specific interactions, and therefore the effectiveness of a legal solution, are not readily transplantable. Geographical context is indispensable to the viability of a law reform proposal based on legal transplant, because as Grossfeld observes:

> [T]he relations between geography and law – though so often overlooked – have far-reaching implications. Any in-depth comparative research must take this factor into account. We have to be aware of the fact that a change in the geographical environment in itself might change the function of a given legal institution.
>
> *(Grossfeld 1984, pp.1518–9)*

A place-sensitive approach to comparative law methodology also accords with Riles' observations that comparative law and law-and-society scholarship can and should co-exist (and that this is particularly true for interdisciplinary projects):

> [T]here is consensus now that scholarship in both fields … needs to be both theoretically informed and empirically grounded – and that different mixes of these two elements should be encouraged and appreciated.
>
> *(Riles 2008, pp.801–2)*

A review of existing literature on methodological fusion of comparative law and legal geography produces scant results. Kedar's argument for "critical comparative legal geography" is the lone methodology paper in this space:

> I have approached comparative law hoping to draw upon its rich tradition of legal comparisons in suggesting an agenda for comparative legal geography … I discovered that legal geography scholarship could contribute to comparative law as much as, if not more than, it has to gain from it, and thus I have found ample space for cross-fertilisation. Yet to date, comparative law and legal geography remain separate academic spheres … legal comparatists and legal geographers virtually ignore each other's work.
>
> *(Kedar 2014, pp.95–6)*

Turton responds to Kedar's call for "cross-fertilisation" in proposing possible comparative avenues for legal geographical research into unconventional gas extraction industries in Australia and overseas (Turton 2015, p.57). Turton's analysis suggests a comparative law approach to "[assessing] the appropriateness of Australia's regulatory regimes" relating to unconventional gas – that is, using comparative legal geographical analysis as a tool to evaluate Australian law (Turton 2015, p.63). This approach has value from a law reform perspective in assessing whether unconventional gas regulation in Australia could be shifted to be better aligned with Australia's cultural and physical geography.

Recent work by Nicolini suggests a "legal analysis of the Cyprus question through the lenses of comparative law and legal geography" (Nicolini 2016, p.301) to "[identify a] legal design capable of solving the territorial and ethnic conflicts" (Nicolini 2016, p.291). The "Cyprus question" refers to the ongoing conflict between Greece and Turkey over occupation and control of the island of Cyprus, inhabited by both Greek Cypriots and Turkish Cypriots. Nicolini explicitly references the issue of legal transplants in suggesting that legal geography can help identify "which types of federalism may be transplanted to Cyprus" and "what Cyprus can learn from other multi-ethnic (con)federations" (Nicolini 2016, p.293). He notes the value of legal geography in this endeavour as "[exposing] the legal and non-legal presuppositions that might allow the transplanted mechanisms of legal geography to work" (Nicolini 2016, p.293). Nicolini, however, does not engage with Kedar's work on comparative legal geography as a legal research methodology, nor any of the extensive comparative law literature on legal transplants. In this respect, Nicolini's work is not in conversation with either legal geography or comparative law scholarship about developing new methodological approaches. His work is nonetheless illustrative of what legal geography has to offer comparative law, as it draws on conceptual tools of territoriality and boundaries to explore the possibilities of importing new constitutional structures into Cyprus to resolve the ongoing conflict.

Kedar's use of comparative legal geography is a critique of legal transplants flowing from colonialism – he directs his hybrid methodology towards "examin[ing] processes of displacement and dispossession in colonial and postcolonial settings" (Kedar 2014, p.101). His work uses comparative legal geography to explain and analyse those processes after the fact, rather than for legal transplant purposes of surveying other jurisdictions for possible future solutions to legal problems. Amongst concluding suggestions as to other possible uses of comparative legal geography, however, Kedar notes it may be of service in "planning reforms and imagining progressive solutions" – the context of his suggestion is the potential for political reform of state–Indigenous relations (Kedar 2014, p.111). This chapter takes up that idea and considers how legal geography, in a hybrid methodology with comparative law, can produce a critical and progressive approach to legal transplants as a law reform mechanism.

The applied context of comparative legal geography in my own field of research, public health law, is a comparative study of welfare laws in Australia and the USA that target food insecurity in vulnerable populations. This comparative enquiry has a law reform purpose, in considering whether ideas from USA food welfare law and policy would improve food security for people receiving welfare benefits if "transplanted" to the Australian context. Developments in Australian social security law since 2007 have included experimentation by successive governments with welfare quarantining, referred to as "income management" and "cashless welfare". This involves people on welfare having all or part of their welfare benefit quarantined to an electronic benefits transfer (EBT) card, called a "BasicsCard", with restrictions on where it can be spent and what can be purchased. My research focuses on food security, as one of the primary purposes of the Australian income management legislation: ss. 123TB and 123TH(1)(a) of the *Social Security (Administration) Act 1999* (Cth).

Official parliamentary material, in support of income management legislation in Australia, cited the USA food welfare system as an exemplar (Buckmaster, Ey & Klapdor 2012). This citation was not accompanied by any elaboration on the efficacy or transferability of USA law and policy to the Australian context (Cox 2011). There was no investigation of the human and environmental contexts in which USA food welfare law operates, nor how those contexts influence the success or otherwise of USA food welfare programmes. Australia's experimentation with the imported concept of welfare quarantining is emblematic of a "legal transplant" approach lacking rigour and validity, and the results to date have been predictably controversial and dysfunctional. My comparative research evaluates the extra-legal factors influencing both the food security outcomes of USA food welfare laws, and of Australia's "botched transplant" of the legal idea of welfare quarantining (Spencer 2018).

The following discussion of comparative law scholarship on legal transplants considers how incorporating legal geography can produce a more nuanced, place-sensitive legal transplant methodology. This is by no means an exhaustive discussion of what legal geography has to offer comparative law methodology in the field of legal transplants. Each comparative law research project will have distinctive requirements: "we need a 'toolbox', not a fixed methodological road map ... it is the aim of the research and the research question that will determine which methods could be useful" (Hoecke 2015, pp.1, 29). What is discussed here is a sample of relevant conceptual tools from legal geography which have been useful to my own comparative project.

Context and place in comparative law methodology

Comparative law scholars have alleged an overall lack of attention in comparative law having been paid to method (De Cruz 2007, p.219; Zweigert & Kötz 1998, p.33; Glanert 2012, p.64). Some specifically articulate their method, while others only reference it obliquely; however, both groups consistently claim a place of pre-eminence for context and background (Siems 2014, p.3; De Cruz 2007, p.18; Gutteridge 1949, p.12; Montesquieu 1989; Menski 2006, p.69; Zweigert & Kötz 1998, p.17) whether discussing legal transplants or other comparative law methods. Arriving at a meaning for what constitutes relevant context and background, and deciding how research question/s frame "the problem" under comparison, lead to "ambiguities of hard and soft law, official and non-official law" (Zumbansen 2005, p.1080) and fundamental questions about whether law and legal institutions can be extricated, or decontextualised as independent objects of study, from surrounding social and material circumstances (Frankenberg 1985, p.454). Menski, for example, cautions against "entrenched assumptions about what is legal and what is not ... drawing one's criteria of what is 'law' too narrowly", with an attendant risk of "impoverish[ed] legal research and understanding" (Menski 2006, p.66).

Demleitner claims that comparative law itself must be viewed in terms of "historical contingency and geographical and cultural situatedness" (Demleitner 1998, p.652). Comparative law originated in France, then Great Britain, Germany and the USA – as a result, Kedar claims, the focus of comparative law has been Eurocentric (Kedar 2014, p.5). Further, the historical timing of the emergence of comparative law coincided with colonialism, and with social Darwinism, classificatory scientific models and legal formalism. A "taxonomy" or "evolutionary scale of social progress" was applied to the comparative analysis of legal systems (Kedar 2014, p.6) with the European system as the high point; thus a justification for colonialism lay in bringing the Western system to more "primitive" societies (Kedar 2014, p.6). In my own research on public health law and food welfare, the selection of "intracultural" (De Cruz 2007, p.236) case studies compares the operation of laws in two Western Anglophone federal democracies. It is obviously far more problematic to propose legal transplants as a mechanism to export the law of wealthy Western countries to poorer non-Western countries. Kedar, for example, explores displacement and dispossession as consequences of colonial legal transplants (Kedar 2014).

Despite consensus in the literature that context must have a pre-eminent place in comparative law, this imperative has not been consistently observed in practice. Kedar claims that two important techniques of comparative law, legal families and legal transplants, are conducted in orthodox comparative law "without much attention to local conditions or 'extra-legal factors'" (Kedar 2014, p.15) or with insight into "the relationship between law, society and culture on a worldwide or regional basis" (Kedar 2014, p.16). ("Legal families" is comparative law research on the classification of legal systems (De Cruz 2007, p.35).) Orthodox comparative law, on this view, has somewhat lost sight of the relationship between law and its social and geographical context.

The first step in legal comparison is defining "the problem" to be studied – and in the case of legal transplants, asking what problem might be remedied in law reform via legal transplant. This step is therefore the first opportunity to integrate social and geographical context into legal transplants methodology. De Cruz asserts the importance of defining problems by reference to context – although he neglects to include geography as a relevant extra-legal factor:

> [W]hat the comparatist is seeking to do is to evaluate the efficacy of a given solution or approach to a legal problem in terms of that particular institution's cultural, economic, political and legal background ... Given a set of priorities, the task is to assess the effectiveness of a solution in terms of achieving those aims and objectives.
>
> *(De Cruz 2007, p.224)*

Context and place in legal transplants research: a case study

In my research on food welfare laws, the purpose of comparative legal analysis is to reconcile commonalities and contextual differences between Australia and the USA, to produce useful knowledge about what effect food welfare laws have on food insecurity and how those effects are conditioned by place-specific factors. The problem of the correlation between poverty, populations receiving welfare benefits, food insecurity and inequitable public health outcomes is not a problem novel to Australia, nor a problem novel to wealthy Western countries. The developing world is now also experiencing increased rates of adverse population health outcomes related to dietary patterns, such as obesity and associated chronic disease, in correlation with poverty and the adoption of a relatively cheaper "Western diet" amongst less privileged members of society (Food and Agriculture Organisation of the United Nations n.d.). There are elements of commonality across the experience of most nations, with these commonalities having contextually diverse expressions. The perishable nature of fresh produce, for example, is a common concern; this concern plays out differently in a wet and cold climate than it does in a desert city, and differently again as between and within urban and rural contexts even in the same climate zone.

Similarly, there are common fundamentals of healthy diet regardless of where in the world a population is located. Adequate intake of fresh vegetables and fruit, for example, is a health need common across human populations; which varieties are available and culturally acceptable is contingent on location, both at the scale of a city and at the nested scales of districts and households within a city. The application of law in different locales affects what food is available. Valverde, for example, observed that the availability of particular cultural types of street food in different locales within a city was contingent on whether local municipal regulation was racially discriminatory or oriented to accommodation of cultural diversity (Valverde 2012, p.150). The absence of legal intervention can also shape access to food security. "Food deserts" or city areas where there are no proximate fresh food outlets illustrate how a regulatory vacuum – the absence of intervention by local planning authorities – can produce places where residents do not have access to a healthy diet (Astell-Burt & Feng 2015; Elias 2013).

The problems of food security may be common, but the contextual expression of the way law and policy interacts with those problems is distinctive in each place. Research questions identifying "the problem" for comparative study can be place-sensitive by acknowledging that the expression of a problem, and therefore the efficacy of possible solutions, is contextually specific. Graham's work on "adaptation/maladaptation" informs the framing of research questions in my food welfare law comparative project (Graham 2011, p.8). Rather than only asking what lessons can Australia learn from USA food welfare law, I consider how those "lessons" or legal ideas from USA food welfare programmes can be appropriately adapted to suit local conditions. The comparative law debate over inadequate attention to context in legal transplants mirrors arguments from legal geography that "placelessness" in law is problematic, for example as Graham says of property law:

> [I]f the paradigm of modern property law is to remain, or rather, become functional, as a paradigm of people-place relations, it needs to acknowledge

that its placelessness and atopic basis is maladapted to diverse and specific places.

(Graham 2011, p.8)

Conceptual tools from legal geography scholarship can be utilised in comparative law methodology to mitigate maladaptation in transplanted laws. My research found that the place-based income management trial of welfare quarantining is maladapted to local conditions in Bankstown and produces adverse food security effects (Spencer 2018), just as English property law has produced adverse environmental effects in the Australian landscape (Graham 2011, pp.89–90). Legal ideas from USA food welfare laws may have something to offer Australian law-and-policymakers, however those ideas have to be adapted to local conditions to achieve the intended aims of reducing food insecurity: "adaptation is necessarily a process of becoming local" (Graham 2011, p.205). The "process of becoming local" in the migration of legal ideas across jurisdictional boundaries arguably underpins scholarly disputes on the topic of legal transplants. Those disputes, discussed below, have implications for what constitutes a rigorous comparative methodology.

Legal transplants as a common-but-controversial mode of law reform

Using legal transplants as a tool for law reform-oriented research accords with long-established views as to the purposes of comparative jurisprudence (De Cruz 2007, p.511; Siems 2014, pp.3–4; Andenas & Fairgreave 2012, pp.16, 25; Von Nessen 2006, p.28; Ehrmann 1976). Sir Henry Maine, the founder of modern comparative law, viewed the "chief function of comparative jurisprudence [as being] to facilitate legislation and the practical improvement of the law" (Maine 1871, p.4). Comparative law as a tool for legislative reform in ancient times dates back to the writings of Cicero and Gaius (De Cruz 2007, p.11), and of Plato and Aristotle (Andenas & Fairgreave 2012, p.25).

Embarking on research that explores the possibility of importing solutions from a foreign jurisdiction into the domestic context, however, invokes the ongoing scholarly debate over legal transplants (Watt 2012, pp.82–4; De Cruz 2007, pp.511–13; Frankenberg 2010, pp.565–70). Watson coined this phrase in 1974 (Watson 1993) and has since produced decades of scholarship arguing that legal transplants are "the most prominent form of legal change" (Watson 2010, p.5). Watson does not claim that "borrowed law [is necessarily] the most satisfactory rules for the social, economic, political conditions of the borrowing state" but rather that borrowed or transplanted law is, as evidenced by history, a function of several factors (Watson 2010, p.11). These factors include imposition by an occupying force, accessibility of a particular foreign law because of shared language or usable form, law-makers borrowing legitimacy where local authority is lacking and chance or historical accident (Watson 2010, p.11). Watson is making a pragmatic observation-after-the-fact that a transplant has, in fact, occurred and that this occurrence was precipitated by identifiable factors. Kahn-Freund argues in a more abstract, normative vein that the transplantability of a rule depends on the alignment between the social and political environments of the transplantor and transplantee jurisdictions, invoking Montesquieu's caution that law is so specific to place that it is only in exceptional cases that the "institutions of one country could serve those of another at all" (Kahn-Freund 1974, pp.6–7; Montesquieu 1989, bk.2 ch.3).

The most vocal critic of Watson's legal transplants scholarship is Legrand, who claims that legal transplants are "impossible" (Legrand 1997, p.111), describing Watson's views as "impoverished" and "crude", on the grounds that law cannot be segregated from society and that, therefore, legal "rules cannot travel" (Legrand 1997, p.113). (Watson retaliated with "I see no substance, just big words, in his article" (Watson 2000).) Much of the dispute between these two positions appears to relate only to differences in the use of language and terminology, as Watson concedes that he and Legrand "might be using the language of 'transplant' in different ways and might both be guilty of misrepresenting the other" (Watt 2012, pp.83–4), by stating that "if I am correct then to a large extent he and I are talking at cross-purposes, using the word 'transplant' in different senses" (Watson 2000). Watson has defended himself against Legrand's criticism by pointing to the objective facts of legal history and the extent of legal "borrowings", and also to his published work on the importance of law in its social and political context (Watson 2000). Watson's objections to Legrand's "impossibility" stance irresistibly recall the title of the essay, folklorically attributed to Kant: "That may be all right in practice, but it doesn't work in theory" (Goldwin 1981, p.39).

There is strong support in the literature for Watson's view, for example de Cruz's observation that "[c]rucially, the 'impossibility' protagonists seem to ignore the facts of legal history and phenomena such as the effects of conquest and colonization" (De Cruz 2007, pp.511–13). Similarly, Örücü's view is that

> scathing approaches to this mode of law reform [legal transplants] suffer from an avoidance syndrome … the movement of legal institutions and ideas is trans-border … this is both a historical and present fact and the future will see more of it.
>
> *(Örücü 2002, p.205)*

Watt contends that "there may be more to Watson's conception of 'legal rules' than Legrand gives him credit for" (Watt 2012, p.8) citing Watson's earlier work:

> Transplanting frequently, perhaps always, involves legal transformation. Even when the transplanted rule remains unchanged, its impact in a new social setting may be different. The insertion of an alien rule into another complex system may cause it to operate in a fresh way. (Watson 1993, p.116)

The terminology "legal transplants" is therefore problematic. Others have expressed this concern – Örücü resolves "disquiet" around the term "transplants" by preferring "legal transposition" (Örücü 2002, pp.205–6; Teubner 1998). It would appear that word choice, language and meaning are largely to blame for the dispute over whether legal transplants are "possible" or not. (This is quite ironic given that the consensus in the literature on comparative law methodology that words, language and meaning have to be treated with extreme caution and that commonality of meaning between jurisdictions cannot be assumed (Curran 2006).) However, naming the process – whether as "legal transplants", "legal borrowings", "legal transposition", "tuning" (Örücü 2002, p.205), "reception", "adaptation", "mutation", "influence", "evolution", "migration" (Frankenberg 2010, p.566) or otherwise – is less important than carefully articulating the content and purpose of the legal transplants methodology to be applied to a given law reform project. This acknowledges the reality that legal institutions and ideas have, past and present, flowed across jurisdictional borders, facilitated by processes and factors inside and outside those borders, and with a variety of outcomes. "Borrowing" legal institutions and ideas from other jurisdictions continues to be a widespread mode of law reform. The differences of opinion between comparatists as to whether legal transplants are "possible" is more accurately a difference in meaning; a distinction needs to be drawn as to where a legal transplant (past, present or proposed in future) lies on a scale of functional versus dysfunctional outcomes in addressing identified problem/s in the transplantee jurisdiction.

It is, therefore, not so much whether a legal transplant can be done that comparatists are arguing about, but *whether it can be done well*. There are two main, and overlapping, considerations in how a legal transplant might be "done well" – that is, by producing a functional law reform. The first consideration is the issue of "comparability" in the selection of the transplantor and transplantee jurisdictions. The second consideration is whether the transplant proposal pays adequate attention to the sociological and geographical context of both jurisdictions. In the following section, I will consider how a legal geographical approach can usefully inform both of these methodological considerations.

Comparability and context in legal transplants

There are two broad "species of comparative study" that may be undertaken, "macro-comparison" and "micro-comparison". Macro-comparison is a study of two or more legal systems in their entirety, whereas micro-comparison is a study of specific aspects or topics within two or more legal systems (De Cruz 2007, p.233). Microcomparison involves "specific legal institutions or problems, that is, with the rules used to solve actual problems or particular conflicts of interests" (Zweigert & Kötz 1998, pp.4–5). Within these groupings, two further permutations emerge, a "cross-cultural comparison" or an "intracultural comparison". A cross-cultural study compares two legal systems, e.g. one belonging to a Western society, and one to a non-Western society; an intracultural study compares two or more legal systems belonging to the same societal type, e.g. Western society

(De Cruz 2007, p.236). My research employs comparative law methodology for the purposes of "intracultural micro-comparison", comparing case studies from within two Western Anglophone federal democracies, Australia and the USA, of social security laws functioning as *de facto* public health laws regulating the risk of food insecurity.

Comparability and context are related considerations in comparative law methodology involving legal transplants. Essentially, the more dissimilar the contexts of the transplantor and transplantee jurisdiction, the less comparable the operation of laws in those jurisdictions. There is some consensus in the literature on comparative law methodology regarding which systems or jurisdictions can be meaningfully compared. Similarity of jurisdictions to be compared goes hand-in-hand with comparability. Comparative law as a vehicle for domestic law reform is claimed to be more effective where the two jurisdictions being compared have minimal social and cultural differences (Von Nessen 2006, pp.6–7; Siems 2014, p.16; Gutteridge 1949, p.12; Kamba 1974, p.1974; De Cruz 2007, p.6; Montesquieu 1989, ch.3). Effectiveness is also claimed to be dependent upon whether the two jurisdictions are at the "same stage of legal, political and economic development" (Gutteridge 1949, p.73), and where there is not "already in place a norm or institution which is functionally equivalent" (Von Nessen 2006, pp.150–1).

Kahn-Freund stresses in respect of legal transplants a need to give adequate consideration to "congruence between the comparative power structure of the transplantee and transplantor countries" lest "failure to take this factor into account [causes] the transplantee country to reject the foreign institution or rule" (Kahn-Freund 1974, p.1). Von Nessen adds that there should be alignment of "socio-political, economic, and legal norms of the borrowing state with those of the state from which the concepts are being borrowed" (Von Nessen 2006, p.115).

Grossfeld views "the cultural, political and economic components of a society, as well as the particular relationship that exists between the State and its citizens, its value system and its particular conception of the individual and the world in general" (Grossfeld 1990, p.73) as determinative of comparability. Other general factors affecting comparability include "a society's 'cultural climate' … resulting from the people's unconscious axioms, collective feelings and prevalent ideas of reality"; also important is the "homogeneity of the society" in question and "whether it has a cultural consensus or not" (Grossfeld 1990, p.73). Importantly, Grossfeld also cites "geographical situation, language and religion" as factors determining comparability, and "stresses that, in order to ascertain the effect of legal solutions in different systems, a multi-disciplinary approach is required" (Grossfeld 1990, p.73).

Similarity of context, then, determines comparability of jurisdictions. Comparative law scholarship claims contemporary comparative law has not paid adequate attention to context. Kedar points to "present-day comparatists' neglect of geography", as a deterioration from early comparative law "scholars [who] compared legal systems while simultaneously attributing determining influences to climate and physical geography" (Kedar 2014, p.97). If context is lacking in comparative law method, geographical context is, according to Kedar, especially lacking:

> The time seems ripe for a 'spatial turn' in comparative law, a perspective found – in an environmental determinist version – in the scholarship of comparative law's precursors but lost with the institutionalisation of the discipline. The reintroduction of spatiality and an engagement with [critical legal geography] scholarship could assist comparative law in recovering from its current malaise and embarking on a new trajectory.
>
> *(Kedar 2014, p.101)*

Grossfeld claims that geography is one of the contextual factors in comparative law method so important as to be potentially determinative of the success or failure of law reform via legal transplant:

> The general trend today is to view the transferability of law as a function of political and economic similarities between legal cultures … The emphasis is on the social and political power structure, on the East-West and North-South conflicts. Other authors emphasize the communality of cultural values … All these aspects certainly have their merits, but they have even more serious limits. They cannot explain why comparative law is so difficult even between Europe and the United States, or between Continental Europe and England. Here we find similar political structures, similar economic standards, shared cultural values – yet still very great difficulties when it comes to comparative law. So far, the present methods in comparative law have not sufficiently explained these tensions. This leads to the suspicion that important factors are often overlooked – probably because they are too apparent. These factors include primarily the natural environment (particularly the geographical situation of a country), the climate, population density, and language and religion. These factors are of utmost importance in comparative law … as legal scholars, we must give them our intensive attention.
>
> *(Grossfeld 1984, pp.1501–11)*

The jurisdictions selected for comparison in my own research are New South Wales (NSW), Australia and California, USA. Within those jurisdictions, I chose two locations for comparative case studies: the urban region of Bankstown in NSW and the city of Oakland in California. Bankstown was selected as a case study because it is the sole NSW site of a pilot programme which further expands trials of welfare quarantining in Australia (Department of Human Services 2018). Oakland, California was selected as a USA case study because it has a degree of demographic and geographic similarity to Bankstown, with a history of racial tension and racial justice activism including early community food welfare programmes, relatively high socioeconomic disadvantage and recent gentrification. Selecting a jurisdiction in the USA as a comparison was based on two important considerations. Firstly, as discussed above, Australian parliamentary material cited the USA food welfare system as an exemplar for trialling welfare quarantining in Australia. Secondly, the USA has distinctive, well-documented and long-established federal legislative schemes using social security law to address food security (for example "food stamps" and the school lunch programme). In contrast to Australia's predominantly cash-based provision of the food security needs of welfare recipients (with the exception of the welfare quarantining trials), the USA makes extensive use of in-kind food welfare via various programmes. This diverse suite of programmes provides a rich source of law and policy ideas for comparison with Australia's trial welfare quarantining schemes, including the pilot place-based income management scheme implemented in Bankstown.

The "process of becoming local" in a legal transplant of ideas borrowed from USA food welfare law turns on the extent to which the borrowed ideas are appropriately adapted to suit local sociological and geographical contexts in Australia. Following is a discussion of how a legal geographical approach can shape a law reform proposal towards being appropriately adapted and place-sensitive.

Sociological and geographical context in legal transplants

As a key theme or leitmotif in legal geography scholarship, "co-constitutivity" – whereby "the impact of law is both felt and made (at least in part) locally" (Bennett & Layard 2015, p.408) – assists in thinking through the place-specific effects of a legal idea and whether those effects are transplantable. In my comparative research, co-constitutivity helps explain connections between food welfare law and "nutrition environments", which are the social and physical context in which people make decisions about food. Nutrition environments have a powerful role in human dietary patterns: it matters where we live, work, shop and eat (Caspi et al. 2012). In the place-based income management trial in Bankstown the legislation profoundly distorts the relationship between welfare recipients participating in the trial, and the local food landscape.

Trial participants in place-based income management have a portion of their welfare benefit quarantined onto an EBT "BasicsCard". The card can only be used at shops which have opted-in to the scheme by applying to the Department of Human Services to become approved merchants. BasicsCard funds cannot be used to purchase "excluded goods and services" which include alcohol, tobacco, pornography or gambling, home-brew kits, gift cards or to obtain cash: *Social Security (Administration) Act 1999* (Cth) s. 123TI and Social Security (Administration) (Excluded Goods — section 123TI) Specification 2010.

At the beginning of the trial, no safeguards were put in place to make sure that food did not become less affordable or less accessible because of these restrictions; nor were safeguards put in place to make sure that Bankstown's diverse population could access foods which met their cultural preferences (Spencer 2018).

Bankstown is one of the most ethnically diverse places in Australia (Bankstown City Council 2007). My empirical research into the effects of the welfare quarantining trial on participants' food security found that Bankstown has a rich and distinctive food landscape, reflecting this diverse ethnic population.

My research found that participants' access to healthy affordable food is severely constrained by the operation of the legislation. Only a small proportion of local shops are licensed to accept the BasicsCard. The list of licensed food outlets is dominated by the more expensive supermarket duopoly of Coles and Woolworths; the less expensive supermarkets and fresh produce outlets are not licensed to accept the card. Most of Bankstown's colourful and diverse local food landscape is "greyed out" for trial participants (Spencer 2018).

The ability of participants to achieve food security in the local nutrition environment is compromised, contrary to the stated purposes of the legislation which include the promotion of food security. The material reality of how the welfare quarantining law operates in Bankstown's nutrition environment, juxtaposed against what are claimed as statutory purposes, recall Graham's assertion (in relation to property law) that "the law is also, partly, what the land says it can be" (Graham 2011, p.7). By failing to take account of the unique nutrition environment of the trial site, the legislation is ineffective in reducing food insecurity; in fact, it compromises participants' food security.

Revealing the "constitutive power of law" (Delaney 2015, p.99) in nutrition environments and health outcomes is one function of legal geographical comparison; the constitutive power of law in human–human relationships is another. As Delaney notes:

> [D]istinctive legal practices are also involved in 'making up' the kinds of persons and non-persons (citizens, consumers, animals, lovers, owners, workers, refugees, children, soldiers, and so on) who live in our world and so how they live … Social relationships of many kinds are also distinctively legal relationships … familial relationships, political relationships … and relations among strangers are all legally inflected in ways that are often determinative of how the relationships are enacted and experienced. (Delaney 2015, p.99)

People subject to welfare quarantining legal schemes are a category of persons whose food security is treated differentially. Their legal denomination as a type of person – "welfare recipient" – changes social relationships with the people from whom they buy food, and with others sharing social spaces who have more autonomy in where and what they buy and eat. Paying for food with a conspicuously different EBT card carries stigma. Being constituted as welfare recipient also potentially alters relationships within a family unit as the law delimits the role of adult "providers".

Blomley points out that legal geography scholarship has thus far delivered a "partial account" of the interaction between law and geography, by focusing on "the production of the public–private divide within judicial decisions and formal state actions" and remaining "confined within statute and case law" (Blomley 2005, p.286). Blomley claims "there has been relatively little empirical research on everyday legal geographies; that is, on the ways people actually navigate and apprehend the spatial dimensions of law" (Blomley 2005, p.286). Legally imposed constraints on the choices that individuals can make about food provisioning for their households very much concern "everyday legal geographies" at the humble scale of people's shopping trolleys, grocery budgets and the food they can place on the family table.

Restricting access to affordable healthy food is the sort of "spatialized inequality" that Blomley and others concerned with power, space and law claim as fertile ground for legal geographical analysis (Braverman et al. 2014, p.13). The only unique aspect of the five trial sites contemplated by the ironically named place-based income management legislation is the relative socioeconomic disadvantage and the proportion of people dependent upon welfare benefits. Imposing an experimental scheme of welfare quarantining on already vulnerable populations illustrates how inattention to the localised effects of legislation can produce and exacerbate spatial inequality. Bankstown has an abundance of healthy food outlets, and fresh produce literally spilling into the streetscape from cultural food vendors, but people on place-based income management are not able to access this nutritious and affordable food because most shops are not signed up to accept the BasicsCard. The law aiming to regulate food insecurity is ineffectual because it does not accommodate the concrete realities of the social and environmental landscape where it is implemented. One of the roles of critical human geography is to "expose the socio-spatial processes that (re)produce inequalities between people and places" (Hubbard et al. 2002, p.62). A legal geographical analysis of the Bankstown trial site frames the place-based income management trial as a form of spatial injustice, where law is complicit in entrenching place-based disadvantage and social exclusion.

Conclusion

A hybrid methodology of legal geography and comparative law places the focus back on the specific geography of each jurisdiction, avoiding the process of orthodox comparative law which Kedar describes as "a de-contextualisation of the laws from their concrete social and geographical settings" (Kedar 2014, pp.8–9). This approach paints mutually constitutive relationships between law, land and people and has the potential to transcend an artificially simplistic understanding of legal transplants as a mode of law reform. Frankenburg claims that the comparative law method is in need of a more sophisticated tolerance and accommodation of ambiguity in research findings (Frankenberg 1985, pp.441, 445). Identifying and addressing ambiguity in a legal transplant proposal is critical to avoiding a dysfunctional imposition of "maladapted" law. The Australian experience of experimentation with the imported legal idea of welfare quarantining is a cautionary tale against disregarding the ambiguities and complexities of the social and environmental contexts in which law operates. A legal geographical approach grounds the evaluation of a proposed law reform in the concrete settings of the "transplantor" and "transplantee" jurisdictions – "[incorporating] the materiality of place into analyses of laws and governance" (Graham, Davies & Godden 2017, pp.17–18).

That is not to say that legal transplants should be removed from the law reform toolkit, only that the methodology must be place-sensitive and appropriately adapted. Other jurisdictions may very well provide useful inspiration and a source of ideas for how to improve the law in the home jurisdiction. As Ehrmann observes:

> Only the analysis of a variety of legal cultures will recognise what is accidental rather than necessary, what is permanent rather than changeable in legal norms and legal agencies, and what characterises the beliefs underlying both. The law of a single culture will take for granted the ethical theory on which it is grounded. (Ehrmann 1976)

This resonates with legal geography scholarship on the illusion of law as a permanent, immutable state of affairs (Blomley 2013, p.47), and with the role of comparative law as a basis for considering a broader range of options in law reform. In looking to other jurisdictions for inspiration, the methodological limitations of legal transplants are mitigated by bringing appropriate conceptual tools from legal geography into the research design.

References

Andenas, M., & Fairgreave, D. 2012, 'Intent on making mischief: Seven ways of using comparative law', in *Methods of Comparative Law*, ed. P.G. Monateri, Edward Elgar Pub, Cheltenham, UK; Northampton, MA, pp. 16, 25.

Astell-Burt, T., & Feng, X. 2015, 'Geographic inequity in healthy food environment and type 2 diabetes: Can we please turn off the tap?', *The Medical Journal of Australia*, vol. 203, no. 6, pp.246–248.

Bankstown City Council 2007, *Multicultural Strategy Bankstown*, Bankstown City Council, Bankstown, http://online.bankstown.nsw.gov.au/Docs/Temp/18C_003EC886.001.pdf.

Bartel, R., Graham, N., Jackson, S., Prior, J. H., Robinson, D. F., Sherval, M., & Williams, S. 2013, 'Legal geography: An Australian perspective', *Geographical Research*, vol. 51, no. 4, pp.339–353.

Bennett, L., & Layard, A. 2015, 'Legal geography: Becoming spatial detectives', *Geography Compass*, vol. 9, no. 7, pp.406–422.

Blomley, N. 2005, 'Flowers in the bathtub: Boundary crossings at the public-private divide', *Geoforum*, vol. 36, no. 3, pp.281–296.

Blomley, N. 2013, 'Performing property, making the world', *Canadian Journal of Law and Jurisprudence*, vol. 26, no. 1, pp.23–48.

Braverman, I., Blomley, N., Delaney, D., & Kedar, A. 2014, 'Introduction: Expanding the spaces of law', in *The Expanding Spaces of Law: A Timely Legal Geography*, eds I. Braverman, N. Blomley, D. Delaney & A. Kedar, Stanford Law Books, Stanford, CA, pp.1–29.

Buckmaster, L., Ey, C., & Klapdor, M. 2012, 'Income management – An overview', Parliamentary Library, Department of Parliamentary Services, viewed 5 June 2019, <http://parlinfo.aph.gov.au/parlInfo/download/library/prspub/1727168/upload_binary/1727168.pdf;fileType=application%2Fpdf>.

Caspi, C., Sorensen, G., Subramanian, S., & Kawachi, I. 2012, 'The local food environment and diet: A systematic review', *Health & Place*, vol. 18, no. 5, pp.1172–1187.

Cox, E. 2011, 'Evidence-free policy making? Part C: Expanding the program', *Journal of Indigenous Policy*, vol. 12, pp.28–84.

Curran, V. G. 2006, 'Comparative law and language', in *The Oxford Handbook of Comparative Law*, eds R. Zimmermann & M. Reimann, Oxford University Press, Oxford, pp.676–704.

Danneman, G. 2008, 'Comparative law: Study of similarities or differences?', in *The Oxford Handbook of Comparative Law*, eds R. Zimmermann & M. Reimann, Oxford University Press, Oxford, p.400.

De Cruz, P. 2007, *Comparative Law in a Changing World*, 3rd ed., Routledge-Cavendish, New York, NY.

Delaney, D. 2015, 'Legal geography I Constitutivities, complexities, and contingencies', *Progress in Human Geography*, vol. 39, no. 1, pp.96–102.

Demleitner, N. V. 1998, 'Challenge, opportunity and risk: An era of change in comparative law', *The American Journal of Comparative Law*, vol. 46, no. 4, pp.647–655.

Department of Human Services 2018, *Income Management and Cashless Debit Card Summary (29 December 2017 – 5 January 2018)*, viewed 16 March 2018, <https://data.gov.au/dataset/income-management-summary-data/resource/3eb4d339-89fb-43ec-adf9-df2e43cf3cd2>.

Ehrmann, H. 1976, *Comparative Legal Cultures*, Prentice-Hall, NJ.

Elias, R. R. 2013, 'Grocery stores: Neighborhood retail or urban panacea? Exploring the intersections of federal policy, community health, and revitalization in bayview hunters point and West Oakland, California', *eScholarship*, viewed 24 May 2017, <http://escholarship.org/uc/item/1d45296j>.

Food and Agriculture Organisation of the United Nations, n.d., *The Developing World's New Burden: Obesity*, viewed 14 March 2018, <http://www.fao.org/focus/e/obesity/obes2.htm>.

Frankenberg, G. 2010, 'Constitutional transfer: The IKEA theory revisited', *International Journal of Constitutional Law*, vol. 8, no. 3, pp.563–579.

Frankenberg, G. 1985, 'Critical comparisons: Re-thinking comparative law', *Harvard International Law Journal*, vol. 26, pp.411–455.

Glanert, S. 2012, 'Method?', in *Methods of Comparative Law*, ed. P. G. Monateri, Edward Elgar Pub, Cheltenham, UK; Northampton, MA, p. 64.

Goldwin, R. 1981, 'That may be all right in practice, but it doesn't work in theory: Comments on "Is federalism compatible with prefectorial administration"?', *Publius*, vol. 11, no. 2, pp.9–45.

Graham, N. 2011, *Lawscape: Property, Environment, Law*, Routledge, Abingdon, Oxon; New York, NY.

Graham, N., Davies, M., & Godden, L. 2017, 'Broadening law's context: Materiality in socio-legal research', *Griffith Law Review*, vol. 26, no. 4, pp.480–510.

Graziadei, M. 2008, 'Comparative law as the study of transplants and receptions', in *The Oxford Handbook of Comparative Law*, eds M. Reimann & R. Zimmermann, Oxford University Press, Oxford, p. 472.

Grossfeld, B. 1984, 'Geography and law', *Michigan Law Review*, vol. 82, no. 5/6, pp.1510–1519.

Grossfeld, B. 1990, *The Strength and Weakness of Comparative Law*, Clarendon Press; Oxford University Press, Oxford, New York.

Gutteridge, H. C. 1949, *Comparative Law: An Introduction to the Comparative Method of Legal Study*, Cambridge University Press, Cambridge.

Hoecke, M. V. 2015, 'Methodology of comparative legal research', *Law and Method*, no. 12, viewed 7 March 2019, <https://www.bjutijdschriften.nl/tijdschrift/lawandmethod/2015/12/RENM-D-14-00001>.

Hubbard, P., Kitchin, R., Bartley, B., & Fuller, D. 2002, *Thinking Geographically: Space, Theory, and Contemporary Human Geography*, Continuum, London ; New York.

Kahn-Freund, O. 1974, 'On uses and misuses of comparative law', *The Modern Law Review*, vol. 37, no. 1, pp.1–27.

Kamba, W. J. 1974, 'Comparative law: A theoretical framework', *The International and Comparative Law Quarterly*, vol. 23, no. 3, pp.485–519.

Kedar, A. 2014, 'Expanding legal geographies: A call for a critical comparative approach', in, *The Expanding Spaces of Law: A Timely Legal Geography*, eds I. Braverman, N. Blomley, D. Delaney, & A. Kedar, Stanford Law Books, Stanford, CA, pp.95–119.

Legrand, P. 1997, 'The impossibility of "legal transplants"', *Maastricht Journal of European and Comparative Law*, vol. 4, pp.111–124.

Maine, H. 1871, *Village Communities in the East and West*, John Murray.

Menski, W. 2006, *Comparative Law in a Global Context: The Legal Systems of Asia and Africa*, Cambridge University Press, New York.

Montesquieu, C. de S. 1989, *The Spirit of the Laws*, Cambridge University Press, Cambridge; New York.

Nicolini, M. 2016, 'Territorial and ethnic divide: A new legal geography for cyprus', in *Law, Territory and Conflict Resolution: Law as a Problem and Law as a Solution*, eds M. Nicolini, F. Palermo, & E. Milano, Brill Nijhoff, Leiden; Boston, pp.285–315.

Örücü, E. 2002, 'Law as transposition', *The International and Comparative Law Quarterly*, vol. 51, no. 2, p.205.

Riles, A. 2008, 'Comparative law and socio-legal studies', in *The Oxford Handbook of Comparative Law*, eds M. Reimann & R. Zimmermann, Oxford University Press, Oxford, pp.775–813.

Siems, M. M. 2014, *Comparative Law*, Cambridge University Press, New York.

Spencer, L. 2018, 'Place-based income management legislation: Impacts on food security', *Flinders Law Journal*, vol. 20, no. 1, pp.1–54.

Teubner, G. 1998, 'Legal irritants: Good faith in British law or how unifying law ends up in new divergences', *The Modern Law Review*, vol. 61, p.11.

Turton, D. 2015, 'Unconventional gas in Australia: Towards a legal geography', *Geographical Research*, vol. 53, no. 1, pp.53–67.

Valverde, M. 2012, *Everyday Law on the Street: City Governance in an Age of Diversity*, The University of Chicago Press, Chicago.

Von Nessen, P. 2006, *The Use of Comparative Law in Australia*, Lawbook Co., Pyrmont, NSW.

Watson, A. 2010, *Comparative Law: Law, Reality and Society*, 3rd ed., Vandeplas Pub, Lake Mary, Florida.

Watson, A. 1993, *Legal Transplants: An Approach to Comparative Law*, University of Georgia Press, Athens.

Watson, A. 2000, 'Legal transplants and European private law', *Electronic Journal of Comparative Law*, vol. 4, no. 4. <https://www.ejcl.org/44/art44-2.html>.

Watt, G. 2012, 'Comparison as deep appreciation', in *Methods of Comparative Law*, ed. P. G. Monateri, Edward Elgar Pub, Cheltenham, UK; Northampton, MA, pp.82–103.

Zumbansen, P. 2005, 'Comparative law's coming of age? Twenty years after critical comparisons†', *German Law Journal*, vol. 6, pp.1074–1084.

Zweigert, K., & Kötz, H. 1998, *Introduction to Comparative Law*, Clarendon Press; Oxford University Press, Oxford; New York.

10

THE OTHER IS US

Conservation, categories and the law

Robyn Bartel

Introduction: categories and privilege

Categories are the product of human attempts to make sense of complexity. They do not overlook the fuzziness of the real world – they are intended as a cure. In so doing, they are somewhat blunt and perverse instruments, as by eliminating nuance and shades of grey they conceal as much as they reveal (see Cloke & Johnston 2005). Misrepresentations result as important features are obscured and alternative perspectives excluded, particularly as categorisation tends to essentialise and universalise. Categories can become normalised and, hence, invisible (see Bowker & Star 2000). The consequences are not just representational, but also matter in a material world where a select few may be favoured. This variant of nominative determinism is both preferential and real. This phenomenon is exacerbated where the number of categories is limited, hierarchical and normative.

It has been observed that "[h]umans have a proclivity for creating good-evil dualities" (Frandsen & McGoun 2010, p.1), and that good is frequently conflated with "us" and evil with "them" (Zimbardo 2008, p.211). Such "othering" becomes particularly powerful when reflected in the law, as legal categories have additional representational and material powers (Godden 2002). The law is fundamentally normative and in addition to moral weight, legal definitions carry "imperial" force (Bowker & Star 2000). The law is backed by a sophisticated political apparatus to support its interpretation and is therefore a powerful world-maker. However, while the law may be informed by the material world and is always to some degree a co-creation of place, this may only be partial (Nagle 2010; Bartel et al. 2013). Even when claiming to be universal, the law makes its world in a preferential way, as it is crafted by the powerful, and favours their interests (Bartel 2017). This may be internalised, as the law acculturates its subjects and is both normalising and self-perpetuating (Nagle

2010). Legal categories can become almost invisible, their epistemological and ontological frames hidden, and therefore their effects pernicious (Fitzpatrick 1984). In Australia, the perspective of the colonisers was and remains privileged within the dominant institutions of the settler state. These institutions are predicated on a human–nature binary. This is a powerful duality, pervasive in many Western and European systems, which separates nature, subordinates it and makes it subject to human control, treating it largely as a resource, and favouring economic production over the environment (Bartel, McFarland & Hearfield 2014). Conservation and natural resource management are also predicated on the view that nature is "other" and humans have the capacity and are entitled to manage the environment (see Hamilton 2010). Similarly arrogant and anthropocentric assumptions underpin the act of categorisation itself – assumptions that humans have the ability and authority to label, and thereby make the world in a particular way.

Binaries of native–non-native and human–nature

Provenance-based categorisations dominate biodiversity conservation and environmental law, dividing plants into a binary of native species, or "native vegetation", and introduced species, or "weeds". Introduced species are demonised – "pest", "noxious" – and the term is treated as inter-changeable with "exotic", "foreign" and "alien" (the Convention on Biological Diversity uses the latter label) (see Shine, Williams & Gündling 2000; Head & Muir 2004; Riley 2009). Such labels are often imbued with moralistic, emotive and political connotations (and declarations) of belonging, and othering. Fearful, glamorising and militant terminology are common, such as "invasion" and "the war on weeds" (see Davis 2009; Gröning & Wolschke-Bulmahn 2003; Olwig 2003; O'Brien 2006; Eskridge & Alderman 2010). There are distinct and troubling parallels with xenophobia (Warren 2007; Riley 2013; Barker 2010).

Eradication of the "other" is deemed essential (see Warren 2007), as harm is conflated with provenance. The harm caused by weeds, in habitat destruction and extinction of other species, is significant (Low 1999; Groves et al. 2005; Kearney et al. 2018). Weeds are considered to pose one of the greatest threats to biodiversity in NSW, second only to land clearance (DPI & OEH 2011; see also NSW Government 2012). However, substantive considerations of harm rarely inform plant categories. Classifications are instead based on provenance. This may appear objective but "nativeness" is not an absolute criterion. It is dependent on time, scale, intervention and reference point, and its determination reliant upon human presence or absence and their values (see Head & Muir 2004; Head 2012; Shine et al. 2000, p.44; Warren 2007; Frawley & McCalman 2014). Even when harm is considered, it is usually viewed via a particular human lens. In Australia, weed management is frequently identified as essential for industrial agriculture (McFadyen 2012; Sinden et al. 2004), which is almost completely reliant on introduced species (Young 1996). Yet crops and introduced pasture are not classified as "weeds", even where they may also cause environmental harm, most obviously by displacing native habitat. Both the categories and their exceptions appear fixed in a rigidity that is inconsistent and exceptional, as priority is given to economic values over environmental (see Pauly 1996).

Environmental law generally fails to challenge such hegemony of production (see Bartel & Graham 2019), and the justification for weed removal on conservation grounds alone is often fragile (see Riley 2009; Nagle 2010; Bartel et al. 2014). Some biologists, are also challenging the idea that provenance should determine plant removal (see Davis et al. 2011; Schlaepfer, Sax & Olden 2011), and that introduced species may not necessarily be invasive or harmful, nor all native species benign. These assessments suggest that desirability of the presence of species, or intervention to remove them, should be based on evaluations of functionality and degrees of "weediness" or "naturalness" (see Hobbs et al. 2006; Dudley 2011). Invasive native species (INS), which are plants that have spread beyond their local ranges or niches (Shine et al. 2000, p.3), have only been intermittently recognised as a modification to the existing categories in biodiversity conservation and environmental law.

The law is instead reliant on the dominant "invasion biology" frame of native–non-native (Davis 2009). Furthermore, it transforms these idealistic definitions into powerful and fixed categories ill-designed for flexibility, irrespective of the agency of the very plants that are intended protection. Non-human agency is rarely considered, including in conservation (Folke et al. 2010; Wallach et al. 2018). This is due to the human–nature binary, which splits concepts that in legal geography are understood to be interrelated. In legal geography, in addition to accepting the geographical self as co-constituted of place and self (see Larsen & Johnson 2016, p.149), there is an appreciation of the co-constitution of place and law (see Bartel et al. 2013). This appreciation demands that greater attention be paid to non-human agency.

Methodology: place-based case study and legal geography

This chapter adopts a place-centric methodology, focusing on the vernacular knowledge held by local landholders, to give voice to the plants and the places in which they grow and to examine binaries in the law. Godden (2002) has observed that particular Western worldviews impose fixed categories that restrict the range of possibilities (see also Fitzpatrick 1984; Watson 2002). New approaches are required to facilitate the realisation of alternative perspectives (see Kahane 2004; Pahl-Wostl 2009; Huntjens et al. 2011). Research in new materialism demonstrates there is a critical need to foreground place and non-human agency in scholarship, management, law and policy (see Brierley, Hillman & Fryirs 2006; Brierley et al. 2018; Wilcock, Brierley & Howitt 2013; Nightingale 2016; Wheeler et al. 2016). Environmental historians have examined the role of place and non-human agency in past events (see Worster 1979; Steinberg 2019;

Morgan 2015), and plant agency has been foregrounded in studies of the influence on culture of grasses (Brooking & Pawson 2007), grapes (Brice 2014) and grains (Head, Atchison & Gates 2012; Scott 2017). Place may also produce its own law, which adds another dimension to legal pluralism scholarship, that foregrounds the diversity of anthropogenic legal systems coincident in one location (Bartel 2018). In legal geography, and for environmental law in particular, there is a growing recognition that these formal legal systems are not the only laws that need to be acknowledged, nor courtrooms the only places in which law belongs (Bartel et al. 2013). There are multiple ways of appreciating place, such as in Indigenous ontologies, and place has been recognised as a collaborator and co-author in seminal work in geography (Bawaka Country et al. 2014). Research design increasingly incorporates multiple ontologies, plural epistemologies (St Pierre, Jackson & Mazzei 2016; Bartel 2018) and new methodologies (Howitt 2019; Lather 2016; Ulmer 2017), including place-centric methodologies, such as the place-based case study (Bartel 2017).

Case studies are designed to garner a deeper appreciation and understanding of particular locales and sites, as well as context, events and circumstances. Therefore, this methodology offers great potential to recognise "the difference that place makes" (Anderson, Adey & Bevan 2010, 590; Scott 2005; Swart 2009; Tuck & McKenzie 2015). Laws are intended to apply universally to all places in a jurisdiction that may have a variety of place laws. In Australia, these jurisdictions are comprised of a heterogeneity of biophysical and socio-economic environments. These may interact to produce a range of different compliance and practical outcomes. These conditions and contingency may be investigated by examining how formal law works "on the ground" and in real-world conditions, similar to how remotely sensed imagery is ground-truthed to examine its veracity. Efficacy may be assessed by whether the law attains self-proclaimed conservation goals (Goates, Hatch & Eggett 2007; Brummitt et al. 2015) and also by analysing the perspectives of the regulated community, as their perspectives will often affect compliance and on-ground outcomes (Hansen 2011; Bartel & Barclay 2011).

As Shine and others (2000) have highlighted, a study of the treatment of weeds will always need to consider the social as well as scientific context in which weeds are understood (see Kahane 2004). Place-based case studies decentre the human, rather than affording them any superiority, exceptionalism or essentialism, and instead foreground the non-human, including place and its inter-relationships. This requires a focus on the apparent "other" that is actually within us (see Bartel 2017; Wilcock et al. 2013). There are several data collection techniques that are sensitive to non-human agency, including walking interviews (Evans & Jones 2011), oral history (Nunn 2018) and analysing vernacular or place-based knowledge which is a co-construction of people and place in particular locales (Bartel 2014). The voice of place may be recognised in the words expressed by human participants, as "something that has already been 'said' by the environment" (Carbaugh 2007, p.68).

The case study: environmental law and plant categories in New South Wales, Australia

Weed lists have been developed by all levels of government in Australia and are defined as "introduced species", meaning those new to Australia since European invasion (Table 10.1). Critical to many definitions is not only the place of the plant's origin but the extant biota in the invaded location. Such provenance-based categorisations, which are common in environmental law, privilege nativeness (Table 10.1). The recognition of the invasiveness of some native species in NSW has been ad hoc and inconsistent. It was not until several years after the *Native Vegetation Act 2003* (NSW) came into force that a scientific investigation of the issue was conducted and the government department responsible concluded:

> Because invasive native scrub is much more extensive and/or much denser than its previous natural condition, clearing it in certain circumstances, and under certain conditions, can improve or maintain environmental outcomes in its own right. (DNR 2005, p.1)

Three adjacent administrative regions, known at the time of the study as catchment management areas (CMAs), were selected for this study due to their agricultural land use profiles, history of land clearance and enduring conflicts over biodiversity management (NRC 2009a; Bartel 2014). All have greater non-native cover and less intact native vegetation than the state average. Their extents follow the catchment boundaries of major river systems (see Figure 10.1).

A random sample of participants was drawn from public telephone directory listings of farmers and graziers residing within each area. The label "farmer" is used to include both croppers and graziers, and nearly all of those interviewed had mixed operations (see Table 10.2). Twelve one hour-long interviews comprising 30 mainly open-ended questions were completed in each CMA (see Bartel 2014). Interviewees were asked to provide basic demographic details, descriptions of farming history and their perspectives on native vegetation legislation and environmental management, particularly regarding biodiversity and weeds. The data were analysed using a modified grounded theory approach in which the answers were transcribed and analysed to identify concepts and themes (after Glaser & Strauss 1967; Corbin & Strauss 2007).

Relationships were characterised from coded responses, and selected quotes are provided to give voice to the participants (Moon & Cocklin 2011). A multi-methods approach was adopted including site visits, and a comprehensive review of the legal and institutional arrangements governing native vegetation and weeds management in the CMAs and NSW was conducted.

Results: disconnection between place and law

Farmers in each CMA described invasive native species as their major environmental issue, as the plants exacerbate soil erosion by impeding the growth

TABLE 10.1 Timeline and Details of Law and Regulation in Australia and New South Wales

Jurisdiction/regulation	Details	
National treatment of weeds and native vegetation		
Environmental Protection and Biodiversity Conservation Act 1999 (Cth)	One of the key threatening processes is identified as the "loss and degradation of *native* plant and animal habitat by invasion of escaped garden plants, including aquatic plants" (emphasis added).	
National Weeds Strategy (Department of the Environment and Energy 2012)	In 2012, the list was expanded to include a further 12 introduced and closely related species.	Eradication is mandatory. There are restrictions on sale and movement, including for entry into the country under quarantine legislation.
National Weeds Strategy (Department of the Environment and Energy 1999)	Defines 20 of the most threatening invasive plant species in Australia as "Weeds of National Significance" (WONS). All are introduced.	There are also multiple extended noxious weed lists in each jurisdiction (Glanznig, McLachlan & Kessal 2004).
New South Wales' treatment of weeds		
Invasive Species Plan 2018–2021 (NSW Department of Primary Industries 2018)	The most serious weeds are classified as "noxious" or "declared" and may also be listed as key threatening processes under the *Threatened Species Conservation Act 1995* (NSW). There are no native species mentioned in the plan.	
New South Wales' treatment of native vegetation and invasive native species		
Biodiversity Conservation Act 2016 (NSW)	Regulates clearance of native vegetation. Clearance requires a permit. Invasive native species are not required to be removed or managed (as for a listed weed) but may be allowed to be removed in some areas under certain conditions, provided that notification has been provided or a certificate obtained from the local government agency (currently the Local Land Services, see the *Land Management (Native Vegetation) Code 2018* (NSW)).	Native vegetation is defined as species that existed in the state prior to European settlement (also referred to as "Indigenous"), comprising trees, understory plants, ground cover and wetland plants (not marine) (also s. 60B of the *Local Land Services Act 2013* (NSW)).

(Continued)

TABLE 10.1 (Continued)

Jurisdiction/regulation	Details	
Native Vegetation Act 2003 (NSW)★	Earlier recognition for invasive native species (INS) was not carried over. A permit requirement to remove INS was later introduced, also stipulating that at least 20% of the extent of the INS on a property remain untreated (see Droulers & Kneipp 2010; INS RPAG 2014).	The status of listed INS varies across the state: in some areas they are not considered invasive and will be protected.
Native Vegetation Conservation Act 1997 (NSW)	INS could be cleared without a permit in the Western Division (a region roughly equivalent to the areas covered by the Western and Lower Murray Darling Catchment Management Areas in Figure 10.1).	INS are defined as a plant species that invades vegetation communities where it has not been known to occur previously, or a species that regenerates densely following natural or artificial disturbance, and the invasion and/or dense regeneration of the species results in change of structure and/or composition of the vegetation community, and the species is within its natural geographic range or distribution.

★ Legislation applicable at the time of the case study.

of other plants. Soil erosion is a major contributor to land and water degradation in Australia (State of the Environment Committee 2011). There was very low regard for government intervention overall, and for the *Native Vegetation Act 2003* in particular, as it was perceived to be unsympathetic to local conditions, as well as overly restrictive and bureaucratic. The approach that had been adopted with regard to management of INS was seen as perverse, particularly given the environmental objectives of the legislation. Landholders expressed a preference for a harms-based approach over a provenance-based demarcation, since this would be more sensitive to impacts in the landscape. Landholders considered that removing INS would be a more effective way to achieve broader environmental objectives and aims related to land degradation, as well as the biodiversity conservation the Act intends to promote:

> I don't think the current regulations do anything to encourage people to get better. The regulation in fact decreases incentives as far as eliminating

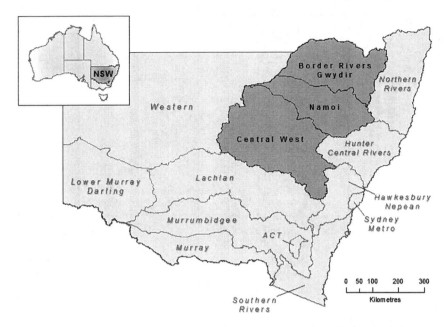

FIGURE 10.1 Catchment management areas in New South Wales (darker shaded region on main map indicates area of case study). Source: author provided.

TABLE 10.2 Description of the CMAs and Participant Profiles

CMA	Area (sq km)	Population	Participants' primary production details and property size (acres)	Age and gender profile
Border Rivers Gwydir	50,000	50,000	Property range: 988–42,007 Average size: 9,757	20% female. Average age 61 (national average farmer age 56).
Central West	92,200	185,515	Nearly all mixed farming operations – including grazing (cattle, sheep, goats) and mostly dryland cropping (wheat, sorghum, canola, barley, oats, cotton and grapes). Limited irrigation due to drought conditions.	Age range: 42–84 years. (40% between 45 and 55, 22% between 55 and 65, 31% between 65 and 75 and 6% between 75 and 85.)
Namoi	42,000	100,000		

Source: ABS 2018; Bartel 2014; NRC 2009a; NRC 2009b; NRC 2009c; NRC 2009d.

the scrub or the excessive timber goes … At this stage the regulations really do zip for encouraging the preservation of ground cover and promoting the multitude of benefits that has. (Central West farmer)

According to this place-based knowledge problematic species could be identified irrespective of their status as native or non-native, as either could be problematic for exacerbating soil erosion.

Our particular area is suffering quite badly from inundation from native and non-native species which is tending to silt the [river] system up. (Border Rivers Gwydir farmer)

While this more nuanced appreciation was commonly expressed by the farmers interviewed, this has not been the approach adopted by the law. The Act protects all native vegetation, irrespective of impact. Even when INS were once again recognised by the Act (similarly although not exactly as they had been under the previous legislation) permits were still required before weeds could be removed (Table 10.1). However, farmers in all CMAs declared that they were aware of certain species that were causing harm, but believed they were unable to remove them because they had become disengaged from the legislation following its introduction and were unaware of the changes (Table 10.3).

Discussion: challenging categories and going beyond categorisation

The way plants are categorised is crucial to the way environmental legislation is devised and definitions are drafted, and this, in turn, affects how law influences human behaviour, including interactions with the environment (see Robards et al. 2011). The environment also affects human behaviour, including through place law. In the case study, landholders have become disengaged with the formal law because they consider it to be unsuited to local conditions. This is more likely to result in poorer environmental outcomes. If left untreated, invasive native species may continue to impede environmental health, as the relevant state department belatedly recognised (DNR 2005). Better environmental outcomes would occur if the law always reflected the characteristics of the places – the place law that the local landholders were familiar with – without risking negative consequences such as disengagement. Conversely, formal laws that are better able to pass a "reality check" may lead to reduced resistance to legal interventions, as well as better environmental outcomes, through engendering social approval and voluntary compliance (Bartel & Barclay 2011; Huntjens et al. 2011). Le Gal (2017, 162) has observed that, "[i]n the invasive species space, place-based knowledge is essential for effective control". The landholders' vernacular knowledge reflects actual instead of assumed harm, due to their place-based (and plant-based) experiences (see also Low 2008). It has previously been observed that regulatory

TABLE 10.3 Invasive Native Species Provisions and Landholder Disengagement

CMA	Illustration	Details
Central West	"it's covered by this Budda and we can't get in and clear it".	Budda, also known as Buddha Wood and False Sandlewood, *Eremophila mitchelii* is a listed invasive native species in the Central West CMA (NSW Government 2006).
Border Rivers Gwydir	"As far as regulation goes, anything that grows from grass to a tree is classified native vegetation, and the tea tree it's growing so fast … You just can't afford to battle against the regulations".	Grey tea tree, *Leptospermum brevipes,* is listed as an invasive native species in the Border Rivers Gwydir CMA (NSW Government 2006).
Namoi	"And the woody weeds they're horrible, the African Boxthorn and Briar bushes they're just the home for feral pigs and kangaroos. I think it's got to be managed properly to try and keep those things [in check] – don't let them have free range. Otherwise you've got such a seed bank and the animals and birds can spread it to pollute the clean and free country".	African boxthorn, *Lycium ferocissimum*, is a declared noxious weed in NSW and it was also included in the 2012 additions to the national WONS list. Briar or briars bush, also known as Mimosa bush, *Vachiella farnesiana* or *Acacia farnesiana*, is listed as an invasive native species in the Namoi CMA (NSW Government 2006).
Namoi	"We don't know if we can touch the roly-poly. We never know what we're allowed to do. That's a real issue for farmers knowing what they can and can't do. And with roly-poly it takes over and there are paddocks and paddocks of it".	Black roly-poly, *Sclerolaena muricate*, is a listed invasive native species in the Namoi CMA (NSW Government 2006).

failures have been due to the disjuncture between law and "real-world" social or geographical conditions (see Sanderson 2002; Home et al. 2014), which may be described as disconnects between place law and "the" (formal) law.

Plant binaries and law

The belated recognition of INS by the formal legal system has occurred only in a fragmented fashion, which unfortunately simply extends rather than reframes pre-existing schema. Exceptions to imperial categories are just that, and they are selective rather than holistic. This merely qualifies and thus reinforces a classification (as in the term "invasive native"), rather than challenging it at a more fundamental level. Given the extent of environmental harm (Brondizio et al. 2019) there may be some understandable reluctance to make concessions to the absolutist idea that native species protection is paramount, and that introduced species have no place in conservation (see discussion of nativism in Wallach et al. 2018). Diversity thus continues to be positioned as an exception rather than the norm, thereby retaining the privileging of the universal and essential over the particular (Bartel 2017). Such limited attempts to account for harm have not troubled key assumptions that threaten to forestall the achievement of environmental objectives, including the production hegemony that is antithetical to sustainable agriculture (Massy 2017), and which threatens even domesticated species (Brondizio et al. 2019). Furthermore, all plants (not just weeds) are naturalised as non-sentient (Bowker & Star 2000; Wallach et al. 2018). The increasing evidence of plant consciousness challenges this categorisation (Hall 2011; Gagliano 2015) and demands even greater consideration of the harms experienced by non-humans (Wallach et al. 2018).

The law is predicated on human harm and human rights and wrongs, while environmental law is similarly biased (Shine et al. 2000). Head (2012) has questioned binarised and binarising policy responses based on classifications that exert dualities of good and bad, native and introduced, and which promote and perpetuate "othering" (see Head & Muir 2004). As Warren (2007, p.432) has observed, "[n]o species is inherently good or evil". All categories are coarse and dualities particularly so. By contrast, biophysical and social environments are complex and dynamic (Low 2008; Orts 1995; Schirmer, Dovers & Clayton 2012), as is the vernacular knowledge of the people living in them (Bartel 2014). Further interrogation of vernacular knowledge, particularly its diversity and relationship with place law in different sites, would be advantageous. For example, certain species of the native genus *Eremophila* may be viewed as invasive, others have medicinal and spiritual uses, and some are considered sacred by Indigenous Australians (Richmond 1993).

Concluding remarks: the other is us

The categorisation of "the" law as superior works in a similar way to the binary categorisation of plants. This approach overlooks diversity, and disregards other laws, including place law, the law of First Nations and Indigenous law (Watson 2012), and ecological and geomorphological laws (see Bartel 2018; Wilcock et al. 2013). Diverse conditions are produced and are governed by these "other" laws that are rarely universal. Instead they are particular, site-specific and place-orientated, due to the unique combinations of soil, weather, climate, landforms, lifeforms and other influences such as culture (see Wilcock et al. 2013). However, non-human nature is predominantly overlooked or appreciated only in limited ways within the dominant institutions of the settler state. Such institutions are imbricated with a human–nature binary that privileges the human over nature (see Bartel 2017). As Howitt (2019, p.5) observes, "[c]olonising societies assumed and assert not only their own superiority but also the inferiority of others".

A more fundamental mistake than plant category errors may be that of the human–nature binary, and the general human inclination towards categorisation, especially the making of coarse categories (Cloke & Johnston 2005). Identifying definitional issues as the problem may be less fruitful than examining the arrogant anthropocentrism underpinning both the human–nature binary and the act of categorisation itself (Robbins 2004; also Farber 1994; Bartel 2017). According to Peterson and others (2007, 75), the "human-nature dualism poses a serious obstacle to conservation generally ... by excluding extrahumans from the community of decision-makers". Recognising non-human agency presents enormous challenges for environmental law (Head & Atchison 2008, p.4; Jones & Cloke 2008; Lorimer 2006, p.541) and its methodologies. Privileged worldviews impose categories of and within law that may be interrogated through designing and pursuing less human-centred methodologies, particularly those that recognise and give voice to non-human agency.

References

ABS (Australian Bureau of Statistics) 2018, *Agricultural Commodities, Australia, 2016–7*, Cat. No. 7121.0. Canberra.

Anderson, J., Adey, P., & Bevan, P. 2010, 'Positioning place: Polylogic approaches to research methodology', *Qualitative Research*, vol. 10, no. 5, pp. 589–604.

Barker, K. 2010, 'Biosecure citizenship: Politicising symbiotic associations and the construction of biological threat', *Transactions of the Institute of British Geographers*, vol. 35, pp. 350–363.

Bartel, R. 2014, 'Vernacular knowledge and environmental law: Cause and cure for regulatory failure', *Local Environment: The International Journal of Justice and Sustainability*, vol. 19, no. 8, pp. 891–914.

Bartel, R. 2017, 'Place-thinking: The hidden geography of environmental law', in *Handbook of Research Methods in Environmental Law*, Philippopoulos-Mihalopoulos, A. & Brooks, V (eds), Edward Elgar Publishing Limited, Cheltenham, pp. 159–183.

Bartel, R. 2018, 'Place-speaking: Attending to the relational, material and governance messages of *Silent Spring*', *The Geographical Journal*, vol. 184, no. 1, pp. 64–74.

Bartel, R., & Barclay, E. 2011, 'Motivational postures and compliance with environmental law in Australian agriculture', *Journal of Rural Studies*, vol. 27, pp. 153–170.

Bartel, R., & Graham, N. 2019, 'Ecological reconciliation on private agricultural land: Moving beyond the human-nature binary in property-environment contests,' in *Ecological Restoration Law: Concepts and Case Studies*, (eds) Richardson, B. & Akhtar-Khavari, A., Routledge, Abingdon, pp. 93–118.

Bartel, R., Graham, N., Jackson, S., Prior, J., Robinson, D., Sherval, M., & Williams, S. 2013, 'Legal geography: An Australian perspective', *Geographical Research*, vol. 51, no. 4, pp. 339–353.

Bartel, R., McFarland, P., & Hearfield, C. 2014, 'Taking a de-binarized envirosocial approach to reconciling the environment vs economy debate: Lessons from climate change litigation for planning in NSW, Australia', *Town Planning Review*, vol. 85, no. 1, pp. 67–96.

Bawaka Country, Wright, S., Suchet-Pearson, S., Lloyd, K., Burarrwanga, L., Ganambarr, R., Ganambarr-Stubbs, M., Ganambarr, B., & Maymuru, D. 2014, 'Working with and learning from Country: Decentring human authority', *Cultural Geographies*, vol. 22, no. 2, pp. 269–283.

Biodiversity Conservation Act 2016 (NSW).

Bowker, G. C. & Star, S. L. 2000, *Sorting Things Out: Classification and Its Consequences*, MIT Press, Cambridge.

Brice, J. 2014, 'Attending to grape vines: Perceptual practices, planty agencies and multiple temporalities in Australian viticulture', *Social & Cultural Geography*, vol. 15, no. 8, pp. 942–965.

Brierley, G., Hillman, M., & Fryirs, K. 2006, 'Knowing your place: An Australasian perspective on catchment-framed approaches to river repair', *Australian Geographer*, vol. 37, no. 2, pp. 131–145.

Brierley, G., Tadaki, M., Hikuroa, D., Blue, B., Šunde, C., Tunnicliffe, J., & Salmond, A. 2018, 'A geomorphic perspective on the rights of the river in Aotearoa New Zealand', Special Issue Paper, *River Research and Applications*, <DOI: 10.1002/rra.3343>.

Brondizio, E. S., Settele, J., Díaz, S., & Ngo, H. T. (eds) 2019, *The Global Assessment on Biodiversity and Ecosystem Services*, Inter-governmental Science-Policy Platform on Biodiversity and Ecosystem Services (IPBES).

Brooking, T., & Pawson, E. 2007, 'Silences of grass: Retrieving the role of pasture plants in the development of New Zealand and the British Empire', *Journal of Imperial and Commonwealth History*, vol. 35, no. 3, pp. 417–435.

Brummitt, N., Bachman, S. P., Aletrari, E., Chadburn, H., Griffiths-Lee, J., Lutz, M., Moat, J., Rivers, M. C., Syfert, M. M., & Lughadha, E. M. 2015, 'The sampled red list index for plants, phase II: Ground-truthing specimen-based conservation assessments', *Philosophical Transactions of the Royal Society of Botany*, vol. 370, <DOI: 10.1098/rstb.2014.0015?>.

Carbaugh, D. 2007, 'Quoting "the environment": Touchstones on earth', *Environmental Communication: A Journal of Nature and Culture*, vol. 1, no. 1, pp. 64–73.

Cloke, P., & Johnston, R. 2005, 'Deconstructing human geography's binaries', in *Spaces of Geographical Thought: Deconstructing Human Geography's Binaries*, (eds) Cloke, P., & Johnston, R., Sage, London, pp. 1–20.

Corbin, J., & Strauss, A. 2007, *Basics of Qualitative Research: Techniques and Procedures for Developing Grounded Theory*, 3rd edition, Sage Publications, Thousand Oaks, CA.

Davis, M. 2009, *Invasion Biology*, Oxford University Press, Oxford.

Davis, M. A., Chew, M. K., Hobbs, R. J., Lugo, A. E., Ewel, J. J., Vermeij, G. J., Brown, J. H., Rosenzweig, M. L., Gardener, M. R., Carroll, S. P., Thompson, K., Pickett, S. T. A., Stromberg, J. C., Del Tredici, P., Suding, K. N., Ehrenfeld, J. G., Grime, J. P., Mascaro, J., & Briggs, J. C. 2011, 'Don't judge species on their origins', *Nature*, vol. 474, pp. 153–154.

Department of the Environment and Energy, 1999, *Weeds of National Significance*, <https://www.environment.gov.au/biodiversity/invasive/weeds/weeds/lists/wons.html>.

Department of the Environment and Energy, 2012, *Weeds of National Significance*, <https://www.environment.gov.au/biodiversity/invasive/weeds/weeds/lists/wons.html>.

Department of Primary Industries, 2018, *NSW Invasive Species Plan 2018–2021*, Author, Orange.

DNR (Department of Natural Resources in New South Wales) 2005, *Clearing/Thinning of Native Vegetation known as Invasive Native Scrub under the* Native Vegetation Act 2003, New South Wales Government, Sydney.

DPI (NSW Department of Primary Industries) & OEH (NSW Office of Environment and Heritage) 2011, *Biodiversity Priorities for Widespread Weeds: Statewide Framework*,

NSW Department of Primary Industries, Orange and NSW Office of Environment and Heritage, Sydney.

Droulers, P., & Kneipp, K. 2010, 'Demonstrating implementation of invasive native scrub property vegetation plans', Paper presented at the 16th Biennial Conference, Australian Rangeland Society, Bourke, New South Wales <https://globalrangelands. org/sites/globalrangelands.org/files/dlio/38151/arsbc-2010-droulers.pdf>.

Dudley, N. 2011, *Authenticity in Nature: Making Choices about the Naturalness of Ecosystems*, Earthscan, Abingdon.

Environmental Protection and Biodiversity Conservation Act 1999 (Cth).

Eskridge, A. E., & Alderman, D. H. 2010, 'Alien invaders, plant thugs and the southern curse: Framing Kudzu as environmental other through discourses of fear', *Southeastern Geographer*, vol. 50, no. 1, pp. 110–129.

Evans, J., & Jones, P. 2011, 'The walking interview: Methodology, mobility and place', *Applied Geography*, vol. 31, pp. 849–858.

Farber, D. A. 1994, 'Environmental protection as a learning experience', *Loyola of Los Angeles Law Review*, vol. 27, no. 3, pp. 791–807.

Fitzpatrick, P. 1984, 'Law and societies', *Osgoode Hall Law Journal*, vol. 22, no. 1, pp. 115–138.

Folke, C., Carpenter, S. R., Walker, B., Scheffer, M., Chapin, T., & Rockstrom, J. 2010, 'Resilience thinking: Integrating resilience, adaptability and transformability', *Ecology and Society*, vol. 15, no. 4. [online] URL: http://www.ecologyandsociety.org/vol15/ iss4/art20/.

Frandsen, A., & McGoun, E. G. 2010, 'To be human is to account?', *Accounting and the Public Interest*, vol. 10, no. 1, pp. 1–12.

Frawley, J., & McCalman, I. (eds.) 2014, *Rethinking Invasion Ecologies from the Environmental Humanities*, Routledge, London.

Gagliano M. (2015) 'In a green frame of mind: Perspectives on the behavioural ecology and cognitive nature of plants', *AoB PLANTS*, vol. 7, DOI: 10.1093/aobpla/plu075.

Glanznig, A., McLachlan, K., & Kessal, O. 2004, *Garden Plants that are Invasive Plants of National Importance: An Overview of their Legal Status, Commercial Availability and Risk Status*, WWF-Australia, Sydney.

Glaser, B., & Strauss, A. 1967, *The Discovery of Grounded Theory*, Aldine, Chicago.

Goates, M. C., Hatch, K. A., & Eggett, D. L. 2007, 'The need to ground truth 30.5m buffers: A case study of the boreal toad (*Bufo boreas*)', *Biological Conservation*, vol. 138, pp. 474–483.

Godden, L. 2002, 'Indigenous heritage and the environment: "Legal categories are only one way of imaging the real"'. *Environmental and Planning Law Journal*, vol. 19, no. 4, pp. 258–266.

Gröning, G., & Wolschke-Bulmahn, J. 2003, 'The native plant enthusiasm: Ecological panacea or xenophobia?' *Landscape Research*, vol. 28, no. 1, pp. 75–78.

Groves, R., Boden, R., & Lonsdale, M. 2005. *Jumping the Garden Fence Invasive Plants in Australia and their Environmental and Agricultural Impacts*, CSIRO and WWF, WWF-Australia.

Hall, M. 2011. *Plants as Persons. A Philosophical Botany*, State University of New York Press, Albany, NY.

Hamilton, C. 2010, *Requiem for a Species: Why We Resist the Truth About Climate Change*, Allen & Unwin, Sydney.

Hansen, C. 2011, 'Forest law compliance and enforcement: The case of on-farm timber extraction in Ghana', *Journal of Environmental Management*, vol. 92, no. 3, pp. 575–586.

Head, L. 2012, 'Decentring 1788: Beyond biotic nativeness', *Geographical Research*, vol. 50, no. 2, pp. 166–178.

Head, L., & Atchison, J. 2008, 'Cultural ecology: Emerging human-plant geographies', *Progress in Human Geography*, vol. 33, no. 2, pp. 236–245.

Head, L., & Muir, P. 2004, 'Nativeness, invasiveness and nation in Australian plants', *Geographical Review*, vol. 94, no. 2, pp. 199–217.

Head, L. M., Atchison, J. M., & Gates, A. 2012, *Ingrained: A Human Bio-geography of Wheat*, Ashgate, Burlington.

Hobbs, R. J., Arico, S., Aronson, J., Baron, J. S., Bridgewater, P., Cramer, V. A., Epstein, P. R., Ewel, J. J., Klink, C. A., Lugo, A. E., Norton, D., Ojima, D., Richardson, D. M., Sanderson, E. W., Valladares, F., Zamora, M. V. R., & Zobel, M. 2006, 'Novel ecosystems: Theoretical and management aspects of the new ecological world order', *Global Ecology and Biogeography*, vol. 15, pp. 1–7.

Home, R., Balmer, O., Jahrl, I., Stolze, M., & Pfiffner, L. 2014, 'Motivations for implementation of ecological compensation areas on Swiss lowland farms', *Journal of Rural Studies*, vol. 34, pp. 26–36.

Howitt, R., 2019, 'Unsettling the taken (-for-granted)', *Progress in Human Geography*, <DOI: 10.1177/0309132518823962>.

Huntjens, P., Pahl-Wostl, C., Rihoux, B., Schlueter, M., Flachner, Z., Neto, S., Koskova, R., Dickens, C., & Kiti, I. N. 2011, 'Adaptive water management and policy learning in a changing climate: A formal comparative analysis of eight water management regimes in Europe, Africa and Asia', *Environmental Policy and Governance*, vol. 21, pp. 145–163.

INS R PAG (Invasive Native Species Research Program Advisory Group) 2014, 'Managing invasive native scrub guide to rehabilitate native pastures and open woodlands: A best management practice guide for the Central West and Western Regions, State of NSW', Local Land Services, *Central West*, <https://centralwest.lls.nsw.gov.au/__dat a/assets/pdf_file/0007/685222/managing-invasive-native-scrub.pdf>.

Jones, O., & Cloke, P. 2008, 'Non-human agencies: Trees in place and time', in *Material Agency: Towards a Non-Anthropocentric Approach*, (eds) Knappett, C. & Malafouris, L., Springer, Berlin, pp. 79–96.

Kahane A. 2004, *Solving Tough Problems: An Open Way of Talking, Listening, and Creating New Realities*, Berrett-Koehler, San Francisco.

Kearney, S. G., Cawardine, J., Reside, A. E., Fisher, D. O., Maron M., Doherty, T. S., Legge, S., Silcock J., Woinarski J. C. Z., Garnett S. T., Wintle B. A., & Watson J. E. M. 2018, 'The threats to Australia's imperilled species and implications for a national conservation response', *Pacific Conservation Biology*, CSIRO Publishing, <DOI: 10.1071/PC18024>.

Land Management (Native Vegetation) Code 2018 (NSW).

Larsen, S. C., & Johnson, J. T. 2016, 'The agency of place: Toward a more-than-human geographical self', *GeoHumanities*, vol. 2, pp. 149–166.

Lather, P. 2016, 'Top Ten+ List: (Re)Thinking ontology in (post)qualitative research', *Cultural Studies ↔ Critical Methodologies*, vol. 16, no. 2, pp. 125–131.

Le Gal, E. 2017, 'Climate change and invasive species law in agriculture', in *Research Handbook on Climate Change and Agricultural Law*, (eds) Angelo, M. J. & Du Plessis, A., Edward Elgar, Cheltenham, pp. 136–166.

Local Land Services Act 2013 (NSW).

Lorimer, J. 2006, 'What about the nematodes? Taxonomic partialities in the scope of UK biodiversity conservation', *Social & Cultural Geography*, vol. 7, no. 4, pp. 539–558.

Low, T. 1999, *Feral Future: The Untold Story of Australia's Exotic Invaders*, Viking, Melbourne.

Low, T. 2008, *Climate Change & Invasive Species: A Review of Interactions, November 2006 Workshop Report*, Department of the Environment, Water, Heritage and the Arts, Commonwealth of Australia, Canberra.

Massy, C. 2017, *Call of the Reed Warbler: A New Agriculture, A New Earth*, University of Queensland Press, Brisbane.

McFadyen, R. 2012, 'Food security for a 9 billion population: More R&D for weed control will be critical', in *Developing Solutions to Evolving Weed Problems*, Proceedings for the 18th Australasian Weeds Conference, Albert Park, Melbourne, Victoria, Australia, 8–11 October 2012, (ed) Eldershaw, V., Weed Society of Victoria Inc., Victoria, pp. 306–309.

Moon, K., & Cocklin, C. 2011, 'A Landholder-Based Approach to the Design of Private-Land Conservation Programs', *Conservation Biology*, vol. 25, no. 3, pp. 493–503.

Morgan, R. 2015, *Running Out? Water in Western Australia*, University of Western Australia Press, Perth.

Nagle, J. C. 2010, *Law's Environment: How Law Shapes the Places We Live*, Yale University Press, New Haven.

Native Vegetation Act 2003 (NSW).

Native Vegetation Conservation Act 1997 (NSW).

Nightingale, A. J. 2016, 'Adaptive scholarship and situated knowledges? Hybrid methodologies and plural epistemologies in climate change adaptation research', *Area*, vol. 48, no. 1, pp. 41–47.

NRC (Natural Resources Commission), 2009a, *Native Vegetation Extent & Condition: Technical Report*, Natural Resources Commission, Sydney.

NRC (Natural Resources Commission), 2009b, *Audit Report Central West Catchment Management Authority*, March 2009, Natural Resources Commission, Sydney.

NRC (Natural Resources Commission), 2009c, *Audit Report Border Rivers-Gwydir Catchment Management Authority*, April 2009, Natural Resources Commission, Sydney.

NRC (Natural Resources Commission), 2009d, *Audit Report Namoi Catchment Management Authority*, October 2009, Natural Resources Commission, Sydney.

NSW Government, 2006 *Native Vegetation Management in NSW: Managing invasive native scrub*, Native Vegetation Information Sheet No. 9, New South Wales Government, Sydney.

NSW Government, 2012. New South Wales State of the Environment, Chapter 5: Biodiversity, New South Wales Government, Sydney.

Nunn, P. 2018, *The Edge of Memory: Ancient Stories, Oral Tradition and the Post-Glacial World*, Bloomsbury, London.

O'Brien, W. 2006, 'Exotic invasions, nativism, and ecological restoration: On the persistence of a contentious debate', *Ethics, Place and Environment*, vol. 9, no. 1, pp. 63–77.

Olwig, K. R. 2003, 'Natives and aliens in the national landscape', *Landscape Research*, vol. 28, no. 1, pp. 61–74.

Orts, E. W. 1995, 'Reflexive environmental law', *Northwestern Law Review*, vol. 89, no. 4, pp. 1227–1340.

Pahl-Wostl, C. 2009, 'A conceptual framework for analysing adaptive capacity and multi-level learning processes in resource governance regimes', *Global Environmental Change*, vol. 19, no. 3, pp. 354–365.

Pauly, P. J. 1996, 'The beauty and menace of Japanese cherry trees: Conflicting visions of American ecological independence', *Isis*, vol. 87, no. 1, pp. 51–73.

Peterson, M. N., Peterson, M. J., & Peterson, T. R. 2007, 'Environmental communication: Why this crisis discipline should facilitate environmental democracy', *Environmental Communication: A Journal of Nature and Culture*, vol. 1, no. 1, pp. 74–86.

Richmond, G. S. 1993, 'A review of the use of *Eremophila* (Myopoiraceae) by Australian aborigines', *Journal of the Adelaide Botanic Gardens*, vol. 15, no. 2, pp. 101–107.

Riley, S. 2009, 'A weed by any other name: Would the rose smell as sweet if it were a threat to biodiversity?' *The Georgetown International Environmental Law Review*, vol. 22, pp. 157–184.

Riley, S. 2013, "Buffalo belong here, as long as he doesn't do too much damage': Indigenous perspectives on the place of alien species in Australia', *Australasian Journal of Natural Resources Law and Policy*, vol. 16, no. 2, pp. 157–196.

Robards, M. D., Schoon, M. L., Meek, C. L., & Engle, N. L. 2011, 'The importance of social drivers in the resilient provision of ecosystem services', *Global Environmental Change*, vol. 21, no. 2, pp. 522–529.

Robbins, P. 2004, 'Comparing invasive networks: Cultural and political biographies of invasive species', *Geographical Review*, vol. 94, no. 2, pp. 139–156.

Sanderson, I. (2002) 'Evaluation, policy learning and evidence-based policy making', *Public Administration*, vol. 80, no. 1, pp. 1–22.

Schirmer, J., Dovers, S., & Clayton, H. 2012, 'Informing conservation policy design through an examination of landholder preferences: A case study of scattered tree conservation in Australia', *Biological Conservation*, vol. 153, pp. 51–63.

Schlaepfer, M. A., Sax, D. F. & Olden, J. D. 2011, 'The potential conservation value of non-native species', *Conservation Biology*, vol. 25, no. 3, pp. 428–437.

Scott, J. C. 2005, 'Afterword to "moral economies, state spaces, and categorical violence"', *American Anthropologist*, vol. 107, no. 3, pp. 395–402.

Scott, J. C. 2017, *Against the Grain: A Deep History of Early States*, Yale University Press, New Haven.

Shine, C., Williams, N., & Gündling, L. 2000, *A Guide to Designing Legal and Institutional Frameworks on Alien Invasive Species*, IUCN Environmental Policy and Law Paper No. 40. IUCN, Gland, Switzerland, Cambridge and Bonn.

Sinden, J., Jones, R., Hester, S., Odom, D., Kalisch, C., James, R., & Cacho, O. 2004, *The Economic Impact of Weeds in Australia*, Technical Series 8, CRC for Australian Weed Management, Adelaide.

St Pierre, E. A., Jackson, A. Y., & Mazzei, L. A. 2016, 'New empiricisms and new materialisms: Conditions for new inquiry', *Cultural Studies ↔ Critical Methodologies*, vol. 16, no. 2, pp. 99–110.

State of the Environment Committee 2011, *Australia: State of the Environment 2011*, Australian Government, Canberra.

Steinberg, T. 2019, *Down to Earth: Nature's Role in American History*, 4th edition, Oxford University Press, Oxford.

Swart, J. A. A. 2009, 'Towards an epistemology of place', in *New Visions of Nature: Complexity and Authenticity*, (eds) Drenthen, M., Keulartz, J., & Proctor, J., Springer, Dordrecht, pp. 197–204.

Threatened Species Conservation Act 1995 (NSW).

Tuck, M., & McKenzie, E. 2015, *Place in Research: Theory, Methodology, and Methods*, Routledge, New York.

Ulmer, J. B. 2017, 'Posthumanism as research methodology: Inquiry in the Anthropocene', *International Journal of Qualitative Studies in Education*, vol. 30, no. 9, pp. 832–848.

Wallach, A. D., Bekoff, M., Batavia, C., Nelson, M. P., & Ramp, D. 2018, 'Summoning compassion to address the challenges of conservation', *Conservation Biology*, vol. 32, no. 6, pp. 1255–1265.

Warren, C. R. 2007, 'Perspectives on the alien versus native species debate: A critique of concepts, language, and practice', *Progress in Human Geography*, vol. 31, pp. 427–446.

Watson, I. 2002, 'Buried alive', *Law and Critique*, vol. 13, pp. 253–269.

Watson, I. 2012 'The future is our past: We once were sovereign and we still are', *Indigenous Law Bulletin*, vol. 8, no. 3, pp. 12–15.

Wheeler, M. J., Sinclair, A. J., Fitzpatrick, P., Diduck, A. P., & Davidson-Hunt, I. J. 2016, 'Place-based inquiry's potential for encouraging public participation: Stories from the common ground land in Kenora, Ontario', *Society & Natural Resources: An International Journal*, vol. 29, no. 10, pp. 1230–1245.

Wilcock, D., Brierley, G., & Howitt, R. 2013, 'Ethnogeomorphology', *Progress in Physical Geography*, vol. 37, no. 5, pp. 573–600.

Worster, D. 1979, *The Dust Bowl: The Southern Plains in the 1930s*, Oxford University Press, Oxford.

Young, A. R. M. 1996, *Environmental Change in Australia Since 1788*, Oxford University Press, Melbourne.

Zimbardo, Philip. 2008, *The Lucifer Effect: How Good People Turn Evil*, Penguin Random House, New York.

11

ASK AN "EXPERT"

Phenomenology and key informant interviews as a research method in legal geography

Paul McFarland

Introduction

Legal geography differentiates itself from other social and natural sciences through the use of relevant theoretical frameworks and methodological processes which analyse contemporary relationships between people and physical concepts. These concepts include things such as space, place, mobility and nature, as well as abstract constructs, such as economics and law (Kitchin & Thrift 2009). These relationships are investigated and analysed using geographically relevant data collection and analysis methods. In studies of contemporary legal geographies, the concepts and constructs often interconnect. This can be particularly noticed in land use planning where the examination of the relationship between people and one construct, such as law, has a relationship to a variety of other concepts and constructs, including economics and the environment (McFarland 2015). As will be discussed in this chapter, key informant interviews are an important method in legal geography.

The interactions in legal geography can be seen in market economies. Operating within a market economy, people seek to exploit the values of land for human benefit. These benefits can be for personal enjoyment of the attributes of the land and its surrounds, or the realisation of the exchange value of the products of that land, such as agriculture or mining (Jessop 2004; Harvey 2006; Brenner, Marcuse & Mayer 2011). The manner in which land use activities are undertaken brings with it benefits for the land owner but can also produce negative externalities, such as water and air pollution (Daly & Farley 2004). These externalities can adversely affect other lands and, therefore, the ability of other land holders or occupiers to achieve enjoyment of their land in their desired manner (Gotham 2009).

The effects of externalities can be regarded as market failures (Poulton 1997). In order to address market failures in relation to land uses legal remedies have been provided (Alexander 2001). This concept has further developed whereby the uses of land are regulated by laws that describe where and how land can be used in a system commonly referred to as land use planning. Zoning and related regulatory frameworks are commonly, but not exclusively, employed for this purpose. Land use regulation often occurs in concert with other regulatory mechanisms controlling specific physical, economic or social issues.

For the purpose of this chapter, legal geography is defined as "a way for enlarged appreciations of relationality, materiality, multiscalarity and agency to be used to interrogate and reform the law" (Bartel et al. 2013, p.339). Legal geography affords an opportunity to examine the operations and effects of land use planning within the framework of planning law, regulation and spatial plans, and thereby reveals the socio-spatial implications of the system. The ultimate outcomes of evaluating land use planning from a legal geography perspective enable an investigation of the performance of the relevant legal system. At the same time, legal geography also provides a context in which to review the implications for reforming the applicable approach to land law and related regulatory systems (Jones 2003).

A key feature of any research is the selection of an appropriate method by which data are collected, so as to address the key research question (Kitchin & Tate 2013). Legal geography research accords with this convention. There are numerous methods that can be utilised to address research questions relating to people, place and the law. If research is specifically being undertaken to examine the processes and procedures relating to land use from the human perspective then an appropriate method would be to employ a phenomenological approach utilising use key informant interviews (Winchester 1996). Key informant interviews provide an opportunity to gain insight into the case being studied from the perspective of those regarded as subject experts. The views provided by key informants can be analysed to expose common themes and key differences from which conclusions can be drawn (Winchester 1996; Dowling, Lloyd & Suchet-Pearson 2016).

Key informant interviews are used in a wide variety of fields of study, including health (for examples see Coreil 1994; Farquhar et al. 2006; Kral et al. 2010), social science (for examples see Marshall 1996; Payne & Payne 2004; Collins 2010) and public administration (for examples see Walker & Enticott 2004; Hermans & Thissen 2009; Rhodes 2014). Key informant interviews are also widely used in geography (see for example Winchester 1999; Ruming 2009; McGuirk & O'Neill 2016).

This chapter focusses on the use of key informant interviews as a research method in legal geography. The chapter begins by defining "key informants". A discussion of the means of selecting key informants is then provided, followed by an examination of the method of analysing key informant interviews. The chapter concludes with a case study using key informant interviews as an example of this method in practice.

Phenomenology and key informant interviews as a methodology in legal geography

Phenomenology concerns the discovery of the meaning underlying human perceptions of the world (Whittemore 2014), and enables the manner in which different frames of reference give meaning to an issue or object to be discovered (Wylie 2013). At the core of phenomenological philosophy is the attempt to reveal the "human lived experience, perception, sensation and understanding" (Wylie 2013, p.54). Key informant interviews provide a technique by which human perceptions of the world can be captured.

Key informant interviews utilise the expertise, experience and insight into organisational issues from the perspective of those with differing roles within a particular social structure. The technique has been used to examine organisations from within, but it can used with equal validity to examine policy and practice of organisational processes and outcomes in a much broader sense (Krannich & Humphrey 1986; Sofaer 1999). For legal geography key informant interviews facilitate different frames of reference about the intersection of place, people and the law.

Key informants are recognised as "opinion leaders" (Barnidge et al. 2013, p.98) who are "judged knowledgeable in their respective areas" (McWilliam et al. 2012, p.756). Key informants are important to research as they bring unique and particular information and perspectives to a topic under investigation. The next section discusses the selection of participants in key informant studies. Before discussing this, it is worth noting that opinion leaders (key informants) must not be self-selecting. There are plenty of people who will advance themselves as opinion leaders in a field; however, self-promotion lacks an objective credibility that can only come from appropriate recognition and identification by others.

Selection of key informants

The selection of appropriate opinion leaders is crucial to a study relying on key informants. How to select the participants and what constitutes an appropriate sample size are key questions that often vex a researcher in undertaking this type of research. In fact, key informant case study research has been criticised for its lack of the same criteria on which empirical research is based: internal and external validity reliability, and objectivity. Key informant interviews can be as valid as other forms of qualitative and quantitative research if they are assessed for confirmability, dependability, credibility and transferability (Faifua 2014).

The identification of key informants can vary depending on the particular study being undertaken. Literature searches, websites and professional networks (see, for example, Barnidge et al. 2013), purposive sampling (Tongco 2007; McWilliam et al. 2012) or the "snowball" method (McWilliam et al. 2012; Barnidge et al. 2013) are among the techniques employed. Regardless of the approach selected, the first step is for the researcher to be clear about what it is they are seeking to know. Having a set of criteria on which to base the selection of participants eases the process (Tongco 2007). It also provides an objective means of verifying the selection of participants. Once the researcher has clarified the purpose of the study and the participant criteria, the search for those who are willing to share their knowledge and experience can begin (Etikan et al. 2016).

The appropriate number of interviews to conduct in order to achieve a reliable sample size is of concern to many researchers. Criticisms of the key informant technique arise due to the sample size seeming relatively small, especially to those who are more used to undertaking research using quantitative methods involving large data sets (Francis et al. 2010). As key informant interviews are in-depth studies that seek opinion leaders, then the main way of determining the appropriate sample size is to ensure that a stratified selection technique has occurred that has filtered potential participants so that it can be objectively agreed that "opinion leaders" have been selected for the study.

In legal geography studies, one can begin by setting the criteria for studying the human perspectives of an issue, such as land use regulation. The next step is to then seek pertinent opinion leaders (key informants). These people can come from diverse backgrounds, such as lawyers, environmentalists, developers, town planners, farmers, elected representatives and key people from the general community. Appropriate diversity improves the capacity of the study to provide a breadth and depth of information for analysis from the most appropriate people. Diversity of participants also presents input from numerous vantage points, thereby providing a means of cross-correlating the information offered.

Structure of key informant interviews

Key informant interviews are designed as semi-structured (Boyce & Neale 2006). Questions of a semi-structured nature are intended to allow the interviewee to explore the concept or premise contained in the question. Semi-structured questions enable the interviewees to express their own views on what they consider to be most important (Del Balso & Lewis 2001). It is crucial to note that participants are asked about "lived experiences, as contrasted with abstract interpretations of experience or opinions about them" (Starks & Trinidad 2007, p.1374).

Interviews provide the research data that will be analysed so as to provide the basis from which informed conclusions will be drawn. As with any data collection in research, the interviews must be conducted in a manner that will remove the opportunity for bias. In order to ensure consistency, all interviewees need to be asked the same questions in a systematic order. This provides transparency in the manner in which the interview process occurs. A systematic process also removes the opportunity for inconsistency in the manner in which the data are collected.

Qualitative data collection methods, such as interviews, are often considered to have significant weaknesses, such as potential biases of the interviewees, or the possibility of responses containing inaccuracies. Participants provide their own perspectives of the overall topic in the specific responses to the questions – there will be areas of similarity and difference. The nature of the questions should draw out these similarities and differences. The sequence of questions should work from the general to the specific. That is, questions should start by seeking broad information about the topic. Each subsequent question should seek to explore the topic in a more focussed manner. In this way the interviewee is encouraged to provide more detailed commentary on the topic as the interview progresses. Introducing techniques into the interviews to overcome these issues, such as closed questions, enables the interviews to be checked for inconsistencies or bias during the analysis phase of the research process.

Interviews are generally recorded and later transcribed. Some researchers also choose to take notes during the interview, or similar means of recording key items, to use as prompts when reviewing the transcripts. Notes made during an interview can also aid in the verification process, enabling the researcher to, for instance, fact-check claims made in the interview through independent, reliable sources.

Analysis of key informant interviews

Thematic analysis of the interviews exposes the meaning underlying what is "said" and enables the interpretation of the past and present to inform the future (Riessman 2008). Analysis of the interviews promotes examination of a topic beyond the superficial, enabling comparisons between interviews not facilitated by the raw interview transcripts.

Analysis of key informant interview transcripts usually occurs through a process of reading and coding to highlight particular concepts, ideas or expressions relevant to the subject of the question, and to the overall research aim and objectives. An iterative process provides the most effective means to become familiar with the transcripts. Various systems can be employed during this process such as highlighting and notations or the use of colour-coded tabs to indicate particular items in each transcript. There are also proprietary software programmes (for example, NVIVO) that facilitate this process.

Independent sources are readily available, including legislation, transcripts of legislative debates, reports of court proceedings, minutes of council meetings, public submissions to planning hearings and exhibitions of planning proposals and so on. In this way a phenomenological approach to legal geography research based on key informant interviews should produce very robust findings.

A legal geography case study using key informant interviews

For over 40 years Oregon's land use planning system has been regarded as a model from which to draw instructive lessons for managing urban growth (Walker & Hurley 2011; Adler 2012). The case study reported in this chapter examined the contemporary relevance of Oregon's planning system in addressing the simultaneous issues of controlling urban growth while protecting and preserving prime agricultural land. The study was particularly concerned with the ways that Oregon's approach to land management may inform the management of land use at the outer edge of urban areas (commonly referred to as the "peri-urban") (Bourne et al. 2003; Busck et al. 2006). The study was undertaken as part of a self-funded overseas study tour. Ethics approval for the Oregon fieldwork was obtained from the University of New England's *Human Ethics Committee* prior to commencement.

According to Gibson et al.'s (2005) discussion, the effects of peri-urban land transformation have remained in the realm of theoretical discourse, and a more direct policy model addressing these issues has yet to emerge. If Oregon's planning system is to assist in the development of a more broadly applicable model for managing the simultaneously competing priorities of accommodating urban growth while retaining prime agricultural land, then an in-depth understanding of Oregon's planning system is necessary. One means of gaining an in-depth understanding was to conduct key informant interviews with subject experts.

The objectives of the Oregon study were to: (a) identify the types of policy interventions being implemented in order to achieve the competing objectives of allowing for urban expansion while preserving prime peri-urban agricultural land; (b) reveal the evolutionary process that led to the simultaneous resolution of the competing objectives; (c) understand the strategies employed to maintain the balance between urban growth and agricultural land preservation; (d) seek opportunities for the transfer of lessons from Oregon to other locations experiencing similar peri-urban pressures. Examination of the policy framework (legislation, regulation, policy and court processes) was key to gaining an understanding of how the management of place and space for social, environmental and economic outcomes operates.

Study method

Fieldwork was conducted in the Portland Metropolitan area in October, 2010, and data were collected using semi-structured interviews. Participants in the study were selected via a purposive sampling procedure (Onwuegbuzie & Collins 2007). This procedure was chosen in order to identify the individuals (key informants) most appropriate to providing maximum understanding of Oregon's land use planning system. While the focus here is specifically on the key informant methodology, the interviews were part of a more structured approach to understanding peri-urban land use planning in Oregon.

The work undertaken in Oregon was used as an international comparative case study for an in-depth analysis of peri-urban land use planning in Australia. The overall study focussed on three key research questions: what are the current paradigms underpinning land use planning regulation in Australia; how do these paradigms relate to and affect peri-urban land use; and how could emerging issues resulting from current paradigms underpinning peri-urban land use planning be appropriately remedied? Initial motivations for the choice of study locations were to enable comparability and transferability of information and findings.

Australia and the USA share many similarities, having originally been colonised by Great Britain. All three countries share common features, such as governance structures, language and social media. Various treaties and agreements relating to items such as trade and defence also continue the close relationship between these countries. Contemporary free-market economic policies, based on neo-liberalism, were initiated in Great Britain and the USA in the late 1970s (Allmendinger 2009). Australia adopted a similar approach to public policy soon after (O'Neill & Argent 2005). Land use planning policy in Australia and the USA also developed from British origins and continues to operate in a similar manner. Thus, the broad range of commonalities makes these countries ideal for comparative analysis studies and for the transferability of outcomes.

Having determined that the USA broadly provided an appropriate location for an international comparative case study the next step was to select the appropriate location for that study within the USA. Analysis of planning literature revealed Oregon as an exemplar of peri-urban land use policy and management. To develop a deep understanding of the overall operation of Oregon's land use planning system and why it is important in analysing peri-urban land use planning policy, the following structured approach was taken: investigation of the historical development of Oregon's land use planning system; examination of the legislative and judicial processes leading to and since the adoption of the land use system; and obtaining personal perspectives (key informant interviews) of the development and operation of land use policy, with particular focus on peri-urban land use management. Overall, the research process employed facilitates a deep and broad understanding of the factors motivating the development, regulation and operation of peri-urban land use policy in Oregon. Ultimately, these insights provide the opportunity to compare similar issues in the Australian context.

As noted earlier in this chapter, key informants are recognised as "opinion leaders" (Barnidge et al. 2013, p.98) who are "judged knowledgeable in their respective areas" (McWilliam et al. 2012, p.756). The first step in identifying the key informants was to send stratified e-mails (n = 120) to publicly listed farming organisations, universities with planning programmes and environmental organisations in Oregon. Responses were analysed for the frequency and expertise of nominated persons. From this a total of 35 semi-structured interviews were conducted. As with the e-mail responses the interviews were analysed for the frequency with which people and their area of expertise were named. This process resulted in seven key informants being identified for formal interview (Table 11.1).

Semi-structured interviews of approximately one hour each were conducted, enabling the exploration of ideas and provision of insights into Oregon's planning system. Questions covered historical, developmental and operational aspects

TABLE 11.1 Expertise of Key Informants in this Study

Sector	Role	Expertise
University	Academic	Land use planning
University	Academic	Land use planning
State government	Senior administrator	State land use planning
Regional government	Councillor	Regional land use planning
Local government	Planning director	Local government land use planning
Non-government sector	Lawyer	Community advocate in regard to land use planning.
Private sector – primary production	Farmer	Community representation re impacts of urban growth on primary production

of land use planning in Oregon. The exploration included the key informants' insights into the positive and negative aspects of the planning system; values of land at the rural/urban fringe; potential effects of urban expansion into the peri-urban; the reasons for protecting farmland; priorities concerning land use; and perspectives on offsetting or controlling the effects of fringe urban expansion.

A phenomenological approach to interview analysis was undertaken. Interview transcripts were coded for themes and sub-themes in an iterative process. This coding process enabled the examination of each individual's interview for common themes (shared concepts or ideas). The results from the analysis revealed, amongst other things, that the legislative framework had achieved the desired outcome of simultaneously managing competing objectives (facilitation of urban growth and protection of prime agricultural land at the urban fringe). The study also revealed that the same legislative framework that had achieved the desired goals had now constrained the planning system in a manner that prevents substantive change to address contemporary issues.

Oregon's planning system was enacted under the Oregon Land Conservation and Development Act 1973 (Oregon, USA), commonly referred to as Senate Bill 100. Oregon's legal system also permits the public to initiate plebiscites on any issue. There have been a number of public plebiscites since 1973 (Table 11.2). Analysis of the plebiscite results facilitated the cross checking of information provided in the key informant interviews. In the subject study the information in Table 11.2 provided confirmation of the issues and perspectives provided by the key informants.

The information provided in Table 11.2 provided a means of cross checking, verifying and validating the interview data.

Study outcomes and the key informant approach

The key informant interviews reinforced the understanding of the key features of Oregon's planning system and why this provides a model for planning in other jurisdictions. First, the integration of economic measures (taxation relief), as a

TABLE 11.2 Public Referenda on Senate Bill 100 Since 1973

Date	Referendum topic	Result	
			For %/against %
1976	Repeal SB 100.	Defeated	57/43
1978	Eliminate state oversight of land use plans.	Defeated	61/39
1982	Repeal SB 100.	Defeated	55/45
2000	Compensate owners for devaluation of property value by zoning change.	Passed. Overturned by the Oregon Supreme Court.	54/46
2004	Ballot Measure 37 – compensation for land use change restricting use after date of ownership. In lieu of compensation the approving authority can remove, modify or not apply the land use change regulation.	Passed. Upheld by Oregon Supreme Court.	61/39
2007	Ballot Measure 49 – modifies Ballot Measure 37 to remove general compensation and other unintended consequences.	Passed. Upheld by the 9th US Circuit Court of Appeals.	62/38

complementary tool to land use regulation for preserving prime agricultural and forestry land, is illustrative of the way in which economic mechanisms and land use regulation can be utilised in concert to achieve land preservation in a market economy.

Second, the development of objective measures of compliance for strategic plans and zoning instruments (called the Statewide Planning Goals), through extensive community consultation, is an example of the manner in which Oregon's planning system accords with Habermasian principles of *Communicative Action* (Innes 1995). This contrasts with planning in other places which is largely considered to be a top-down process where strategies and regulations are imposed as a *fait accompli* (see, for example, Mees 2011; McFarland 2013). That Oregon also has rigorous on-going public debates over property rights is also an indication of strong community involvement in land use planning.

Third, the appointment of a tenured panel of planning experts to oversight draft plans and to consider planning appeals provides independent oversight of land use planning. The panel members have strict tenures which cannot be renewed, are drawn from experts working in fields related to land use planning from across the USA, panel members are non-remunerated, decision-making is transparent at all stages and the panel operates independently of the political system. Also, the state government provides resource support (personnel, mapping, demographic analysis) to assist local councils in preparing plans and reviewing Urban Growth Boundaries (UGBs). These are examples of features that may be of benefit to land use planning in other locations.

The key informant interviews also revealed limitations in the planning system. The first is that the system has been unable to accommodate even relatively small changes so as to better achieve preservation of forestry and farm land. Adhering to exclusive use zoning and prohibiting any ancillary economic activity on these lands have led to increasing tension between those who defend the existing system and those who advocate private property interests. The adoption of Measure 37 and the response in Measure 49 demonstrate this.

The second is the inconsistent manner in which development occurs in some local council areas around the high-growth city of Portland. In some counties development has been allowed by local councils using "exceptions" powers in the planning legislation. Exceptions powers provide an unintended means of circumventing the development constraints of an approved UGB. These exceptions produce *ad hoc* urban development in peri-urban areas. This growth belies the apparently considered, controlled manner in which SB100, the SPGs and UGBs are supposed to protect land from unplanned urban sprawl. The immediate effects are urban sprawl, loss of farmland, erosion of the planning principles and encouragement of others to follow suit. The long-term effect is to leave future generations with the direct financial costs for infrastructure maintenance and erosion of the values on which the current planning system is based.

The overall conclusion from the study was that, since its outset, Oregon's land use system has been in tension with those who believe most strongly in centralised control and regulation to achieve common goods, and those who argue in favour of private property rights to achieve these goods. Despite a number of challenges since its inception, the planning system has been able to retain its core values. Paradoxically, this has resulted in the system being anchored at its point of origin. It is now unable to accommodate complex issues that have arisen in the four decades since its adoption, such as globalisation and climate change. Escape from the current limits reached by Oregon's planning system will require undergoing a reform process capable of crossing the existing limitations of the system in which it is currently bounded.

This case study reveals that Oregon's land use planning approach still has lessons from which others can learn. The manner in which productive peri-urban land can be retained while accommodating urban fringe growth is of relevance to jurisdictions elsewhere. A transformative process, such as that utilised when Oregon's planning system was created, may provide the means to achieving such revisions in a manner acceptable to both conservative and liberal perspectives of land use.

Summary and conclusions

Legal geography provides the opportunity to examine issues that draw together concepts related to people, place and the law. There are many different types of research that respond to this, utilising a wide variety of methods. The choice of method is that which those undertaking the research regard as most applicable to use in order to address the research aim.

If the focus of a particular study is on the manner in which regulatory systems apply to land then legal geography affords a framework for this. Again, there are many methods that can be utilised for examining land use regulation. If a research project in legal geography is specifically focussed on a single case study with the intent of understanding the social processes, power relationships and organisational motivations, then a phenomenological approach utilising key informant interviews is an appropriate method.

Being concerned with the understanding of human direct experience with place, phenomenology provides the philosophical standpoint from which to examine people's experience with the regulatory framework applying to land and its uses. For single case studies, in-depth analysis of the system can be achieved by obtaining the views of those people recognised as opinion leaders. The method employed to identify, collect, and analyse information from the opinion leaders is known as the "key informant interview" process.

This chapter has demonstrated that selection of opinion leaders, interview structure and analysis of interviews can be usefully deployed in legal geography research. Careful and critical attention to the research process and using key informants can provide the research with credibility, transparency and certainty in relation to the manner in which each step is conducted. Ensuring the thematic discussion of the interviews is consistent with the analysis phase, and supported by independently verifiable information from secondary sources, completes the study and provides convincing evidence from which to draw any findings.

The case study summary completing the body of this chapter illustrates the manner in which phenomenological research has been applied in a legal geography research project. There were readily available secondary sources upon which to draw to verify interviewees' perceptions and recollections. The conclusions revealed the strengths and weaknesses of the land use planning system, and pathways for reform.

Phenomenological research using key informant interviews is eminently suitable to case studies in legal geography. The method is appropriate to the examination of policy development and implementation for land use. It reveals the strengths and weaknesses of the regulatory system based on multiple perspectives. It exposes the human factor in the development and operation of law in relation to place and space. Finally, on a more personal note, fieldwork of this nature is quite rewarding as participants are both knowledgeable about the subject and very generous with their time.

References

Adler, S. 2012, *Oregon Plans: The Making of an Unquiet Land-use Revolution*, Oregon State University Press, Corvallis.

Alexander, E. R. 2001, 'A transaction-cost theory of land use planning and development control: Towards the institutional analysis of public planning', *The Town Planning Review*, vol. 72, no. 1, pp. 45–75.

Allmendinger, P. 2009, *Planning Theory*, second edition, Palgrave Macmillan, Basingstoke.

Barnidge, E. K., Radvanyi, C., Duggan, K., Motton, F., Wiggs, I., Baker, E. A., & Brownson, R. C. 2013, 'Understanding and addressing barriers to implementation of environmental and policy interventions to support physical activity and healthy eating in rural communities', *The Journal of Rural Health*, vol. 29, no. 1, pp. 97–105.

Bartel, R., Graham, N., Jackson, S., Prior, J. H., Robinson, D. F., Sherval, M., & Williams, S. 2013, 'Legal Geography: An Australian Perspective', *Geographical Research*, vol. 51, no. 4, pp. 339–353.

Bourne, L. S., Bunce, M., Taylor, L., Luka, N. & Maurer, J. 2003, 'Contested ground: The dynamics of peri-urban growth in the Toronto Region', *Canadian Journal of Regional Science*, vol. 26, no. 2 & 3, pp. 251–270.

Boyce, C., & Neale, P. 2006, *Conducting In-depth Interviews: A Guide for Designing and Conducting In-depth Interviews for Evaluation Input*, Pathfinder International tool Series: Monitoring and Evaluation –2, Pathfinder International, MA.

Brenner, N., Marcuse, P., & Mayer, M. (eds) 2011, *Cities for People, not for Profit: Critical Urban Theory and the Right to the City*, Routledge, London.

Busck, A. G., Kristensen, S. P., Praestholm, S., Reenbenrg & Primdah, J. 2006, 'Land system changes in the context of urbanisation: Examples from the peri-urban area of greater copenhagen', *Danish Journal of Geography*, vol. 106, no. 2, pp. 21–34.

Collins, K. 2010, 'Advanced sampling designs in mixed research: Current practices and eemerging trends in the social and behavioral sciences', in *SAGE Handbook of Mixed Methods in Social & Behavioural Research*, second edition, Tashakkori, A. & Teddlie, C. (eds), Sage, London, pp. 353–377.

Coreil, J. 1994, 'Group interview methods in community health research', *Medical Anthropology*, vol. 16, no. 1–4, pp. 193–210.

Daly, H. E. & Farley, J. 2004, *Ecological Economics: Principles and Applications*, Island Press, Washington DC, USA.

Del Balso, M. & Lewis, A. 2001, *First Steps: A Guide to Social Research*, second edition, Nelson Thomson Learning, Scarborough, Ontario.

Dowling, R., Lloyd, K., & Suchet-Pearson, S. 2016, 'Qualitative methods 1: Enriching the interview', *Progress in Human Geography*, vol. 40, no. 5, pp. 679–686.

Etikan, I., Musa, S. A., & Alkassim, R. S. 2016, 'Comparison of Convenience Sampling and Purposive Sampling', *American Journal of Theoretical and Applied Statistics*, vol. 5, no. 1, pp. 1–4.

Farquhar, S. A., Parker, E. A., Schulz, A. J., & Israel, B. A. 2006, 'Application of qualitative methods in program planning for health promotion interventions', *Health Promotion Practice*, vol. 7, no. 2, pp. 234–242.

Faifua, D. 2014, 'The key informant technique in qualitative research', *Sage Research Method Cases*, Sage Publications, London.

Francis, J. J., Johnston, M., Robertson, C., Glidewell, L., Entwistle, V., Eccles, M. P., & Grimshaw, J. M. 2010, 'What is an adequate sample size? Operationalising data saturation for theory-based interview studies', *Psychology & Health*, vol. 25, no. 10, pp. 1229–1245.

Gibson, C., Dufty, R., & Drozdzewski, D. 2005, 'Resident attitudes to farmland protection measures in the Northern Rivers Region, New South Wales', *Australian Geographer*, vol. 36, no. 3, pp. 369–383.

Gotham, K. F. 2009, 'Creating liquidity out of the spatial fixity: The secondary circuit of capital and the subprime mortgage crisis', *International Journal of Urban and Regional Research*, vol. 33, no. 2, pp. 355–371.

Harvey, D. 2006, *The Limits to Capital (New and Fully Updated Edition)*. Verso, New York.

Hermans, L. M., & Thissen, W. A. H. 2009, 'Actor analysis methods and their use for public policy analysts', *European Journal of Operational Research*, vol. 196, pp. 808–818.

Innes, J. 1995, 'Planning theory's emerging paradigm: Communicative action and interactive practice', *Journal of Planning Education and Research*, vol. 14, no. 3, pp. 183–189.

Jessop, B. 2004, 'On the limits of *The Limits to Capital*', *Antipode*, vol. 36, no. 3, pp. 480–496.

Jones, G. A. 2003, 'Camels, chameleons, and coyotes: Problematizing the 'histories' of land law reform', in *Law and Geography*, Holder, J. & Harrison, C. (eds), Oxford University Press, Oxford, pp. 169–190.

Kitchin, R., & Tate, N. 2013, *Conducting Research in Human Geography: Theory, Methodology and Practice*, Routledge, London.

Kitchin, R., & Thrift, N. 2009, *International Encyclopedia of Human Geography*, Elsevier, London.

Kral, A. H., Malekinejad, M., Vaudrey, J., Martinez, A. N., Lorvick, J., McFarland, W. & Raymond, H. F. 2010, 'Comparing respondent-driven sampling and targeted sampling methods of recruiting injection drug users in San Francisco', *Journal of Urban Health*, vol. 87, no. 5, pp. 839–850.

Krannich, R. S., & Humphrey, C. R. 1986, 'Using key informant data in comparative community research: An empirical assessment', *Sociological Methods & Research*, vol. 14, no. 4, pp. 473–493.

Marshall, M. N. 1996, 'The key informant technique', *Family Practice*, vol. 13, no. 1, pp. 92–97.

McFarland, P. 2013, 'A tale of two cities: Sydney and Melbourne's growth strategies and the flawed city-centric approach', Paper presented at the State of Australian Cities Conference, Sydney, 29 November–2 December.

McFarland, P. 2015, 'The Peri-urban land use planning tangle: An Australian perspective', *International Planning Studies*, vol. 20, no. 3, pp. 161–179.

McGuirk, P. M., & O'Neill, P. 2016, 'Using questionnaires in qualitative human geography,' in *Qualitative Research Methods in Human Geography*, I. Hay (eds), Oxford University Press, Don Mills, Canada, pp. 246–273.

McWilliam, W., Eagles, P., Seasons, M. L., & Brown, R. D. 2012, 'Evaluation of planning and management approaches for limiting residential encroachment impacts within forest edges: A Southern Ontario case study,' *Urban Ecosystems*, vol. 15, no. 3, pp. 753–772.

Mees, P. 2011, 'Who killed Melbourne 2030?' Paper presented at the State of Australian Cities Conference, University of Melbourne, Melbourne, 9 November–2 December.

O'Neill, P., & Argent, N. 2005, 'Neoliberalism in antipodean spaces and times: An introduction to the special theme issue', *Geographical Research*, vol. 43, no. 1, pp. 2–8.

Onwuegbuzie, A. J., & Collins, K. M. T. 2007, 'A typology of mixed methods sampling designs in social science research', *The Qualitative Report*, vol. 12, no. 2, pp. 281–316.

Payne, G., & Payne, J. 2004, *Key Concepts in Social Research*, Sage, London.

Poulton, M. 1997, 'Externalities, transaction costs, public choice and the appeal of zoning: A response to Lai Wai Chung and Sorensen', *The Town Planng Review*, vol. 68, no. 1, pp. 81–92.

Rhodes, R. A. 2014, 'Recovering the "Craft" of pubilc administration in network governance', Keynote address, IPAA International Conference, Perth, 30 October 2014.

Riessman, C. K. 2008, *Narrative Methods for the Human Sciences*, Sage Publications, California.

Ruming, K. 2009, 'Following the actors: Mobilising an actor-network theory methodology in geography', *Australian Geographer*, vol. 40, no. 4, pp. 451–469.

Sofaer, S. 1999, 'Qualitative methods: What are they and why use them?' *Health Services Research*, vol. 34, no. 5, Pt 2, pp. 1101–1118.

Starks, H., & Trinidad, S. B. 2007, 'Choose your method: A comparison of phenomenology, discourse analysis and grounded theory', *Qualitative Health Research*, vol. 17, no. 10, pp. 1372–1380.

Oregon Land Conservation and Development Act 1973 (Oregon, USA).

Tongco, D. 2007, 'Purposive sampling as a tool for informant selection', *Ethnobotany Research and Applications*, vol. 5, pp. 147–158.

Walker, R. M., & Enticott, G. 2004, 'Using multiple informants in public administration: Revisiting the managerial values and actions debate,' *Journal of Public Administration Research and Theory*, vol. 14, no. 3, pp. 417–434.

Walker, P., & Hurley, P. 2011, *Planning Paradise: Politics and Visioning of Land Use in Oregon*, The University of Arizona Press, Arizona.

Whittemore, A. H. 2014, 'Phenomenology and city planning', *Journal of Planning Education and Research*, vol. 34, no. 3, pp. 301–308.

Winchester, H. P. M. 1996, 'Ethical issues in interviewing as a research method in human geography', *The Australian Geographer*, vol. 27, no. 1, pp. 117–131.

Winchester, H. P. M. 1999, 'Interviews and questionnaires as mixed methods in population geography: The case of lone fathers in Newcastle, Australia', *The Professional Geographer*, vol. 51, no. 1, pp. 60–67.

Wylie, J. 2013, 'Landscape and phenomenology', in *The Routledge Companion to Landscape Studies*, I. Thompson, Howard, P. & Watson, E. (eds), Routledge, New York, pp. 54–65.

PART 4

Investigating the legal geographies of extractive industries

12

SYDNEY'S DRINKING WATER CATCHMENT

A legal geographical analysis of coal mining and water security

Nicole Graham[1]

Introduction

The Greater Sydney drinking water catchment is located within the Sydney Basin bioregion on the east coast of Australia. The Sydney Basin is filled with "near horizontal sandstones and shales of Permian to Triassic age" which were uplifted "during the formation of the Great Dividing Range", leaving "deep cliffed gorges and remnant plateaus across which an east-west rainfall gradient and differences in soil control the vegetation of eucalypt forests, woodlands and heaths" (NSW National Parks and Wildlife 2003, p.186). The Basin is one of the "most species diverse [areas] in Australia" resulting from its "variety of rock types, topography and climates" (NSW National Parks and Wildlife 2003, p.186). Within the Sydney Basin, the Sydney drinking water catchment comprises the catchments of the Hawkesbury-Nepean, Shoalhaven and Woronora Rivers, draining into 21 storage dams and reservoirs (which also capture rainfall across the catchment). There are several historic and active coal mines beneath it. The Greater Sydney catchment is currently the source of drinking water for 5 million people, a number which will increase by 1.74 million people by 2036 (Greater Sydney Commission 2017, p. 9). In December 2018, 99.8% of New South Wales (NSW) was in drought, and the Upper Nepean and Woronora catchments were assessed as being in "intense drought", adversely effecting dam levels (IEPMC 2018, pp.12–13).

The Greater Sydney drinking water catchment is also a lawscape: a landscape constructed in part by legally prescribed and prohibited land use practices. Analysing the catchment as a lawscape reveals to us the interaction of human laws with the laws of Earth's systems, as a complex network of relationships rather than as a two separate spheres: human and non-human, culture and nature (Graham 2011). Numerous state agencies within two government departments are responsible for managing Sydney's drinking water catchment: (a) the NSW Department of Planning and Environment (which includes the National Parks and Wildlife Service, Environmental Protection Authority and Office of Environment and Heritage); and (b) the NSW Department of Industry (which includes the Office of Water in NSW (including the Sydney Catchment Authority), the Office of Trade and Investment and the Dams Safety Committee). A declared catchment area (*Water NSW Act 2014 No 74* (NSW), s. 40) covering a surface area of approximately 16,000 square kilometres, the lawscape of the Greater Sydney drinking water catchment comprises a network of five major declared catchments, each containing several dams (*Water NSW Act 2014* (NSW); *Dams Safety Act 2015 No 26* (NSW); *Mining Act 1992* (NSW) s. 369), drinking water filtration plants (*Independent Pricing and Regulatory Tribunal Act 1992* (NSW); *Water Industry Competition (General) Regulation 2008* (NSW); NHMRC 2011) and canal and pipe networks (*Water NSW Act 2014* (NSW)) from which drinking water is supplied. The largest dams are situated within certain landscapes designated as Special Areas (*Water NSW Act 2014 No 74*, s. 47) and National Park reserves (*National Parks and Wildlife Act 1974* (NSW)) totalling approximately 3,700 square kilometres in which human access and activities are restricted to "protect the quality of stored waters" (*Water NSW Act 2014 No 74*, s. 47(2)(a)) and maintain "the ecological integrity" of the surrounding landscape (*Water NSW Act 2014 No 74*, s. 47(2)(b)). Beneath the catchment are several long-wall coal mines (*Mining Act 1992* (NSW); *Environmental Planning and Assessment Act 1979* (NSW)), including three mines in current operation within the Special Areas (two of which are operated as Dendrobium Mine, *Consolidated Coal Lease 768, Mining Lease 1510, Mining Lease 1566, Groundwater Licence 10BL161946*; and Metropolitan Mine, *Consolidated Coal Lease 703, Mining Lease 1610, Groundwater Licence 10BL603595*).

Thus, the Greater Sydney drinking water catchment does not fit neatly into the categories of "natural" nor "cultural" places. It may be more helpfully understood as a lawscape, a place created, in part, by laws and which has created, in part, the local laws. Lawscapes defy conventional categories of natural and cultural places because they disrupt the human/nature binary and there reveal the "mutual constitution and embeddedness" (Head & Gibson 2012, p.702) of human laws in the world (Graham 2011). Far beyond a set of coordinates on a map, the source of drinking water for over 5 million people and the site of valuable coal deposits, Sydney's drinking water catchment is a place of complex spatial, temporal and legal connectivities. A legal geographical analysis of the catchment subverts atopic conceptualisations of its legal, economic, geological and hydrological processes, activities and agents. The legal taxonomy of the catchment into categories of industry, environment, property, health and utility has more than administrative significance – it abstracts laws that regulate the catchment and fragments it into component parts, including surface water, groundwater, dams, coal, biodiversity, swamps. The atomistic approach of law has become physically manifest in significant and adverse changes in the catchment, including those caused by colonisation, "Big Water" projects (Allon & Sofoulis 2006, p.48) and longwall mining subsidence impacts. These changes in Sydney's "sandstone country" (Karskens 2009) underline the disconnectedness of law from the catchment while paradoxically demonstrating the inextricable connection between law and place. They point to the need to rethink law relationally and to tailor "[t] he shape of the human community, the shape of its laws and responsibilities ... to the watershed. They must be specific and as local as the watershed is specific and local" (Worster 2006, p.7).

This chapter contends that current catchment laws are incongruent and unsound because they are insufficiently responsive to the material entanglement and co-constitutivity of human and more-than-human ecologies, economies and laws. The chapter begins by introducing the agency of place in the human histories and ecologies of what is now the Sydney drinking water catchment, focusing on its history of colonisation. The chapter then considers the ways in which the catchment, as a place, has been regarded as a "resource frontier" (Tsing 2003, p.5100) revealing the bonds of Sydney's people to water and to coal. The chapter then reviews the increasing tension within the interaction of laws that regulate activity within the catchment. Against the atomistic and abstract status quo, the chapter adopts a legal geography perspective to conclude that the Greater Sydney drinking water catchment, "should be understood as relationally as possible, as a product of relationships rather than as a container within which the world proceeds" (Layard 2019, p.2) and thus, that its laws should be, correspondingly, more relational.

A colonised catchment

For tens of thousands of years, the diverse lands of what is now the Sydney drinking water catchment belonged to the Gandangara (Blue Mountains including Nepean River), Dharug (northern Basin from east coast to western plains) and Dharawal (from Wollongong up to Georges River) nations (Goodall & Cadzow 2009, p.31). British colonial settlement was initially limited to coastal and inland "islands" adjacent to rivers where agriculture depended on riparian soils, and "stock grazed over the woodlands and pastures created by Aboriginal burning" (Karskens 2009, p.20) on the Cumberland Plain (see Figure 12.1). The elevated sandstone country that surrounded the colony of Sydney, including the Blue Mountains, was avoided by the colonists at first because of its steep and rocky terrain (Karskens 2009, p.20; Haworth 2003, p.41).

When the armed settlers took, cleared, farmed and populated the so-called *terra nullius* of the Cumberland Plain in the early 19th century, the Aboriginal people retreated into bushland and responded by burning down newly built huts and crops (Karskens 2009, p.457). The "sustained attacks and raids by Aboriginal people" on colonial settlers were referred to as "a war" (Karskens 2009, p.449). Violence and killing continued at ever greater scales in the Sydney catchment, the landscape of which became part of the strategy of war itself. For example, in

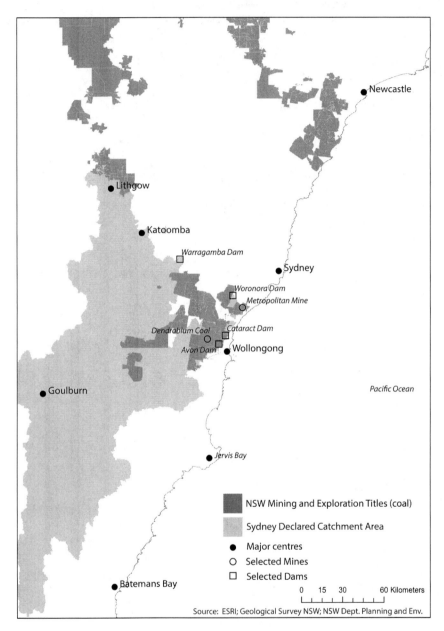

FIGURE 12.1 Geospatial mapping of the Sydney water catchment dam system. Source: author provided.

April 1816, Governor Macquarie sent a "tremendous procession" of soldiers out to the furthest frontiers of Sydney under orders to "capture or kill all Aborigines" and hang up the bodies of the slain "in order to strike the greater terror into the survivors", and drive them "across the mountains" (Karskens 2009, p.507). At 1am on 17 April, the soldiers "formed line rank entire" and "pushed through thick brush" towards the 60-metre-deep gorge of the Cataract River. The soldiers opened fire on the retreating clan of Aborigines who "fled over the cliffs" and were smashed to death in the gorge. Others were wounded or shot dead (Karskens 2009, p.510). The "Appin massacre" became known by its landscape-based strategy, referred to as "the drive" because it was "tied to local landmarks like cliffs, bluffs and gorges" and was later repeated elsewhere across the lands of what is now the Sydney drinking water catchment. In July 1816, Aborigines also applied the "drive" strategy to 200 sheep at Mulgoa (Karskens 2009, p.514), again underlining the agency of the landscape of the Sydney Basin in frontier conflict.

Aboriginal resistance to the occupation and dispossession of their lands by the British also included sustained efforts to defend and retain those lands by using the colonial legal categories of private title and its processes, including both grants and purchase, in relation to land ownership. Numerous and repeated individual claims to various kinds of ownership through title in land using colonial law were refused. The NSW Aborigines Protection Board later declared and controlled six Aboriginal Reserves in the Burragorang Valley (under what is now Lake Burragorang, created by the construction of Warragamba Dam). When the Reserves were revoked towards the end of the 19th century, the Gandangara people moved to live in the Gully on the edge of West Katoomba from 1894 until the local government developed the area into a racetrack 60 years later (Water NSW 2019a).

The British dispossession of the Gandangara, Darug and Tharawal nations from their lands (Johnson 2006, pp.34–56, 84) and the war between them was accompanied by the rejection of Aboriginal laws that articulated and were informed by long-standing knowledge and experience of local geomorphological, hydrological and climate conditions. Traditional knowledge of Sydney's sandstone country is dynamic because the catchment itself is the product and part of ongoing change (Goodall & Cadzow 2009, p.11; Karskens 2009, p.31). The "mutability" and "mercurial" nature of life in the Sydney catchment "confronted and confounded" the British colonists (Karskens 2009, p.31) who struggled for decades to establish reliable food and water supplies.

Bonded to water

The most critical resource that sustained the British colony in Sydney was water. Thus, bonded to water, the locations of colonial settlement were necessarily determined by its presence and movement across what is now the Sydney drinking water catchment (Karskens 2009, p.105). Despite the agency of Sydney's waterscape to life in the colony, early law and regulation failed to prevent health crises through the contamination of water sources (Davies & Wright 2014, p.451) and failed also to provide water security. The source of the colony's drinking water supply changed multiple times throughout the 18th and 19th centuries because of repeated instances of contamination of surface water and shallow groundwater from human and industrial waste, droughts and insufficient levels to meet the demands of the ever-expanding colony. What the Gandangara, Darug and Tharawal peoples "had long found sufficient", the British settlers "perceived as scarce" (Morgan 2015, p.38). From the Tank Stream, to the Lachlan Swamps, and then the Botany Swamps, the drinking water supplies in the settlement were progressively polluted and exhausted (Karskens 2009, p.21), causing environmental degradation and several human health crises (Davies & Wright 2014, p.451). By the mid-1800s, the colonists were determined to establish a safe and reliable water infrastructure on a grand scale. To this end, in 1867 they established a Royal Commission into Sydney's water supply (Davies & Wright 2014, p.451). "Engineering works from the late-nineteenth century spoke to increasing imperial confidence in technocratic solutions able to transform 'unproductive' places and peoples" (Beattie & Morgan 2017, p.62).

The geohydrological features of Sydney's sandstone country and its high-quality water and flow through "deeply dissected sandstone gorges ... proved ideal for building deep water storage reservoirs" (Davies & Wright 2014, p.451). From 1869 to 1888, the Upper Nepean Scheme was built to capture the headwaters of the Nepean River by building a weir at the junction of the Cordeaux and Avon Rivers and connecting to another weir on the Cataract River. Learning from past mistakes and hoping to reduce the risk of contaminating another water supply, the first Special Area was declared in 1880 to protect the land of the Upper Nepean catchment. The scheme was completed with Prospect Reservoir in 1888, but within a short time, water levels were too low to remain viable leading to the construction of four dams in the Upper Nepean Scheme: the Cataract Dam (1902–8), the Cordeaux Dam (1918–26), the Avon Dam (1921–8) and the Woronora Dam (1927–41) (see Figure 12.1).

Since the late 19th century, protection of the quality of Sydney's drinking water supply has been achieved through the creation of the legal category of "Special Areas" that was applied to land surrounding Sydney's dams by successive colonial and then state governments in NSW. While this legal development addressed the importance of water quality and the risk of contamination, the laws of Sydney's catchment had not yet adapted to the dynamism of its climatic conditions and hydrological cycles. By the 1930s, Sydney's drinking water supply had again become inadequate to meet demand. Water restrictions were introduced during the eight-year drought (1934–42), and the Warragamba Emergency Scheme was introduced, for which a weir on the Warragamba River and a 52-kilometre pipeline to Prospect Reservoir were constructed. It began supplying water in 1940 and "temporarily saved Sydney" (Water NSW 2019b). Construction of the Warragamba Dam began in 1946 with the compulsory acquisition of private property, eviction of residents, relocation of settlers' graves (but not the sacred sites of the Gandangara people) and a programme of intensive land-clearing. Warragamba Dam was completed in 1960, but the dam wall was raised by 5 metres between 1987 and 1989 in response to risks posed by unforeseen flood events, and a Spillway was constructed between 1998 and 2002. Legislation was recently passed by the NSW Parliament to raise this dam wall, again, in the name of flood mitigation (Water NSW Amendment (Warragamba Dam) Bill 2018 (NSW)). But critics of the Bill point to negative impacts to the environment, Gandangara cultural and burial sites, rock art, and artefacts, caused by inundating areas upstream of the dam in the Blue Mountains World Heritage site (Dalberger 2019). There is also concern that the underlying objective of this legislation is to open up to property developers land that is currently zoned as flood prone (Dalberger 2019).

In addition to these grand projects, the Sydney drinking water catchment was subject to further substantial infrastructure development. The Shoalhaven Scheme, constructed between 1971 and 1977, transfers water through an extensive pipe system up to the Sydney metropolitan area at substantial cost. And one of the biggest industrial water recycling schemes in Australia was constructed in Wollongong in 2003 to supply BlueScope Steel at Port Kembla, whose annual 7.3 billion litre water use had been drawn from the Wollongong drinking water supply via the Avon Dam. Life in Sydney today has "become dependent on the uninterrupted provision of these technologies" (Beattie & Morgan 2017, p.62) that are commonly believed to exist beyond limit.

Sydney's rapidly expanding urban population and multiple, sustained and severe dryness events did not lead to adaptive measures in relation to water consumption. Conversely and "counter-rationally" (Allon & Sofoulis 2006, p.48), these problems were consistently cited as the reason for the ongoing and ever grander development of water infrastructure in the Sydney catchment. The building of more and larger dams throughout the 20th century, the raising of walls on existing dams, the introduction of water recycling for the water-intensive mining industry and seawater desalination to supplement Sydney's drinking water in the 21st century are adjustments and variations of colonial-scale "metabolism" (Karskens 2007, p.118) rather than a process of adaptation to the geographical conditions of the Sydney Basin. Big Water (Allon & Sofoulis 2006, p.48) technological "solutions" to problems of supply do not address problems of limitless demand (Troy 2008, p.193). They underline the continuing dominance of an abstract, old world resource frontier (Tsing 2003) logic despite abundant evidence of the vulnerability it continues to create. Sydney's reliance on water engineering measures (notwithstanding energy costs) to "predict, protect and provide" (Morgan 2015, p.206) a water supply in the face of both ever-growing demand and climate change is integral to both its colonial history and foreseeable future. The human bond to water has substantially changed the ecology and landscape of the catchment.

The enduring place-based knowledge and laws that successfully sustained the Gandangara, Darug and Tharawal peoples for tens of thousands of years were replaced with a patchwork of colonial laws that regulate immature water infrastructure, degraded landscapes and waterways and increased health crises and water insecurity. The settlers' ongoing search for drinking water, that persists into the 21st century, reflects an atomistic view of the landscape as a set of separable resources, rather than as a material whole comprised of intimately connected cycles and systems. Absent geographically based knowledge and laws, the co-construction of the catchment's landscapes and laws since colonisation is characterised by a low-level of "hydroresilience" (Morgan 2015, p.206). The vulnerability of human life in Sydney is predicated on the inextricable bond of its people to the dynamic hydrological and climatic conditions of the Sydney catchment, including regular cycles of dryness, to which it has yet to learn and adapt. As Morgan observes, the definition of water scarcity is historically and culturally contingent (Morgan 2015, p.4). Life in colonial Sydney was vulnerable to water scarcity, but it was also vulnerable to the effects of diminished water quality.

Two centuries later, both water scarcity and quality remain sources of vulnerability to life in Sydney. Despite the long-standing status of the legal category of Special Areas, the landscapes they protect are now literally and increasingly undermined by the extractive industry. In law, the sub-surface of the catchment, the coal that is mined, is a "shadow place" (Plumwood 2008, p.139), somewhere that is "isolated, detached and devoid of sentiment or worthiness outside of [its] economic implications" (McLean et al. 2018, p.621). The coal is economically valuable, but underground longwall coal mining within the Sydney catchment presents significant risks to its geohydrological cycles and processes. Rather than valuing the groundwater of the Sydney catchment and its connectivity to the surface waters, geology and biodiversity on which the catchment as a system depends, these subterranean waters may be regarded as "shadow waters" whose "uses and values have been ignored or under-valued" (McLean et al. 2018, p.616). The risk to water and the landscape it creates, from coal mining within the Sydney catchment, is a risk made real by law. The following section explores the risk to the waters and lands of the Sydney catchment caused by the effects of subsidence impacts created by underground, longwall coal mining.

Bonded to coal

Today, longwall coal mining is the "most common method of underground coal mining in Australia, as it is safer and more efficient in extracting coal than other extraction techniques" (Department of the Environment 2014, p.2). Longwall coal mining accounts for about half of all coal mine operations in NSW, occurs predominantly in the Hunter and Illawarra regions of the Sydney Basin and contributes significantly to the economy of NSW, notwithstanding substantial government subsidies and low royalties (Baer 2016, p.195). Most of the coal mined in NSW is exported (about 88%). Less than a fifth is used for domestic energy needs. If the NSW Government is bonded, or indeed "addicted" (Pearse 2010, p.26), to coal, it is because of its function as revenue, not energy.

The ongoing and expanding Big Water programme and the extraction of coal from the Sydney catchment "are fundamentally matters of territorialisation – the expression of social power in geographical form" (Bridge 2010, p.825) – that have "alienated" the "material and cultural attachments of existing resource users" (Bridge 2010, p.824).

Like water, coal is predominantly regarded as an isolated component of Sydney's resource frontier (Tsing 2003), rather than as an integral part of the Sydney catchment. Resource frontiers are placeless sites of abstract economic output, comprising "non-human assets that can be owned and exchanged on some market" (Picketty 2014, p.46). As Tsing has argued, the colonial land-scape, imagined to be unowned, was thought to be "ready to be dismembered and packaged for export" (2003, pp. 5001–6). But coal does not exist in the abstract; it is the bedrock and environment of the Sydney catchment on which life in Sydney depends. Everything is the environment of something (Conacher & Conacher 2000, p.6; Brown 2004). In other words, it is not only social power that is bonded to coal – the landscape of Sydney's drinking water catchment is also bonded to the coal seams that support it.

The compacted and solidified plants that now constitute rocks of coal in the Sydney Basin once converted UV radiation from the sun into energy stored in the chemical bonds of organic carbon molecules. Buried through glacier activ-ity, these plants retained their organic carbon and energy content (Freese 2003). Viewed atomistically, solely as a source of energy, the extraction of these coal rocks is a necessary but otherwise insignificant step towards the release of their energy (which is created later by heating and burning them). Viewed geographi-cally, however, as part of a complex and particular geohydrological system, the extraction of coal rocks within the Sydney catchment is a highly significant event. Longwall coal mining uses hydraulic supports to hold up the ceiling, so the entire coal seam can be cut out. It takes place at a depth of 200 to 600 metres under the Earth's surface and is used to extract large panels of coal from 150 to 400 metres wide and 1 to 4 kilometres long (Department of the Environment 2014, p.3). When the coal is removed, the hydraulic supports are also removed, resulting in the fracture of the geological strata above, generating subsidence of the overlaying sub-surface, groundwater and surface. "Rocks are typically 10 to 30 times weaker in tension than compression" which means that surfaces in tension are more susceptible to fracturing (IEPMC 2018, p.29). Subsidence from longwall mining is thus regarded as predictable and immediate (Darmody et al. 2014, p.207). The scale of the fracturing and subsidence varies, depending on particular geology and mining dimensions (IEPMC 2018, p16). Subsidence can impact groundwater systems by causing drainage into mined spaces, change the connectivity between surface and groundwater and change the quality of groundwater. "Water can flow through the subsidence, dissolve the contami-nants, transporting them through surface and subsurface flows and subsequently deposit in sediments downstream from the mine" (Ali et al. 2018, p.696). The contaminated water can also re-emerge at another point at the surface, as is the case on the Waratah Rivulet (and hence, potentially, in Woronora Dam) (Jankowski 2007).

Underground coal mining within what is now the Sydney drinking water catchment began in the Illawarra Region with the Coal Cliff Colliery in 1878. In 1880, the Metropolitan Special Area was declared to protect the Upper Nepean catchment. The Metropolitan Colliery opened at Helensburgh in 1888. Several mines operated in the Illawarra area throughout the second half of the 19th century. Until the mid-20th century, most coal mines were governmed-owned, and the quantum of coal extracted was modest (Pearse 2010, p.22). Mechanised longwall mining was introduced to NSW in 1972 and progressed from contributing 4.3% of all NSW coal extraction in 1979 to 39.6% in 1988 (Wilkinson 1995, p.15). The substantial expansion of existing coal mines and approvals of new mines arose in the late 20th century, with a further 20 mines opened in the catchment. Nineteen historic mines are located beneath the Avon, Cordeaux, Cataract and Woronora Dams. Three mines under current operation are located beneath the Avon, Cordeaux and Woronora Dams: the Dendrobium Mines Areas 2, 3A and 3B and the Metropolitan Mine (Water NSW 2016).

Mining-induced subsidence is thus not a newly encountered consequence of underground coal mining. Subsidence from all underground mining (not only longwall mining) has a long history in NSW, dating back to at least the 1880s (Subsidence Advisory NSW 2019). In the 19th and early 20th centuries, the effects of mining-induced subsidence included the loss of human life, landslip and damage to a tramway, houses, a hotel, Council chambers, a hospital, water pipes, gas services and a soccer ground in Newcastle in the Hunter region at the north of the Sydney Basin (Mine Subsidence Board 2015). The major adverse environmental consequences of underground mining include disturbance, alteration and possible contamination of groundwater, and surface water processes and flows in "streams and rivers flowing on landscapes above the mines' underground workings" (Morrison et al. 2018, p.695). Some changes can cause "long-term damage to stream channels, hydrology and water quality" (Morrison et al. 2018, p.695). Recent studies of river sediments up and downstream from coal mining sites in the Sydney Basin reveal "that the sediment quality, specifically for arsenic, nickel and zinc in downstream coal mining discharge locations exceeded the ANZECC guideline limits for sediments" (Ali et al. 2018, p.701).

The effects of the impacts of subsidence from underground coal mining in the Sydney Basin led the NSW Government in the late 20th and early 21st centuries to review its bond to coal not only as a source of revenue, and as a risk to human safety (through subsidence in urban areas), but also as a part of the geographical integrity of the Sydney drinking water catchment (see, for example, Reynolds 1976; McNally & Evans 2007). An ever-growing quantum of scientific research (see for example, Morrison et al. 2019) and independent panel recommendations (IEPMC 2018) indicates that the current regulatory framework is overly confident in evaluating and regulating both the immediate and cumulative risks of longwall coal mining within the catchment, because it relies on outdated, and largely proponent-generated, risk assessment data.

A lawscape of subsidence

Law has long played a key role in the creation of mines in the Sydney Basin and, more recently, also in the regulation of the effects of mining subsidence. Conventionally, mining law and laws related to mining were concerned with the socio-economic aspects of mining, including: various questions of acquisition, proprietorship, access and use; revenue, health, safety and labour conditions; and compensation for adverse effects of mining on improved lands, whether private or public (Hepburn 2015, p.81). In NSW, the *Mining Act 1992* (NSW) imposes restrictions on mining to protect privately owned and agricultural land generally. It prohibits the mining of the surface of such "improved" lands without express written consent of the landowner (s. 62(1)(c) & sch 1, cl 22), unless the Minister is satisfied that specific conditions attaching to the mining lease would minimise any damage to the surface of agricultural land (sch 1, cl 23(3)).

The concept of "improvement" reaches back to the 17th century, when it was used to rationalise the enclosure of the common lands of Britain and the contemporaneous colonisation of foreign lands around the world (Graham 2011, pp.32–6, 60–6, 100–4). The concept of improvement is predicated on an anthroparchic model of the world that separates and privileges human over non-human life in the landscape. It also places a higher value on landscapes corresponding to the degree of certain anthropogenic landscape change associated with particular cultural practices.

The vocabulary and logic of improvement continue to be used today to differentiate legal remedies for the effects of longwall subsidence impacts corresponding to the classification and consequent economic value of land as either improved or unimproved. Improvements on land in mining law are synonymous with human structures and infrastructure. Early subsidence-related law was generated by concern about effects of the impacts of subsidence to the safety of human life and urban settlement. Repeated and serious subsidence events in the Hunter region of the Sydney Basin (see above) led to the enactment of the *Mine Subsidence Act 1928* (NSW). Replaced with the *Mine Subsidence Compensation Act 1961* (NSW) and now the *Coal Mine Subsidence Compensation Act 2017 No37* (NSW), the scope of this legislation is also limited to the category of "improved" lands which Section 4 defines as (a) "any building or work erected or constructed on land; (b) infrastructure, whether or above or below the surface of the land, including railways, roads, electricity transmission or distribution networks, pipelines, ports, wharf or boating facilities, telecommunications, sewerage systems, etc.". The NSW Court of Appeal in *Ulan Coal Mines v Minister for Mineral Resources* [2007] NSWSC 1299 reiterated the importance of protecting "improved" land subject to an application for a mining lease such as "important agricultural infrastructure or valuable historical buildings" from the adverse effects of mining (Hepburn 2015, p.82).

The case demonstrates the overreach of 17th-century European thought into 21st-century Australian mining law, when human structures and infrastructure are legally protected in ways that the more-than-human world is not. This legal distinction is exacerbated by the potential application of Ministerial exceptions to the limited legal protections for the more-than-human world.

Some environmental and planning laws have broadened the scope of mining regulation to address environmental risks and effects of mining operations. The *Environmental Planning and Assessment Act 1979* (NSW) requires all major mining project applications to include an Environmental Impact Assessment (EIA). This important reform is compromised, however, by the fact that it is the mine proponent who prepares the EIA and, because it is an "anatomised" approach to a landscape, focusing on its parts, rather than on the place as a whole. This approach "fractures landscapes and knowledge into a dozen or more expert reports" (Karskens 2007, p.144). Since 2011, Section 91 of the *Water Management Act 2000* (NSW), together with the Aquifer Interference Policy, "require traditionally licence-exempt extractive activities … to hold regular water licences … [and] aquifer interference approval" (Nelson 2018, p.247). This suite of reforms defines the thresholds of harm in terms of cumulative impacts within a long temporal scope (Nelson 2018, p.247). However, the reforms are impressive on paper only, as these provisions have not yet come into effect several years after their appearance on the statute books (Nelson 2018, p.247). This leaves the most practically meaningful conditions attached to the approval of a mining project application as its Extraction Plan (EP) (replacing Subsidence Management Plans in 2014).

EPs predict subsidence impacts in the foreseeable future and outline strategies to mitigate and rehabilitate their effects. They are not directed to consider long-term or cumulative impacts. The non-regulated method and vocabulary of EPs are from subsidence research funded mostly by the Australian Coal Association Research Program (ACARP) (Department of the Environment 2014, p.42). ACARP is a body, established in 1992, comprising the NSW Minister for Primary Industries and Energy and the Australian Coal Association. The ACARP system for assessing the risks of the effects of subsidence is an attempt by industry to regulate itself in the absence of legally prescriptive criteria for the preparation of EPs. It involves a formula that "scores" the risks presented by a given proposal (Pells et al. 2014). The ACARP system of "scoring" incorporates several diverse factors into a single assessment process, but it is not a scientific methodology. The weighting of subjective and qualitative factors (for example, "aesthetics"), combined with the conflict of interest created by the proponent providing the risk assessment, renders the "calculation" less reliable than an independent scientific assessment standard or process (Seedsman & Pells 2014, p.64). Independently engaged studies produce remarkably different results to those engaged by mining proponents. Water NSW recommends "that selected future impact assessment reports should be engaged by government but funded by the mining company" (Water NSW 2018, p.44). EPs manage and monitor the risks of subsidence to public health and safety and built features, and thus articulate anthropocentric

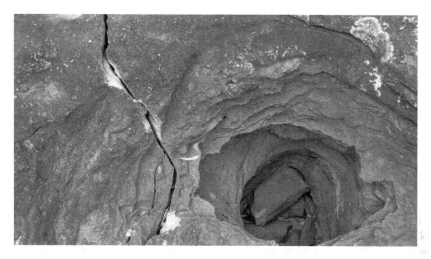

FIGURE 12.2   Dried-out bedrock in Waratah Rivulet (an inflow to Woronora Dam) caused by subsidence. Source: author provided.

concerns to protect only human health and "improved" lands. The protection of the category of "unimproved" lands from the effects of subsidence is still in its legal nascence. Critics have long contended that the EP approval process is "failing to protect the environment from subsidence damage" because "they offer no accurate assessment of the damage that may occur" (Total Environment Centre 2007, p.4) (Figure 12.2)

Unimproved lawscapes

A category of "unimproved" land known as a "State Conservation Area" was introduced in NSW in 2002, specifically "to reserve lands 'only where conservation values and mineral values do not allow for reservation under any other reserve category, such as national park or nature reserve'" (Department of Environment and Climate Change 2008, p.4). State Conservation Areas "protect natural and cultural heritage values and provide recreational opportunities" while simultaneously providing for "mineral exploration and mining, and petroleum exploration and production" (Department of Environment and Climate Change 2008, p.4). There is no restriction of mining in the lands of state conservation areas (Environmental Defenders Office 2012, pp.39–40). National Parks, on the other hand, enjoy protection from mining of both their surface and the earth beneath them under the *National Parks and Wildlife Act 1974* (NSW), unless an Act of Parliament authorises mining (ss. 41(1) & (3)). However, this protection is subject to the exception of Ministerial approval or "alternatively, the Minister may redraw the boundaries or revoke this status in order to allow for mining to occur" (Lyster et al. 2016, p.460). Mining also happens beneath other categories of land surrounding water storage and infrastructure in the Sydney catchment: Special Areas, Controlled Areas and Dam Safety Notification Areas. Mining within these categories of landscapes foregrounds the inter-connected geography of the Sydney drinking water catchment and invites us to rethink the legal categorisation of landscape in a more relational way.

Special Areas (*Water NSW Act 2014 No 74,* s. 47) comprise a quarter of the Sydney catchment and are "dominated" by sclerophyll forest, rainforests, freshwater and forested wetlands and upland swamps (NSW Chief Scientist and Engineer 2014, p.4). Upland swamps "act like sponges, storing surface water and in some cases accessing groundwater storage to contribute to base flow. In times of drought, they are critical in maintaining stream flow" (NSW Chief Scientist and Engineer 2014, p.4). This "necklace of waterholes" (Karskens 2009, p.105), once lost, cannot be rehabilitated (Lock the Gate 2018, p.7). It is classified as an Endangered Ecological Community under the *Biodiversity Conservation Act 2016* (NSW). Upland swamps are in severe decline (Krogh 2007), with both water drainage and contamination caused by subsidence impacts occurring up to 900 metres away from a mine site and soil moisture falling below baseline levels in all swamp monitoring sites (Lock the Gate 2018, p.7; IEPMC 2018, p.105). This damage has been, in the case of the Dendrobium Mine, subject to "offset" arrangements under its Subsidence Management Plan approval which permits the damage to the swamps in return for the purchase, by the mining company, of other land for donation to the NSW National Parks Estate (IEPMC 2018, pp.115–16). This regulatory approach to managing the effects of subsidence demonstrates acceptance of damage to landscapes undermined in the Sydney catchment that contradicts the rationale of the category of Special Areas, which were conceptualised and declared for the sole purpose of protecting water quality and maintaining the ecological integrity of these landscapes (*Water NSW Act 2014,* s. 47(2)(a) & (b)).

Of great concern to researchers working on this issue is the failure to connect the effects of subsidence impacts today to effects in the future: "[c]umulative impacts from past and continuing mining need to be considered as a potential factor impacting water supply from the Special Areas" (IEPMC 2018, p.13). Legally, cumulative impacts are technically subject to the assessment of the environmental impact of proposed projects (Environmental Planning and Assessment Regulation 2000, cl228(2)(o)). However, that process is overseen (in some instances) by the Independent Planning Commission (formerly Planning Assessment Commission) whose approval of the extension of an existing mine was recently overturned by the NSW Court of Appeal for having applied the wrong standard to assess the mines' effects on water quality in the Sydney catchment (*4nature Incorporated v Centennial Springvale Pty Ltd* [2017] NSWCA 191). Section 34B (now s. 3.26) of the *Environmental Planning and Assessment Act 1979* (NSW) (and applied in the relevant planning approval provision) provides that "consent to a development application relating to any part of the Sydney drinking water catchment" should be refused "unless the consent authority is satisfied that the carrying out of the proposed development would have a neutral or beneficial effect on the quality of the water" (s. 34B). The NSW Parliament reversed the Court's decision by amending the statute not long after this case in late 2017, so that it now applies only to prospective mine applications, not to proposed extensions of existing mines (Environmental Planning and Assessment Amendment (Sydney Drinking Water Catchment) Bill 2017 (NSW)). In late 2018, the Independent Expert Panel for Mining in the Catchment released its Initial Report indicating that proponent-generated risk assessments relating to water flow and quality from the effects of subsidence have likely under-estimated the diversion of water from dams within the catchment into the mines (IEPMC 2018, p.iii). The report led a spokesperson for WaterNSW to state that "no further longwall mining should be approved within the Special Areas with dimensions of the size currently undertaken at the Dendrobium mine" and a scientist for the National Parks Association to state that "Special Areas are no place for mining development by trial and error" (Hannam 2019).

The regulation of the effects of subsidence impacts (current and cumulative) in the Sydney catchment is thus contradictory and contested, with some government agencies granting approvals that other government agencies dispute. For example, in 2013, the Sydney Catchment Authority (SCA) expressed its concern to the Department of Planning and Infrastructure that a longwall mining project expansion, located entirely beneath land owned by the SCA that it managed as a declared "Special Area", was within 1 kilometre of the Cataract Dam and could impact the dam wall and reservoir. They feared that the expansion could induce leakage from the reservoir, affect water quality and that it was based on a questionable risk assessment of subsidence impacts (Sydney Catchment Authority 2013). Soon after this submission was made, the contract of the CEO of the SCA was terminated less than one year into his five-year term (Nicholls 2013). The regulation of the risks and effects of subsidence impacts on "improved lands" is well-established, and the tolerance for risk to human life and private property is very low (Pells et al. 2014). However, the regulation relating to "unimproved lands" such as National Parks, State Conservation Areas and Special Areas is equivocal, inconsistent and exhibits a very high tolerance for very serious risk. At the basis of the assessment process is the preparation and submission of an EP by the proponent of the new longwall project. All EPs are thus premised on the acceptability of the (unavoidable) effects of subsidence impacts. Unlike damage to "improved lands" for which compensation is payable and rebuilding available, repair and compensation for irreversible damage to "unimproved" lands would seem to be perversely abstract remedies (see Graham 2015).

The objective of laws regulating subsidence impacts is to maximise the coal extracted from the Earth while simultaneously maximising human safety through relevant infrastructure features (Robinson 2007, p.13). In the process, these laws uncritically reproduce the long-outdated distinction between improved and unimproved lands, or cultural and natural places. The Greater Sydney drinking water catchment is neither.

Conclusion

Legal geography is concerned with the "conjoined and co-constituted" (Braverman et al. 2014, p.1) worlds of law and geography, people and place, "culture" and "nature". Analysing the Sydney drinking water catchment using legal geography is important because it disrupts the uncritical reproduction of anachronistic legal categories of land in terms of "improvement" and challenges the atomistic separation of water and coal as unrelated component parts. In isolation, each law may appear fit for purpose, but in concert the laws seem less viable given the adverse effects of mining within the catchment that highlight their collective incongruity. The impacts of subsidence from longwall coal mining within the catchment underline the laws' abstractness: decreased water flow and water contamination, the loss of swamp ecologies and the unquantifiable risk to the drinking water supply to the largest city in the country. The Greater Sydney drinking water catchment "will be at risk unless humans can be reimagined and co-opted as active co-constructors" of it (Head & Muir 2006, p.90). Legal geography provides one such method by which we might begin this reimagining of our place in the world.

Note

1 The author thanks Dr Kate Owens and Professor Stuart Khan for their feedback on earlier versions of this chapter. This research was funded by the Australian Government through the Australian Research Council's Discovery Projects funding scheme (DP190101373). The views expressed herein are those of the author and are not necessarily those of the Australian Government or the Australian Research Council.

References

Ali, A. E., Strezov, V., Davies, P. J., & Wright, I., 2018, 'River sediment quality assessment using sediment quality indices for the Sydney Basin, Australia affected by coal and coal seam gas mining', *Science of the Total Environment*, vol. 616–617, pp. 695–702.

Allon, F., & Sofoulis, Z. 2006, 'Everyday water: Cultures in transition', *Australian Geographer*, vol. 37, no. 1, pp. 45–55.

Baer, H. A. 2016, 'The nexus of the coal industry and the state in Australia: Historical dimensions and contemporary challenges', *Energy Policy*, vol. 99, pp. 194–202.

Beattie, J., & Morgan, R. 2017, 'Engineering Edens in this "rivered earth"? A review article on water management and hydro-resilience in the British Empire, 1860s–1940s', *Environment and History*, vol. 23, no. 1, pp. 39–63.

Biodiversity Conservation Act 2016 (NSW).

Braverman, I. Blomley, N., Delaney, D., & Kedar, A. (eds) 2014, *The Expanding Spaces of Law: A Timely Legal Geography*, Stanford University Press, California.

Bridge, G. 2010, 'Resource Geographies 1: Making carbon economies, old and new', *Progress in Human Geography*, vol. 35, no. 6, pp. 820–834.

Brown, P. 2004, 'Are there any natural resources?' *Politics and the Life Sciences*, vol. 23, no. 1, pp. 12–21.

Coal Mine Subsidence Compensation Act 2017 No 37 (NSW).

Conacher, A., & Conacher, J. 2000, *Environmental Planning and Management in Australia*, Oxford University Press, South Melbourne.

Dalberger, J. 2019, 'Warragamba Dam: Damned if you do, damned if you don't', *The Fifth Estate*, viewed 12 April 2019, <https://www.thefifthestate.com.au/urbanism/environ ment/warragamba-dam-damned-if-you-do-damned-if-you-dont/>.

Dams Safety Act 2015 No 26 (NSW).

Darmody, R. G., Bauer, R., Barkley, D., Clarke, S., & Hamilton, D. et al. 2014, 'Agricultural impacts of longwall mine subsidence: The experience in Illinois, USA and Queensland, Australia', *International Journal of Coal Science and Technology*, vol. 1, no. 2, pp. 207–212.

Davies, P. J., Wright, I. A. 2014, 'A review of policy, legal, land use and social change in the management of urban water resources in Sydney, Australia: A brief reflection of challenges and lessons from the last 200 years', *Land Use Policy*, vol. 36, pp. 450–460.

Department of Environment & Climate Change, 2008, *Review of State Conservation Areas: Report of the First Five-year Review of State Conservation Areas under the National Parks and Wildlife Act 1974*, Sydney, viewed 12 April 2019 <https://www.environment.nsw.go v.au/-/media/OEH/Corporate-Site/Documents/Parks-reserves-and-protected-area s/Types-of-protected-areas/review-state-conservation-areas-first-five-year-revie w-080516.pdf>.

Department of the Environment 2014, *Subsidence from Coal Mining Activities: Background Review*, Australian Government.

Environmental Defenders Office 2012, *Mining Law in New South Wales: A Guide for the Community*, viewed 12 April 2019, <https://www.edonsw.org.au/mining_law:in_ nsw:a_guide_for_the_community>.

Environmental Planning and Assessment Act 1979 (NSW).

Environmental Planning and Assessment Amendment (Sydney Drinking Water Catchment) Bill 2017.

Environmental Planning and Assessment Regulation 2000 (NSW).

Freese, B. 2003, *Coal: A Human History*, Perseus Publishing, USA.

Goodall, H., & Cadzow, A. 2009, *Rivers and Resilience: Aboriginal People on Sydney's Georges River*, University New South Wales Press, Sydney.

Graham, N. 2011, *Lawscape: Property, Environment, Law*, Routledge, Abingdon.

Graham, N., 2015, 'Improving on Sugarloaf: The regulation of longwall subsidence impacts on 'unimproved' lands in New South Wales', *Australasian Journal of Natural Resources Law and Policy*, vol. 18, no. 2, pp. 125–144.

Hannam, P. 2019, 'No place for mining': coal mines drain water from dams', *Sydney Morning Herald*, 7 January 2019, viewed 12 April 2019, <https://www.smh.com.au/en vironment/conservation/no-place-for-mining-coal-mines-drain-water-from-dams -20190106-p50pu3.html>.

Haworth, R. J. 2003, 'The shaping of Sydney by its urban geology', *Quaternary International*, vol. 103, pp. 41–55.

Head, L., & Gibson, C. 2012, 'Becoming differently modern: Geographic contributions to a generative climate politics', *Progress in Human Geography*, vol. 36, no. 6, pp. 699–714.

Head, L., & Muir, P. 2006, 'Edges of Connection: Reconceptualising the human role in urban biogeography', *Australian Geographer*, vol. 37, no.1, pp. 87–101.

Hepburn, S. 2015, *Mining and Energy Law*, Cambridge University Press, Cambridge.

Independent Expert Panel for Mining in the Catchment (IEPMC) 2018, *Initial Report on Specific Mining Activities at the Metropolitan and Dendrobium Coal Mines*, Office of the

Chief Scientist and Engineer, NSW Department of Planning and the Environment, New South Wales.

Independent Pricing and Regulatory Tribunal Act 1992 No 39 (NSW).

Jankowski, J. 2007, 'Changes in water quality in a stream impacted by longwall mining subsidence', in *7th Triennial Conference on Mine Subsidence*, Li & Kay (eds), 26–27 November 2007, Wollongong, Australia, Mine Subsidence Technological Society, pp. 241–251.

Johnson, D. 2006, *Sacred Waters: The Story of the Blue Mountains Gully Traditional Owners*, Halstead Press, Rushcutters Bay, NSW.

Karskens, G. 2007, 'Water dreams, earthen histories: Exploring urban environmental history at the Penrith Lakes Scheme and Castlereagh, Sydney', *Environment and History*, vol. 13, no. 2, pp. 115–154.

Karskens, G. 2009, *The Colony: A History of Early Sydney*, Allen & Unwin, Sydney.

Krogh, M. 2007, 'Management of longwall coal mining impacts in Sydney's southern drinking water catchments', *Australasian Journal of Environmental Management*, vol. 14, no. 3, pp. 155–165.

Layard, A. 2019, 'Reading law spatially', in *Routledge Handbook on Socio-Legal Theory and Methods*, Creutzfeldt, N., Mason, M., & McConnachie, K. (eds), Routledge, Abingdon, 232–243.

Lock the Gate Alliance 2018, *Submission to Independent Expert Panel for Mining in the Catchment*, viewed 12 April 2019, <https://www.chiefscientist.nsw.gov.au/__data/assets/pdf_file/0019/180271/Lock-the-Gate-Submission.pdf>.

Lyster, R., Lipman, Z., Franklin, N., Pearson, L., & Wiffen, G. 2016, *Environmental and Planning Law in New South Wales*, fourth edition, Federation Press, Alexandria, New South Wales.

McLean, J., Lonsdale, A., Hammersley, L., O'Gorman, E., & Miller, F. 2018, 'Shadow waters: Making Australian water cultures visible', *Transactions of the Institute of British Geographers*, vol. 43, no. 4, pp. 615–629.

McNally, G., & Evans, R. 2007, *Impacts of Longwall Mining on Surface Water and Groundwater, Southern Coalfield NSW*, report prepared by eWater Cooperative Research Centre for NSW Department of Environment and Climate Change, Canberra.

Mine Subsidence Board 2015, viewed 19 February 2015, <http:www.minesub.nsw.gov.au/templates/mine_subsidence_board.aspx?edit=false&pageID=3870>.

Mining Act 1992 (NSW).

Mine Subsidence Act 1928 (NSW).

Mine Subsidence Compensation Act 1961 (NSW).

Morgan, R. A. 2015, *Running Out?: Water in Western Australia*, University of Western Australia Publishing, Perth.

Morrison, K. G., Reynolds, J. K., & Wright, I. A. 2019, 'Subsidence fracturing of stream channel from longwall coal mining causing upwelling saline groundwater and metal-enriched contamination of surface waterway', *Water Air and Soil Pollution*, vol. 230, no. 2, <https://doi-org.virtual.anu.edu.au/10.1007/s11270-019-4082-4>.

Morrison, K. G., Reynolds, J. K., & Wright, I. A. 2018, 'Underground coal mining and subsidence, channel fracturing and water pollution: A five-year investigation', in *Proceedings of the 9th Australian Stream Management Conference*, 12–15 August 2018, Hobart, Tasmania, River Basin Management Society, Victoria, pp. 689–696.

National Parks and Wildlife Act 1974 (NSW).

Nelson, R. L. 2018, 'Regulating cumulative impacts in groundwater systems: Global lessons from the Australian experience', in Holley, C. & Sinclair, D. (eds), *Reforming Water Law and Governance*, Springer, Singapore, 237–256.

NHMRC 2011, *Australian Drinking Water Guidelines,* Paper 6 National Water Quality Management Strategy, National Health and Medical Research Council, Commonwealth of Australia, Canberra.

Nicholls, S. 2013, 'Mystery behind water chief's sudden departure', *Sydney Morning Herald,* 5 October 2013, viewed 30 April 2019 <https://www.smh.com.au/national/nsw/mystery-behind-water-chiefs-sudden-departure-20131004-2uzra.html>.

NSW Chief Scientist & Engineer 2014, *On Measuring the Cumulative Impacts of Activities Which Impact Ground and Surface Water in the Sydney Water Catchment,* NSW Government, New South Wales.

NSW National Parks and Wildlife Service 2003, *The Bioregions of New South Wales: Their Biodiversity, Conservation and History,* NSW National Parks and Wildlife Service, Hurstville.

Pearse, G. 2010, 'King Coal', *The Monthly,* May 2010, pp. 20–26.

Pells, P. J. N., Young, A., & Turner, P. 2014, 'On the establishment of acceptability criteria for subsidence impacts on the natural environment', Paper presented at the 9th Triennial Conference on Mine Subsidence, Sebel Kirkton Park, Pokolbin, 11–13 May 2014, Mine Subsidence Technological Society.

Picketty, T. 2014, *Capital in the Twenty-First Century,* Harvard University Press, Massachusetts.

Plumwood, V. 2008, 'Shadow places and the politics of dwelling', *Ecological Humanities Review,* vol. 44, pp. 329–333.

Reynolds, R. G. 1976, *Coal Mining Under Stored Waters; Report on an Inquiry into Coal Mining under or in the Vicinity of the Stored Waters of the Nepean, Avon, Cordeaux, Cataract and Woronora Reservoirs,* New South Wales, Australia, Department of Public Works, Sydney.

Robinson, M. 2007, 'West Wallsend Colliery: A coordinated approach to managing subsidence impacts on multiple high risk sensitive surface features: LW27 case study', Paper presented at *7th Triennial Conference on Mine Subsidence,* Mine Subsidence Technological Society, University of Wollongong, 26–27 November 2007.

Seedsman, R., & Pells, P. J. N. 2014, 'On the deception in requiring and providing singular accurate predictions for surface subsidence, tilt and strain', Paper presented at the 9th Triennial Conference on Mine Subsidence, Sebel Kirkton Park, Pokolbin, 11–13 May 2014, Mine Subsidence Technological Society.

Sydney Catchment Authority 2013, *Submission on the Environmental Assessment NRE No.1 Colliery, Stage 2 Underground Expansion Project Application No.* MP 09-0013, viewed 30 April 2019, *<https://cdn.fairfaxregional.com.au/storypad-bMiDV7PLuBggjrMfH7qr6f/SCA%20Submission.pdf>.*

Total Environment Centre 2007, *Impacts of Longwall Coal Mining on the Environment in New South Wales',* Sydney, viewed 12 April 2019, <https://d3n8a8pro7vhmx.cloudfront.net/boomerangalliance/pages/606/attachments/original/1486609983/tec_report_final.pdf?1486609983>.

Troy, P. (ed) 2008, *Troubled Waters: Confronting the Water Crisis in Australia's Cities,* ANU ePress, Canberra.

Tsing, A. L. 2003, 'Natural resources and capitalist frontiers', *Economic and Political Weekly,* vol. 38, no. 48, pp. 5100–5106.

Ulan Coal Mines v Minister for Mineral Resources [2007] NSWSC 1299.

Water Industry Competition (General) Regulation 2008 (NSW).

Water Management Act 2000 (NSW).

Water NSW 2016, *Literature Review of Underground Mining Beneath Catchments and Water Bodies,* viewed 12 April 2019 <https://www.waternsw.com.au/__data/assets/pdf_file

/0011/127559/20161223-WaterNSW-Literature-Review-Underground-Mining-V3 .pdf>.

Water NSW 2018, *Submission to the Independent Expert Panel on Mining Sydney Catchment*, Paramatta, viewed 12 April 2019 < https://www.chiefscientist.nsw.gov.au/__data/ assets/pdf_file/0006/225168/Submission-WaterNSW-2.pdf>.

Water NSW Act 2014 No 74. (NSW).

Water NSW Amendment (Warragamba Dam) Bill 2018 (NSW).

Water NSW2019a, Sydney, viewed 12 April 2019, <https://www.waternsw.com.au/su pply/heritage/water-schemes/warragamba-supply-scheme>.

Water NSW2019b, *Warragamba Supply Scheme, Sydney*, viewed 12 April 2019, <https ://www.waternsw.com.au/supply/heritage/water-schemes/warragamba-supply -scheme>.

Wilkinson, J. 1995, 'Coal Production in New South Wales', Briefing Paper No 10/95, NSW Parliamentary Library Research Service.

Worster, D. 2006, 'Watershed Democracy: Recovering the lost vision of John Wesley Powell', in *Water: Histories, Cultures, Ecologies*, Leybourne, M. & Gaynor, A. (eds), University of Western Australia Press, Perth, 3–14.

4nature Incorporated v Centennial Springvale Pty Ltd [2017] NSWCA 191.

13

LAWYERS IN LEGAL GEOGRAPHY

Parliamentary submissions and coal seam gas in Australia

David J. Turton

Introduction

Almost a decade ago, Martin et al. observed that lawyers remained "largely unseen" by legal geography scholars, despite the profession's "strong shaping influence over … the construction of space" (2010, pp.177, 179). While there have been sporadic efforts to redress this silence (Kocher 2017), this has not been without difficulty for researchers. This chapter surveys some of the ways in which scholars have taken up the task of bringing lawyers into legal geography. This chapter seeks to achieve two goals. First, it will delve into the unconventional gas and legal geography literatures, providing a brief overview of research efforts that have so far employed lawyer perspectives. It will pose some ideas to future researchers about how to increase the presence of lawyers in their work. Second, this chapter explores the contribution of lawyers to Australia's coal seam gas (CSG) debate through an empirical case study of submissions prepared by lawyers for the Queensland State Development, Natural Resources and Agricultural Industry Development Committee's ("Committee") inquiry into the then Mineral, Water and Other Legislation Amendment Bill 2018 (Qld) (Queensland State Development, Natural Resources and Agricultural Industry Development Committee Report 2018). A brief explanation of the debate is appropriate for context and to explain the methodological approach to the case study.

CSG is an unconventional fossil fuel (primarily composed of methane) and one of a number of contested land uses in Australia. Technological innovation in extraction techniques has transformed CSG from a danger to underground coal miners into a profitable industry. The socio-economic effects (positive and negative) of CSG have been fiercely debated, as have environmental concerns – for example, weeds, damage to aquifers and the merits of CSG as a transition fuel to lower carbon emission forms of energy (Measham et al. 2016). Tension has emerged in the eastern Australian states of New South Wales and Queensland in recent years, as CSG expands into more heavily populated rural areas traditionally associated with intensive agricultural and pastoral production. This led in part to the creation of a national anti-CSG social movement known as "Lock the Gate" in 2010. It is an alliance of farmers and environmentalists committed to obstructing CSG and coal mining developments on the grounds of water and food security (Cronshaw & Grafton 2016; Sherval & Hardiman 2014; see also Sherval's chapter in this volume). Extensive media commentary, political debate, litigation and government inquiries are all features of this debate. Due to CSG's ongoing presence in the public mind, lawyers and the range of community interests they represent are capable of being explored, in this case, through the analysis of submissions made to the Bill. Indeed, as actors in the policy-making process, lawyers may seek to demand and/or draft new laws to create and maintain statutory arrangements for the benefit of CSG operators, landowners, government and other interested parties (Apple 2014). By focusing on parliamentary submissions written by lawyers prior to the enactment of legislation, it is possible to explore multiple lawyer perspectives in a formal setting beyond the courts, opening up research possibilities for legal geography. In this case, lawyers and community groups drew attention to a spatially infused clause of the Bill relating to CSG and the geographical scope of compensation for landholders affected by gas extraction. Arguments presented in submissions by lawyers at the inquiry prior to the passing of the Bill in the Queensland Parliament on 18 October 2018 showcase the potential for further analysis of lawyers in legal geography research, while also highlighting compensation – a crucial aspect of CSG research (Measham et al. 2016). To appreciate some of the ways in which lawyers' voices can be obtained by legal geographers investigating human–environment interactions, including unconventional gas disputes, an overview of past scholarly efforts follows. Inquiry submissions that form the basis of the empirical case study will then be examined.

Lawyers, legal geography and methods

Legal geography is breaking down disciplinary binaries, with its practitioners seeking to investigate the "interconnections between law and spatiality, and especially their reciprocal construction" (Braverman et al. 2014, p.1). Property, the legal-spatial nexus of urban life, public space and the environment are key themes in the field (Pruitt 2014; Graham et al. 2017). Legal geographers might have "arguably paid too little attention to the environment" (Andrews & McCarthy 2014, p.7) in the past, but this is changing rapidly (e.g. Bartel et al. 2013; Salgo & Gillespie 2018). Extractive industries have likewise received increasing attention from scholars, due in part to a variety of contemporary resource conflicts involving mining (e.g. Ey & Sherval 2016; Bosca & Gillespie 2019; Sherval & Hardiman 2014). Although forays into the legal geography of unconventional gas are few (Hesse et al. 2016; Turton 2015a, 2015b, 2017;

Andrews & McCarthy 2014), different approaches to human and legal geography can offer important viewpoints on extractive industries (Delaney 2016; Bennett & Layard 2015). There is also merit in drawing on different disciplinary viewpoints to enrich legal geography investigations, with political ecology being one example of this (e.g. Andrews & McCarthy 2014; Salgo & Gillespie 2018). In considering the research methods of legal geographers operating in the extractive industries space, including CSG, it should be acknowledged that, "the way in which geography has approached the extractive sector is far from static" (Ey et al. 2017, p.156). For example, there is increasing emphasis on impacts beyond the physical sites of resource extraction (Ey & Sherval 2016). This has implications for the ways in which legal actors – lawyers, judges, regulators, legislators and others – associated with CSG are studied. Therefore, it is useful to briefly survey the extant social science literature to gain an appreciation for how lawyers have been incorporated into CSG and broader international unconventional gas research, before turning to consider methodological approaches deployed by legal geographers in particular.

Lawyers and judges are by no means invisible in either legal geography or Australian CSG social science literature (e.g. Curnow et al. 2017; Kennedy 2017; Turton 2015a, 2015b, 2017). There is, however, a tendency for lawyers' viewpoints to be scattered throughout wider qualitative interviewee and submission samples, community-level case studies and broader research questions (e.g. de Rijke 2013; Farrugia et al. 2018). Various international studies also make use of lawyer opinions (e.g. Fry & Brannstrom 2017; Walsh & Haggerty 2019). Beyond the words of lawyers, the literature reveals some of the views held by those who have interacted with lawyers while negotiating unconventional gas development, both within Australia and overseas (e.g. Malin & DeMaster 2016; Curnow et al. 2017). Studies hint at the diversity of this profession's involvement in CSG, from lawyers interviewed in their capacity as local residents within a CSG-impacted community (Farrugia et al. 2018), to lawyers serving as key players in private landholder–CSG company access agreement negotiations in Queensland (Curnow et al. 2017). Lawyers also provide specialist legal services relating to CSG (such as water and planning law) in rural and regional communities (Scott 2016; Mundy & Kennedy 2017), participate in community education events about CSG (Turton 2015b) and, in common with other CSG stakeholders, prepare submissions to government inquiries. These activities present opportunities for researchers to examine lawyers, whether through interviews, questionnaires, focus groups, participant observation or other forms of data collection and analysis (Dunn 2016; Cameron 2016; McGuirk & O'Neill 2016; Kearns 2016). The extent of a lawyer's involvement in CSG issues may also dictate their capacity to comment on the subject, whether in a professional or personal context.

This can be methodologically significant, especially when a lawyer draws upon their professional expertise – or simply states that they are a legal professional – when answering research questions. For example, one lawyer who identified as being pro-CSG, when asked for their opinion on CSG in the New South Wales

Camden region, stated that: "I do, as a lawyer, have a fundamental problem, and that is I've always believed that it should ultimately be up to the people that live on that property as to whether a well is drilled" (Bennetto 2013, p.44). The above quote is further complicated by the presumption that lawyers are aware of the state's ownership of CSG resources, namely that CSG drilling decisions are not beholden to an individual landholder's preferences (e.g. Turton 2015b). Lawyers' personal values and professional expertise, and the ways in which these are invoked, matter and should be carefully considered by legal geographers.

Before examining this chapter's case study, the methods used by legal geographers and other socio-legal researchers to obtain lawyer perspectives will be discussed. One reason for doing so is to reflect upon methodological choices in order to make greater use of lawyers in legal geography research (Martin et al. 2010). Lawyers appear in various empirically based studies, some of these making express use of legal geography (e.g. Kocher 2017; Terry 2009). An important obstacle sometimes encountered by researchers seeking to make use of lawyers is access (consider O'Donnell's chapter in this volume). I recall the challenges of attempting to secure both the interest and consent of lawyer participants for what was intended to be an interview-based examination of legal practitioners and their engagement with Australia's CSG debate. Although direct approaches to government, industry and non-government sector lawyers proved fruitless (Turton 2015b), this lack of response became a catalyst for the adoption of a different methodological approach. My decision to focus on community forums which made use of lawyers as speakers on CSG legal issues arose from a two-fold realisation that lawyers were prominent in public CSG discussions, as were audio-visual materials. But for my lack of success in obtaining interviews with a sufficient target population of lawyers, it is possible that I might have dismissed, or remained ignorant of, the significance of community forums. As it was, refusals to be interviewed from lawyers who were contacted by phone and e-mail spurred on a search for YouTube videos with lawyers discussing CSG, in addition to documentary evidence that indicated lawyer involvement in CSG. Differing approaches to public engagement by lawyers, including their receptiveness to university researchers, has ramifications for source material availability. Some lawyers seem very willing to grant interviews to researchers and/or contribute their views through a plethora of media (e.g. Jessup & McIlwraith 2015; Kennedy 2017).

Informant recruitment challenges are not limited to lawyers, remaining an ongoing issue for CSG researchers for a variety of reasons, including interviewee fatigue, the real or imagined sensitivity of information being sought and informant perceptions (accurate or not) of CSG industry funding for research at both university and government levels (Espig & de Rijke 2016). Stakeholder reaction to the publication of academic research is also relevant (Carrington 2013; Hardie et al. 2016). In view of the reality that different researchers can obtain varying degrees of access to diverse actors in the CSG debate, some perspective on so-called controversial research partnerships is beneficial here. Notwithstanding contemporary debate on this issue, collaboration with the CSG industry is not new for Australian universities (Oldroyd 1993); nor are the research alliances between CSG companies and the Australian Government's Commonwealth Scientific and Industrial Research Organisation (CSIRO) (Enever & Jeffrey 1991). Beyond blurred public–private research agendas and funding, anthropologists Espig and de Rijke alert CSG researchers to the need to remain conscious of the "politics of representation" (2018, p.220) and the methodological implications of legislative measures that include and exclude stakeholders from consideration in legal and policy processes. Who constitutes a "local" stakeholder? What determines their legitimacy for research purposes? Who is in and out of scope and why? In the case of arbitration in CSG disputes in New South Wales, for example, the very definition of a lawyer may determine which legal professionals are permitted into the room as representatives for landholders (Turton 2015b; Kennedy 2017). Ethical concerns are also a necessary consideration.

In a study of Queensland vegetation management legislation encompassing court cases, documentary materials and interviews with lawyers and regulators, Kehoe (2013) noted that adherence to human research ethics protocols is crucial when conducting research involving lawyers and their clients. Equally important was the need for consent from all parties to share information, while still observing lawyer–client privilege. This accords with standard research practice in human geography and the need to negotiate access, consent and power dynamics between the subject of the research and the researcher (Dowling 2016). Some legal geographers have also identified strategies that could assist scholars attempting to recruit lawyer informants for their research.

Personal networks and mutual friendships can be an invaluable tool for recruiting lawyer informants (Keenan 2009; Kehoe 2013). The extent of cooperation can also influence research findings, for example, where initial government approval to interview lawyers was withdrawn, requiring a change in focus from "active-duty" to former military lawyers in one study of targeting legal advice and the Israeli military (Jones 2015, p.678). Conversely, cooperation from some law firms has enabled researchers to better understand points of law raised in particular cases (e.g. Atkins et al. 2006; Nikiforuk 2015). However, the exact nature of this cooperation is unclear, making it difficult to know the extent to which lawyers informed and perhaps influenced the resulting publication. On this point, Irus Braverman has discussed the challenges and ultimate rewards of "intense revision" in preparing her zoo ethnography book manuscript following her decision to enter "into detailed communications about the contents of the manuscript" with interviewees prior to publication (2014, p.133). In any case, both refusal and cooperation from prospective interviewees have value for a researcher. What informants reveal is also worth reflecting on, as just like "other advocates [and sources of all kinds] … lawyers tell different versions of stories to suit their purposes" (Pruitt & Sobczynski 2016, p.328). Embedded within such stories are the spaces in which these narratives are told.

To this end, legal geographers can benefit from paying attention to the sites in which interviews are conducted, for example, by incorporating the location of legal firms and the division of legal labour into their analyses (Kocher 2017). This may be an important consideration when approaching law firms who perform CSG-related services, as some have established themselves in rural centres to "take advantage of the business on offer" (Scott 2016, p.21). Outside their offices, lawyers can also be analysed by legal geographers using ethnographic research techniques in the courtroom itself, promoting a more holistic understanding of the intersection between law, space and everyday litigation practices (Walenta in press). Although interviews are an important means of engaging with the views of lawyers, documentary sources and fieldwork, potentially in combination, have considerable usefulness in exploring how lawyers go about translating and transforming legal meanings, in addition to acting as agents and exerting power over others (Martin et al. 2010).

Court cases are regularly explored by legal geographers (e.g. Turton 2017; O'Donnell 2016; Jessup & McIlwraith 2015; Prior et al. 2013), but historical research into water governance litigation in the United States of America also reveals the utility of court documents and lawyer correspondence as tools to explain how legal processes can forge new geographies (Jepson 2012; see also Kay 2016). As a general proposition, there is certainly a great deal of potential for archival material to be used in combination with oral histories of legal actors (e.g. Turton 2015c), greatly enriching text-only perspectives (George & Stratford 2016). That said, archival records are not immune from their own access requirements, as well as gaps, silences and biases (Roche 2016). The type of lawyers who actively contribute empirical data to research projects can also be telling. Indeed, there appears to be a research gap in the range of interests that lawyers represent in CSG scholarship, with an emphasis on representatives from community legal centres and staff from environmental advocacy organisations (Turton 2015b; Kennedy 2017). Legal practitioners who represent government and CSG industry stakeholders, while still present, are seemingly less noticeable in the public domain. To partially remedy this silence, and in recognition of the fact that observing lawyers can involve collecting documentary sources of their statements and activities (Martin et al. 2010), submissions from lawyers and stakeholder organisations who employ lawyers are used in the following case study. Drawing on a Queensland Government inquiry into the then Mineral, Water and Other Legislation Amendment Bill (2018), this case study delves into the issue of compensation and spatial arguments presented by lawyers when attending to legislative change on behalf of different stakeholders.

Situating submissions: CSG case study

Background

The Mineral, Water and Other Legislation Amendment Bill 2018 (Qld) was introduced into the Queensland Legislative Assembly on 15 February 2018 and referred to the Queensland Parliament's State Development, Natural Resources and Agricultural Industry Development Committee the same day (Queensland State Development, Natural Resources and Agricultural Industry Development Committee Report 2018). The Committee itself built upon a predecessor inquiry into a 2017 version of the Bill, which had lapsed when Parliament dissolved at the end of 2017 (Queensland Infrastructure, Planning and Natural Resources Committee 2017). Addressing a number of issues, including arbitration, the payment of professional costs for landholders negotiating land access agreements and the compensation entitlement of affected landholders impacted by resource activities, the Bill was referred by the Minister for Natural Resources, Mines and Energy, Dr Anthony Lynham, to the State Development, Natural Resources and Agricultural Industry Development Committee, with a report expected by 9 April 2018 (Queensland Hansard 15 February 2018). This Committee was charged with the task of determining "whether the Bill … [had] sufficient regard to the rights and liberties of individuals, and to the institution of Parliament" (Queensland State Development, Natural Resources and Agricultural Industry Development Committee Report 2018, p.iv). While it is beyond the scope of this chapter to examine each of the proposed amendments to what would become the *Mineral, Water and Other Legislation Amendment Act 2018* (Qld), the chapter will focus on the Bill's amendment to a provision for general liability to compensate affected landholders under section 81 of the then *Mineral and Energy Resources (Common Provisions) Act 2014* (Qld) who had suffered a "compensatable" effect (costs and impacts) due to authorised resource activities. Proposed changes to this section provoked significant disagreement between the Queensland Government's Department of Natural Resources, Mines and Energy, as well as mining and non-mining interests who had prepared submissions for the Committee's inquiry into the Bill. As the Committee explained:

> Landholder and legal groups raised their significant concerns regarding cl[ause] 38 [of the Bill] … and the proposed amendment to s[ection] 81 of the Mineral and Energy Resources (Common Provisions) Act 2014 … which establishes the general liability of resource authority holders to compensate the owners and occupiers of public and private land. Non-mining sector stakeholders did not support amendments to s[ection] 81(4)(a) that state the compensatable effects only apply where authorised activities have been carried out 'on the eligible claimant's land'. It was argued that the amendment removes the rights of neighbours within the resource authority area, but without authorised activities on their land, to claim compensation for the impacts of nearby activities. (Queensland State Development, Natural Resources and Agricultural Industry Development Committee Report 2018, p.7)

Disagreement over the spatial and compensation aspects of the proposed amendment, and ultimately the Act itself, provides an insight into the ways in which CSG can be approached by a wide spectrum of stakeholders and their lawyers.

Many of the submissions made to the Committee either specifically identified as being lawyers, relied upon written advice from legal practitioners or employed lawyers in the course of their business. Given the challenges of seeking lawyer input through interviews (discussed above), I took the view that formal submissions from a wide cross-section of interested legal parties on a significant topic for all involved in CSG development could offer a fresh vantage point from which to comprehend the place of lawyers in Australia's CSG debate and their capacity to influence both their clients and others (Martin et al. 2010). Some explanation for my choice of case study is therefore appropriate.

Methods

Case study research can be a combination of a researcher's specific purpose and serendipity (Stratford & Bradshaw 2016). In this case, I made the conscious choice to attempt a case study that differed from my previous work, which involved CSG court cases and recorded community forums (Turton 2015a, 2015b, 2017). Drawing upon earlier fieldwork involving community meetings that had lawyer participants, as well as previous knowledge of key lawyers and advocates in stakeholder groups who had appeared at CSG-related inquiries, I sought to focus on submissions to government inquiries into CSG which offered both a relatively small number of submissions (for practical reasons, Stratford & Bradshaw 2016) and a reasonable number of submissions from lawyers. Keywords to this effect ("CSG", "coal seam gas", "unconventional gas", "solicitor", "lawyer", "barrister" and "legal practitioner") were entered into all Australian parliamentary websites to obtain relevant submissions, resulting in the above case study. In addition to a written response from the Department of Natural Resources, Mines and Energy, the Committee received public submissions ($n = 17$) from environmental organisations, private law firms, a community legal centre, mining and rural industry lobby groups, as well as private citizens. All submissions are available online (Queensland Parliament 2018). One non-lawyer, George Houen (a veteran of Queensland agricultural–mining disputes since the 1980s, see Turton 2014), obtained legal advice from barrister Darlene Skennar QC to support his submission (Skennar 2018).

The Bill addressed a number of amendments across a range of Acts. To narrow the case study further, I determined that a focus on section 81 of the *Mineral and Energy Resources (Common Provisions) Act 2014* (Qld) (Clause 38 of the Bill) would be beneficial as an exercise in exploring stakeholder understandings of CSG compensation in a legal-spatial context. Of the 17 written submissions received by the Committee, seven were removed from analysis due to no reference being made to the subject of the case study. This left ten submissions remaining. As noted above, the Committee resumed the earlier work of the 2017 Infrastructure, Planning and Natural Resources Committee. Thus, several of the 2018 submitters repeated verbatim or re-attached their original 2017 submission for the reinvigorated parliamentary committee process in 2018. The 2018 Bill was substantially the same as the 2017 version; however, the researcher cross-referenced 2017 submissions against those put forward in 2018 to ensure the content remained unchanged. Where differences occur, acknowledgment is made. Additionally, verbal submissions at public hearings of the Committee's inquiry into the Bill were used to supplement written submissions (State Development, Natural Resources and Agricultural Industry Development Committee, Transcript of Proceedings on 5 March 2018; State Development, Natural Resources and Agricultural Industry Development Committee, Transcript of Proceedings on 9 March 2018). Before investigating these, analysis of stakeholder submissions is an accepted research method in CSG social science, used by a range of authors from anthropologists to environmental scientists alike (de Rijke et al. 2016; Fenton 2013). Record numbers of submissions are a hallmark of CSG inquiries and reviews (Witt et al. 2018b). Queensland and its CSG industry are also "unique in the Australian regulatory context as the government did not conduct any pre-development inquiries … [I]nstead, legislation has evolved … as the industry has progressed" (Witt et al. 2018a, p.423).

Compensation, compensatable effects and space

The issue of "compensatable effects" arose from what representatives of the Department of Natural Resources, Mines and Energy described as a "minor wording change" to section 81 of the *Mineral and Energy Resources (Common Provisions) Act 2014* (Qld) (State Development, Natural Resources and Agricultural Industry Development Committee, Transcript of Proceedings on 5 March 2018, p.2). As the Department explained:

> Section 81 [of the Act] imposes a liability on the resource tenure holder to compensate each owner or occupier of private or public land that is in an area of the resource authority for any compensatable effect caused by the authorised activities. Neighbours whose land is not being accessed to carry out these activities are not entitled to have a conduct and compensation agreement [CCA] under the land access framework … There is no change to the obligation to compensate neighbouring landholders as a result of changes to section 81. That section has always been about compensation for landholders upon whose land advanced activities are being conducted … This is clear when the existing section 81 is read in context with the Mineral and Energy Resources (Common Provisions) Act as a whole … [A] CCA is only required where a resource authority intends to enter private land to carry out an advanced activity. There is no requirement for a CCA on private land where no advanced activities are proposed. (State Development, Natural Resources and Agricultural Industry Development Committee, Transcript of Proceedings on 5 March 2018, pp.2, 4, 5)

The Queensland Law Society, informed through its Mining and Resources Law Committee (whose membership have substantial expertise and experience in this area of the law), agreed with the above assessment (Krulin & Grossberg 2017; Carter Newell Lawyers 2019; Queensland Law Society 2018), as did national and state industry lobby groups for the mining and CSG sectors (Australian Petroleum Production and Exploration Association 2018; Queensland Resources Council 2018). A divergent view was expressed by the anti-CSG social movement Lock the Gate, rural law firms and landholder advocates, as well as community and environmental organisations (Lock the Gate Alliance 2018; Marland Law 2018; Shay Dougal 2018; George Houen 2018; Lee McNicholl 2018; Protect the Bush Alliance 2018; Shine Lawyers 2018; Shine Lawyers 2017). In articulating their concerns to the Committee in 2018, Shine Lawyers also referred back to their 2017 submission, pointing out that: "[W]e, and others, asserted that the change to section 81 meant that 'neighbouring' landholders to extensive gas activity that impacted on them could [only] apply for compensation if they were in the relevant [resource] tenement" (Shine Lawyers 2018, p.6). In their 2017 submission, Shine Lawyers had observed that: "[T] proposed clause … removes the right of neighbours who are within the tenement area to claim compensation for the impacts of activities carried on next door to them" (Shine Lawyers 2017, p.1). In an effort to underline the impact of the proposed provision, Shine Lawyers referenced the industry's scale:

> [G]as activity, typically has a huge and very widespread impact and does not only affect the landowner on whose land the [resource] activity is conducted. Given its nature it sometimes affects neighbouring landowners – and sometimes very significantly. (Shine Lawyers 2017, p.1)

Another rural law firm endorsed this view (Marland Law 2018). Prior to the passing of the Bill on 18 October 2018, the Act read as follows:

> A resource authority holder is liable to compensate each owner and occupier of private land or public land that is in the authorised area of, or is access land for, the resource authority (each an eligible claimant) for any compensatable effect the eligible claimant suffers caused by authorised activities carried out by the holder or a person authorised by the holder. (*Mineral and Energy Resources (Common Provisions) Act 2014* (Qld), current as at 6 December 2016, s. 81(1))

For the purpose of section 81 of the Act, "compensatable effects" referred to: any deprivation of possession of the land's surface, diminution of its value or use, severance of any part of the land, any cost damage or loss arising from the carrying out of activities under the resource authority on the land and finally, necessary and reasonable accounting, legal or valuation costs the claimant incurred in order to negotiate or prepare a CCA (*Mineral and Energy Resources (Common Provisions) Act 2014* (Qld), current as at 25 October 2018, s. 81(4)). Minister Lynham acknowledged what appeared to be a conflict over the statutory interpretation of section 81 in the Bill's first reading speech to the Queensland Legislative Assembly on 15 February 2018, but stated that: "The policy intent of section 81 is that compensation should be payable for any compensatable effect suffered by a landholder on the land on which the resource activities are being carried out" (Queensland Hansard 15 February 2018, p.116). An explanation as to why a rewording of the section was necessary came in the form of verbal submissions from the Department to the Committee's predecessor in late 2017:

> The amendment to section 81 was driven by the issue of the compensation associated with the cost of developing the CCA. Right now it is defined in that existing section as compensation due to activities. Activities can only occur when you have a CCA, so the costs leading up to the CCA were outside of that framework. In giving effect to Professor Scott's recommendation [(Scott 2016)], Parliamentary Counsel basically subdivided that section, creating a new section to cover those necessary and reasonably incurred costs associated with developing the agreement. They then, in restructuring that provision, provided words that, in their view, were designed to provide clarity. Now obviously that in itself has created some concern. (Infrastructure, Planning and Natural Resources Committee, Transcript of Proceedings on 11 October 2017, p.35)

As has been noted elsewhere (Turton 2015b), the opinions of lawyers can be used by others to craft arguments both for and against the CSG industry, reversing the "lawyers as translators" concept suggested by Martin et al. (2010). There was clear evidence during the Committee's deliberations that submitters had read the views of others, incorporating these arguments to support their own. For example, Lock the Gate identified that:

> Shine Lawyers and the Queensland Law Society have both drawn attention to a major problem with the Bill. Clause 38 of the Bill amends section 81 (4) (a) of the Mineral and Energy Resources (Common Provisions) Act 2014. Those amendments will limit compensation such that it will only apply to compensatable effects from resource activities which happen *on the claimants own land*. Currently, under that section, a landholder is liable for compensation from resource activities '*in relation to the eligible claimants land*' … Gas activities frequently cause … noise and air pollution impacts … [L]andholders can be heavily affected by noise and air pollution, but if it is occurring just outside their land, they will have no claim for compensatable effects. (Lock the Gate Alliance 2018, p.1 with original emphasis; see also Protect the Bush Alliance 2018)

Beyond drawing on the legal advice of others to advance their own positions, submitters made various assumptions about rural sociospatiality, a phenomenon observed in judges, legislators, lawyers and laypersons (Pruitt 2014). One example of this can be seen in both the written and verbal submissions of Tom Marland, a self-described "regional solicitor" (State Development, Natural Resources and Agricultural Industry Development Committee, Transcript of Proceedings on 9 March 2018, p.4) and "one of a very small number of rural lawyers who act for landholders" (Marland Law Submission 2018, p.1). In providing critique on the proposed amendment to compensatable effects, Marland also remarked upon rural sociospatiality as it applied, in his view, to this provision of compensation:

> My experience acting in this area is that landowners who sometimes do not have any infrastructure on their properties are often the ones that are most vulnerable ... Obviously, if you have a gas well on your property you can measure the area, you can measure the roads and you can measure the pipelines. If you are a property next to a coal seam gas well you have to measure the noise, the dust, the stress, the vibration, the sleepless nights where you wake up or you wake up to your kid crying because of the overblast pressure from a drill rig. That is very hard to measure. (State Development, Natural Resources and Agricultural Industry Development Committee, Transcript of Proceedings on 9 March 2018, p.5)

The absence of infrastructure, of materiality, in rural landscapes has been described as the "slate on which ... [legal institutions] impose their will and write their story" (Pruitt 2014, p.197). Yet it would be simplistic to suggest that corporate entities, in this case CSG companies, exert their authority upon blank spaces devoid of power and knowledge. Instead, in drawing a distinction between the negotiation experiences of landholders with and without pre-existing infrastructure, Marland's comments to the Committee highlight the role of rural spatiality in "producing power and knowledge [and in the context of submissions to a parliamentary committee in particular,] ... the relationship between individuals and the state" (Pruitt 2014, p.197). In urging the Committee not to amend section 81 of the Act through the Bill, Marland referred again to spatial impacts:

> There is a difference between a petroleum lease and a mining lease ... [A] mining lease can only occur where the coal is. They will be site-specific. They will go in and purchase those properties and they will compensate those properties ... With a petroleum lease, what you find is that they are over a huge swathe of area ... I think the intent of the original drafters of the legislation was to reflect the fact that a gas field can cover a far larger area and so there might be people stuck in the middle of it who are going to be impacted. That is where you start to see the difference in how the current legislation is drafted and I think any change to that is going to have significant impacts on that very small percentage of people who are ultimately impacted. (State Development, Natural Resources and Agricultural Industry Development Committee, Transcript of Proceedings on 9 March 2018, p.6)

For its part, the Australian Petroleum Production and Exploration Association, having checked with its members, was:

> [N]ot aware of any neighbours being compensated under ... [section 81 of the Act. Nor was the Association] aware of any claims for compensation under [section 81], but acknowledged that there were obligations to neighbours under the *Environmental Protection Act 1994* (Qld) in terms of environmental approvals, with parties to enter into arrangements to alleviate concerns. (State Development, Natural Resources and Agricultural Industry Development Committee, Transcript of Proceedings on 9 March 2018, p.19)

The Department recognised that apparent stakeholder confusion around the interpretation of section 81 seemed to arise from a reading of the section in isolation from the remaining provisions in the land access framework in Chapter 3 of the *Mineral and Energy Resources (Common Provisions) Act 2014* (Qld). In the Department's view, this could not be sustained:

> It is also not a feasible outcome to allow this confusion to continue ... The minor change to section 81 clarifies ... that it is landholders who have resource activities being carried out on their land who are entitled to compensation under the land access framework for the effects of those activities. (State Development, Natural Resources and Agricultural Industry Development Committee, Transcript of Proceedings on 9 March 2018, p.30)

Ultimately, section 81 was amended to reflect the above remarks (*Mineral and Energy Resources (Common Provisions) Act 2014* (Qld), current as at 25 October 2018), but the disagreement created by the modification of a few words serves to highlight how proposed legislation can rouse rural sociospatial anxieties among stakeholders, including lawyers and legal organisations. By studying the engagement of lawyers in the legislative process, it is possible to gain some insight into the ways in which lawyers seek to interpret the law surrounding CSG for different stakeholders. In turn, the submissions considered above reveal contested practitioner viewpoints, different personal values, client experiences and spatial understandings (Martin et al. 2010). Submissions can be helpful in understanding how practitioners present arguments about legislative changes that have legal-spatial consequences for regional and rural Australia.

Conclusion

CSG is a controversial land use linked with a variety of societal concerns. Research challenges, as identified in this chapter, may in fact provide opportunities for researchers to seek out new perspectives and to generate different questions from new sources. This chapter has sought to explore some of the ways in which lawyers have been included in past legal geography research efforts, while also identifying their current place in CSG and unconventional gas literature. Acknowledging the challenges of obtaining lawyer viewpoints in the form of interviews, such obstacles need not preclude lawyers from analysis, provided a flexible approach is taken to gathering source material. Although lawyers tend to be found as one stakeholder among many others in qualitative research, the case study above also shows that it is possible to isolate their contribution so as to make this profession the focus of study. However, other stakeholders were noted, where they made use of lawyer perspectives. As agents for others and officers of the legal system, lawyers have much to offer legal geographers as multifaceted contributors to CSG discussions and other contested uses of the Australian landscape. In speaking for others and themselves, they give shape to significant conversations about CSG which can assist other research being conducted in Australia and overseas. While the confidential nature of their work will likely always be a part of any researcher–subject negotiation, legal geographers are more than capable of answering the call made by Martin et al. (2010) to bring the practice of law – and by extension its practitioners – to bear in their scholarship.

References

Andrews, E., & McCarthy, J. 2014, 'Scale, shale, and the state: Political ecologies and legal geographies of shale gas development in Pennsylvania', *Journal of Environmental Studies and Sciences*, vol. 4, no. 1, pp. 7–16.

Apple, B. E. 2014, 'Mapping fracking: An analysis of law, power, and regional distribution in the United States', *Harvard Environmental Law Review*, vol. 38, pp. 217–244.

Atkins, P. J., Hassan, M. M., & Dunn, C. E. 2006, 'Toxic torts: Arsenic poisoning in Bangladesh and the legal geographies of responsibility', *Transactions of the Institute of British Geographers*, vol. 31, no. 3, pp. 272–285.

Australian Petroleum Production and Exploration Association 2 March 2018, *Submission on the Mineral, Water and Other Legislation Amendment Bill 2018*, viewed 3 January 2019 https://www.parliament.qld.gov.au/documents/committees/SDNRAIDC/2018/3MinWatOLAB2018/submissions/017.pdf.

Bartel, R., Graham, N., Jackson, S., Prior, J. H., Robinson, D. F., Sherval, M., & Williams, S. 2013, 'Legal geography: An Australian perspective', *Geographical Research*, vol. 51, no. 4, pp. 339–353.

Bennett, L., & Layard, A. 2015, 'Legal geography: Becoming spatial detectives', *Geography Compass*, vol. 9, no. 7, pp. 406–422.

Bennetto, J. A. 2013, 'Taking the public seriously: Factors affecting trust in coal seam gas development', unpublished BSc. (Hons) thesis, Australian National University, available from the Fenner School of Environment and Society, Australian National University, Canberra.

Bosca, H. D., & Gillespie, J. 2019, 'The construction of 'local' interest in New South Wales environmental planning processes', *Australian Geographer*, vol. 50, no. 1, pp. 49–68.

Braverman, I. 2014, 'Who's afraid of methodology? Advocating a methodological turn in legal geography', in *The Expanding Spaces of Law: A Timely Legal Geography,*

Braverman, I. Blomley, N., Delaney, D., & Kedar, A. (eds), Stanford University Press, California, pp. 120–141.

Braverman, I., Blomley, N., Delaney, D., & Kedar, A. 2014, 'Introduction: Expanding the spaces of law', in *The Expanding Spaces of Law: A Timely Legal Geography*, Braverman, I. Blomley, N., Delaney, D., & Kedar, A. (eds), Stanford University Press, California, pp. 1–29.

Cameron, J. 2016, 'Focussing on the focus group', in *Qualitative Research Methods in Human Geography*, fourth edition, Hay, I. (ed), Oxford University Press, Melbourne, pp. 203–244.

Carrington, K. 2013, 'Corporate risk, mining camps and knowledge/power', in *Crime, Justice and Social Democracy: International Perspectives*, Carrington, K., Ball, M., O'Brien, E., & Tauri, J. M. (eds), Palgrave Macmillan, London, pp. 295–314.

Carter Newell Lawyers 2019, *Team: James Plumb*, viewed 5 January 2019, https://www.carternewell.com/page/Team/Our_partners/James_Plumb/.

Cronshaw, I., & Grafton, R. Q. 2016, 'A tale of two states: Development and regulation of coal bed methane extraction in Queensland and New South Wales, Australia', *Resources Policy*, vol. 50, pp. 253–263.

Curnow, K., Hunter, T., Weir, M., & Boulle, L. 2017, 'Negotiation and regulation of land access agreements: Lessons from Queensland', *The Journal of World Energy Law and Business*, vol. 10, no. 2, pp. 117–135.

Delaney, D. 2016, 'Legal geography II: Discerning injustice', *Progress in Human Geography*, vol. 40, no. 2, pp. 267–274.

Shay Dougal 2018, *Submission on the Mineral, Water and Other Legislation Bill 2018*, 26 February 2018, viewed 3 January 2019, <https://www.parliament.qld.gov.au/documents/committees/SDNRAIDC/2018/3MinWatOLAB2018/submissions/001.pdf>.

Dowling, R. 2016, 'Power, subjectivity, and ethics in qualitative research', in *Qualitative Research Methods in Human Geography*, fourth edition, Hay, I. (ed), Oxford University Press, Don Mills, pp. 29–44.

Dunn, K. 2016, 'Interviewing', in *Qualitative Research Methods in Human Geography*, fourth edition, Hay, I. (ed), Oxford University Press, Don Mills, pp. 149–188.

Enever, J. R., & Jeffrey, R. 1991, 'Coal bed methane – A new source of energy', in *Seminar on New Technologies for Electricity Generation*, Energy Research Development and Information Centre and New South Wales Electricity Commission, Sydney, pp. 59–70.

Espig, M., & de Rijke, K. 2018, 'Energy, anthropology and ethnography: On the challenges of studying unconventional gas developments in Australia', *Energy Research and Social Science*, vol. 45, pp. 214–223.

Espig, M., & de Rijke, K. 2016, 'Navigating coal seam gas fields: Ethnographic challenges in Queensland, Australia', *Practicing Anthropology*, vol. 38, no. 3, pp. 44–45.

Ey, M., Sherval, M., & Hodge, P. 2017, 'Value, identity and place: Unearthing the emotional geographies of the extractive sector', *Australian Geographer*, vol. 48, no. 2, pp. 153–168.

Ey, M., & Sherval, M. 2016, 'Exploring the minescape: Engaging with the complexity of the extractive sector', *Area*, vol. 48, no. 2, pp. 176–182.

Farrugia, D., Hanley, J., Sherval, M., Askland H., Askew, M., Coffey, J., & Threadgold, S. 2018, 'The local politics of rural land use: Place, extraction industries and narratives of contemporary rurality', *Journal of Sociology*, vol. 55 no. 2, pp. 306–322.

Fenton, R. 2013, 'Demystifying science – Communication of complex science to reduce community fear of industry', *Journal of the Australian Petroleum Production & Exploration Association (APPEA)*, vol. 53, no. 1, pp. 295–300.

Fry, M., & Brannstrom, C. 2017, 'Emergent patterns and processes in urban hydrocarbon governance', *Energy Policy*, vol. 111, pp. 383–393.

George, K., & Stratford, E. 2016, 'Oral history and human geography', in *Qualitative Research Methods in Human Geography*, fourth edition, Hay, I. (ed), Oxford University Press, Don Mills, pp. 189–202.

Graham, N., Davies, M., & Godden, L. 2017, 'Broadening law's context: Materiality in socio-legal research', *Griffith Law Review*, vol. 26, no. 4, pp. 480–510.

Hardie, L., Devetak, N. S., & Rifkin, W. 2016, 'Universities in contentious energy debates – Science, democracy and coal seam gas in Australia', *Energy Research & Social Science*, vol. 20, pp. 105–116.

Hesse, A., Baka, J., & Calvert, K. 2016, 'Enclosure and exclusion within emerging forms of energy resource extraction: Shale fuels and biofuels', in *The Palgrave Handbook of the International Political Economy of Energy*, van de Graaf, T., Sovacool, B. K., Ghosh, A., Kern, F., & Klare, M. T. (eds), Palgrave Macmillan, London, pp. 641–660.

George Houen 2018, *Submission to Inquiry into Mineral, Water and Other Legislation Bill*, 27 February 2018, viewed 3 January 2019, https://www.parliament.qld.gov.au/documents/committees/SDNRAIDC/2018/3MinWatOLAB2018/submissions/012.pdf.

Infrastructure, Planning and Natural Resources Committee, 11 October 2017, Transcript of Proceedings, *Public Hearing – Inquiry into the Mineral, Water and Other Legislation Amendment Bill 2017*, viewed 2 January 2019, https://www.parliament.qld.gov.au/documents/committees/IPNRC/2017/MWOLA2017/trns-ph-11Oct2017-MWOLA.pdf.

Jepson, W. 2012, 'Claiming space, claiming water: Contested legal geographies of water in South Texas', *Annals of the Association of American Geographers*, vol. 102, no. 3, pp. 614–631.

Jessup, B., & McIlwraith, C. 2015, 'The sexual legal geography in *Comcare v PVYW*', *University of New South Wales Law Journal*, vol. 38, no. 4, pp. 1484–1506.

Jones, C. A. 2015, 'Frames of law: Targeting advice and operational law in the Israeli military', *Environment and Planning D: Society and Space*, vol. 33, no. 4, pp. 676–696.

Kay, K. 2016, 'Breaking the bundle of rights: Conservation easements and the legal geographies of individuating nature', *Environment and Planning A: Economy and Space*, vol. 48, no. 3, pp. 504–522.

Kearns, R. 2016, 'Placing observation in the research toolkit', in *Qualitative Research Methods in Human Geography*, fourth edition, Hay, I. (ed), Oxford University Press, Don Mills, pp. 313–333.

Keenan, S. 2009, 'Australian legal geography and the search for postcolonial space in Chloe Hooper's *The Tall Man: Death and Life on Palm Island*', *Australian Feminist Law Journal*, vol. 30, no. 1, pp. 173–199.

Kehoe, J. A. 2013, 'The making and implementation of environmental laws in Queensland: the *Vegetation Management Act 1999* (Qld) and the *Land Act 1994* (Qld)', unpublished PhD thesis, Australian National University, viewed 10 January 2019, https://openresearch-repository.anu.edu.au/handle/1885/109352.

Kennedy, A. 2017, *Environmental Justice and Land Use Conflict: The Governance of Mineral and Gas Resource Development*. Routledge, London.

Kocher, A. C. 2017, 'Notice to appear: Immigration courts and the legal production of illegalized immigrants', unpublished PhD thesis, Ohio State University, viewed 10 January 2019, https://etd.ohiolink.edu/pg_10?::NO:10:P10_ETD_SUBID:150691.

Krulin, V., & Grossberg, H. 2017, 'Mining for good law: The QLS mining and resources law committee', *Proctor*, vol. 37, no. 4, pp. 10–11.

Lock the Gate Alliance 2018, *Submission by Lock the Gate Alliance to the State Development, Natural Resources and Agricultural Industry Development Committee*, viewed 3 January 2019, https://www.parliament.qld.gov.au/documents/committees/SDNRAIDC/2018/3MinWatOLAB2018/submissions/006.pdf.

Malin, S. A., & DeMaster, K. T. 2016, 'A devil's bargain: Rural environmental injustices and hydraulic fracturing on Pennsylvania's farms', *Journal of Rural Studies*, vol. 47, pp. 278–290.

Marland Law 2018, *Submission on Mineral, Water and Other Legislation Amendment Bill*, viewed 3 January 2019, https://www.parliament.qld.gov.au/documents/committees/SDNRAIDC/2018/3MinWatOLAB2018/submissions/016.pdf.

Martin, D. G., Scherr, A. W., & City, C. 2010, 'Making law, making place: Lawyers and the production of space', *Progress in Human Geography*, vol. 34, no. 2, pp. 175–192.

Measham, T. G., Fleming, D. A., & Schandl, H. 2016, 'A conceptual model of the socioeconomic impacts of unconventional fossil fuel extraction', *Global Environmental Change*, vol. 36, pp. 101–110.

McGuirk, P. M., & O'Neill, P. 2016, 'Using questionnaires in qualitative human geography' in *Qualitative Research Methods in Human Geography*, fourth edition, Hay, I. (ed), Oxford University Press, Don Mills, pp. 246–273.

Lee McNicholl, 2018, *Submission to Mineral, Water and Other Legislation Amendment Bill 2018*, 27 February 2018, viewed 3 January 2019, https://www.parliament.qld.gov.au/documents/committees/SDNRAIDC/2018/3MinWatOLAB2018/submissions/010.pdf.

Mineral and Energy Resources (Common Provisions) Act 2014 (Qld), current as at 6 December 2016, viewed 2 January 2019, https://www.legislation.qld.gov.au/view/pdf/inforce/2016-12-06/act-2014-047.

Mineral and Energy Resources (Common Provisions) Act 2014 (Qld), current as at 25 October 2018, viewed 5 January 2019, https://www.legislation.qld.gov.au/view/pdf/inforce/2018-10-25/act-2014-047.

Mineral, Water and Other Legislation Amendment Bill 2018 (Qld), viewed 24 December 2018, https://www.legislation.qld.gov.au/view/pdf/bill.first/bill-2018-027.

Mundy, T., & Kennedy, A. 2017, 'Rural and regional legal practice', in *The Place of Practice: Lawyering in Rural and Regional Australia*, Mundy, T., Kennedy, A. & Nielsen, J. (eds), Federation Press, Sydney, pp. 46–63.

Nikiforuk, A. 2015, *Slick Water: Fracking and One Insider's Stand Against the World's Most Powerful Industry*, Greystone Books, Vancouver.

O'Donnell, T. 2016, 'Legal geography and coastal climate change litigation: The Vaughan litigation', *Geographical Research*, vol. 54, no. 3, pp. 301–312.

Oldroyd, G. C. 1993, *Prediction of Natural Gas Production from Coal Seams*, Energy Research and Development Corporation, Canberra.

Prior, J., Crofts, P., & Hubbard, P. 2013, 'Planning, law, and sexuality: Hiding immorality in plain view', *Geographical Research*, vol. 51, no. 4, pp. 354–363.

Protect the Bush Alliance 2018, *Submission to the Mineral, Water and Other Legislation Amendment Bill 2018*, 24 February 2018, viewed 3 January 2019, https://www.parliament.qld.gov.au/documents/committees/SDNRAIDC/2018/3MinWatOLAB2018/submissions/015.pdf.

Pruitt, L. R., & Sobczynski, L. T. 2016, 'Protecting people, protecting places: What environmental litigation conceals and reveals about rurality', *Journal of Rural Studies*, vol. 47, pp. 326–336.

Pruitt, L. R. 2014, 'The rural lawscape: Space tames law tames space', in *The Expanding Spaces of Law: A Timely Legal Geography*, Braverman, I., Blomley, N., Delaney, D., & Kedar, A. (eds), Stanford University Press, California, pp. 190–214.

Queensland Hansard, 15 February 2018, *Record of Proceedings*, viewed 3 January 2019, https://www.parliament.qld.gov.au/documents/hansard/2018/2018_02_15_WEEKLY.pdf.

Queensland Infrastructure, Planning and Natural Resources Committee 2017, *Mineral Water and Other Legislation Amendment Bill 2017*, viewed 3 January 2019 https://www.parliament.qld.gov.au/work-of-committees/former-committees/IPNRC/inquiries/past-inquiries/MWOLA2017.

Queensland Law Society 2018, *Submission for Mineral, Water and Other Legislation Amendment Bill*, 26 February 2018, viewed 3 January 2019, https://www.parliament.qld.gov.au/documents/committees/SDNRAIDC/2018/3MinWatOLAB2018/submissions/005.pdf.

Queensland Parliament 2018, *Report No. 4, 56th Parliament – Mineral, Water and Other Legislation Amendment Bill 2018*, viewed 24 December 2019, https://www.parliament.qld.gov.au/work-of-committees/committees/SDNRAIDC/inquiries/past-inquiries/3MinWatOLAB2018.

Queensland Resources Council 2018, *Submission for Mineral, Water and Other Legislation Amendment Bill*, 27 February 2018, viewed 3 January 2019, https://www.parliament.qld.gov.au/documents/committees/SDNRAIDC/2018/3MinWatOLAB2018/submissions/008.pdf.

Queensland State Development, Natural Resources and Agricultural Industry Development Committee Report 2018, *Mineral, Water and Other Legislation Amendment Bill 2018*, April 2018, Report No. 4, 56th Parliament, April 2018, viewed 24 December 2018, https://www.parliament.qld.gov.au/Documents/TableOffice/TabledPapers/2018/5618T466.pdf.

de Rijke, K., Munro, P., & de Lourdes Melo Zurita, M. 2016, 'The Great Artisan Basin: A contested resource environment of subterranean water and coal seam gas in Australia', *Society and Natural Resources*, vol. 29, no. 6, pp. 696–710.

de Rijke, K. 2013, 'The agri-gas fields of Australia: Black soil, food, and unconventional gas', *The Journal of Culture & Agriculture: Culture, Agriculture, Food and Environment*, vol. 35, no. 1, pp. 41–53.

Roche, M. 2016, 'Historical research and archival sources', in *Qualitative Research Methods in Human Geography*, fourth edition, Hay, I. (ed), Oxford University Press, Don Mills, pp. 225–245.

Salgo, M., & Gillespie, J. 2018, 'Cracking the code: A legal geography and political ecological perspective on vegetation clearing regulations', *Australian Geographer*, vol. 49, no. 4, pp. 483–496.

Scott, R. P. 2016, *Independent Review of the Gasfields Commission Queensland and Associated Matters*, State of Queensland, Department of State Development, Brisbane, July 2016, viewed 7 January 2019, https://cabinet.qld.gov.au/documents/2016/Oct/RevGasComm/Attachments/Report.PDF.

Sherval, M., & Hardiman, K. 2014, 'Competing perceptions of the rural idyll: Responses to threats from coal seam gas development in Gloucester, NSW, Australia', *Australian Geographer*, vol. 45, no. 2, pp. 185–203.

Shine Lawyers 2018, *Submission for Mineral, Water and Other Legislation Amendment Bill 2018*, 27 February 2018, viewed 3 January 2019 https://www.parliament.qld.gov.au/documents/committees/SDNRAIDC/2018/3MinWatOLAB2018/submissions/011.pdf.

Shine Lawyers 2017, *Submission for Mineral, Water and Other Legislation Amendment Bill 2017*, 13 September 2017, viewed 3 January 2019 https://www.parliament.qld.gov.au/

documents/committees/IPNRC/2017/MWOLA2017/submissions/001-Shine%20 Lawyers.pdf.

Skennar, D. 2018, *Advice RE: Proposed Amendments to the Mineral and Energy Resources (Common Provisions) Act 2014*, Chancery Barristers and Mediators, Brisbane, 8 March 2018, 5 January 2019, https://www.parliament.qld.gov.au/documents/committees/ SDNRAIDC/2018/3MinWatOLAB2018/3-tp-9Mar2018-1.pdf.

State Development, Natural Resources and Agricultural Industry Development Committee, 5 March 2018, Transcript of Proceedings, *Public Briefing – Inquiry into the Mineral, Water and Other Legislation Amendment Bill 2018*, Brisbane, viewed 3 January 2019, https://www.parliament.qld.gov.au/documents/committees/SDNRAIDC/ 2018/3MinWatOLAB2018/3-trns-ph5Mar2018.pdf.

State Development, Natural Resources and Agricultural Industry Development Committee, 9 March 2018, Transcript of Proceedings, *Public Hearing – Inquiry into the Mineral, Water and Other Legislation Amendment Bill 2018*, Brisbane, viewed 3 Janauary 2019, https://www.parliament.qld.gov.au/documents/committees/SDNRAIDC/ 2018/3MinWatOLAB2018/3-trns-ph9Mar2018.pdf.

Stratford, E., & Bradshaw, M. 2016, 'Qualitative research design and rigour', in *Qualitative Research Methods in Human Geography*, fourth edition, Hay, I. (ed), Oxford University Press, Don Mills, pp. 117–129.

Terry, W. C. 2009, 'Working on the water: On legal space and seafarer protection in the cruise industry', *Economic Geography*, vol. 85, no. 4, pp. 463–482.

Turton, D. J. 2017, 'Legal determinations, geography and justice in Australia's coal seam gas debate', in *Natural Resources and Environmental Justice: Australian Perspectives*, Lukasiewicz, A., Dovers, S., Robin, L., McKay, J., Schilizzi, S., & Graham, S. (eds), CSIRO Publishing, Clayton South, Victoria, pp. 155–168.

Turton, D. J. 2015a, 'Unconventional gas in Australia: Towards a legal geography', *Geographical Research*, vol. 53, no. 1, pp. 53–67.

Turton, D. J. 2015b, 'Lawyers in Australia's coal seam gas debate: A study of participation in recorded community forums', *The Extractive Industries and Society*, vol. 2, no. 4, pp. 802–812.

Turton, D. J. 2015c, 'Delivering a "new administrative law": Commonwealth-Queensland ombudsman cooperation, 1976–1981', *Journal of Australian Studies*, vol. 39, no. 2, pp. 216–234.

Turton, D. J. 2014, 'Codifying coexistence: Land access frameworks for Queensland mining and agriculture in 1982 and 2010', *Journal of Australasian Mining History*, vol. 12, pp. 172–192.

Walenta, J., in press, 'Courtroom ethnography: Researching the intersection of law, space, and everyday practices', *The Professional Geographer*, https://doi.org/10.1080/ 00330124.2019.1622427.

Walsh, K. B., & Haggerty, J. H. 2019, 'I'd do it again in a heartbeat: Coalbed methane development and satisfied surface owners in Sheridan County, Wyoming', *The Extractive Industries and Society*, vol. 6, no. 1, pp. 85–93.

Witt, K., Kelemen, S., Schultz, H., & Vivoda, V. 2018a, 'Industry and government responses to unconventional gas development in Australia', *The Extractive Industries and Society*, vol. 5, no. 4, pp. 422–426.

Witt, K., Whitton, J., & Rifkin, W. 2018b, 'Is the gas industry a good neighbour? A comparison of UK and Australia experiences in terms of procedural fairness and distributive justice', *The Extractive Industries and Society*, vol. 5, no. 4, pp. 547–556.

14

ENERGISING THE LAW

Greening of fossil fuels and the rise of gendered political subjects

Meg Sherval[1]

Introduction

There are few areas in which the intersection between law and geography is more profound than in the material pursuit of energy. Energy in all its forms has occupied centre stage in the world's evolution for millennia. Globally, it has been deployed in the negotiation of new geographies of exploitation and production, and in the subsequent reworking of economies, infrastructure, governance and communities. It has long been an agent of environmental change as well as an instigator of land use change around which many communities have mobilised. In short, its impacts are complex and multifaceted and have implications for future land, water, food, migration and climate security (Neville et al. 2017).

Over the past 20 years, many nations have actively engaged in the exploration of unconventional or "extreme" energy sources such as tight oil, shale gas and coal bed methane (also known as coal seam gas). Through technological advances, what was once thought economically non-viable has now become feasible and standard practice. So much so, that these newer forms of hydrocarbon are now offering renewed hope to governments in the ongoing search for long-term energy security (Sherval 2015; Sherval 2013). While this is cause for celebration by many in government and industry circles, with shale gas and coal bed methane in particular being framed as "bridging fuels" in the global transition towards a decarbonised future (see: Pierce 2012; Obama 2014; Leadsom 2016; Frydenberg 2016), in the rural areas where these new energy projects are to be situated, a vastly different response is emerging (Sherval et al. 2018; Turton 2015; Kuch & Titus 2014).

One of the problems that exists in the development of energy is quite simply, as Ghosen (2009, p.7) notes, that it "needs space". Thus, spaces or landscapes of extraction not only require access to large tracts of land and water resources, but they are also located in parts of the world that are increasingly valued for their biodiversity (Bridge 2009; Sultana 2011). Likewise, rural areas that have traditionally been associated with other land uses, such as agricultural production or livestock management, are now faced with new challenges as external powers seek to control not only what resources are to be developed, but how they are to be developed and by whom. As such, many rural regions today are becoming sites of opposition to energy development and privatisation as well as places where power imbalances between government, industry and community are made visible (Willow & Keefer 2015; Neville et al. 2017; Malin et al. 2018).

There has long been a complex and contested relationship between energy and the power associated with its ownership. In part, this is due to ancient British property laws that have promoted the separation of people from place and the natural environment, whilst also maintaining ownership and control over nature and the means of production. Even the language of property law is designed to deprive nature of any meaning or value beyond its utility to the human economy (Graham 2012, p.101). Cognisant of this, in 1983, Ivan Illich noted that the concept of energy had been appropriated by economists and others as the ability "to make nature do work" (Illich 1983, p.13). Prior to this, Heidegger also noted poignantly that nature had become "one vast gasoline station for human exploitation" (Heidegger cited in Graham 2012, p.101). Transplanted to colonial Australia, these laws have successfully erased other ways of knowing and understanding nature and being "in place" (Graham 2011; Bawaka et al. 2015; Ey, Sherval & Hodge 2017). As such, "new" energy and its ongoing advancement has a contradictory relationship with the law and civil society especially in debates surrounding land use, individual rights, justice and the altering of place and its associated environment.

This is unsurprising though, as even in its earliest forms, energy (attained through the exploitation of wood, biogas, peat and coal) has historically been entangled with power and has remained the purview of the Crown and/or associated polity. In the UK and Australia today, this remains the case with all land or "property" assumed to be owned by or leased from the Crown (Reed 2011). Equally, this control of land tenure also extends to "free-hold" titles where private ownership is not absolute, with the Crown's representatives empowered to withhold certain rights, such as the right to any mineral or petroleum found on or under the land (Reed 2011). In more recent times, this legacy of separation and detachment has provided governments with the power not only to licence state intervention in land use practices such as establishing energy infrastructure, but also to determine the path this intervention will take. As such, social and environmental justice issues loom large in the development of unconventional energy sources globally (Schlosberg, Rickards & Byrne 2018; Turton 2017; Witt, Whitton & Rifkin 2018).

Given this, and the fact that legal geography advocates for a reconfigured relationship between situated society and the natural world (Graham 2011; Turton 2015), this chapter explores the mobilisation of women as political subjects in the energy debates currently taking place in the UK and in New South Wales (NSW), Australia. It examines some of the motivations driving action within communities and explores the types of themes to emerge from each location. The chapter begins by setting the scene in both an Australian and UK context in terms of the discourse that governments are using in the promotion of shale and coal seam gas as "bridging fuels". It then moves on to discuss why women were the focus of this research and how different qualitative methods were used to inform the research. It then moves to explore the common themes to emerge from the different case study sites. Overall, the chapter serves to highlight that when it comes to decisions about the future of rural land use, communities will not easily be silenced nor give up their rights to peaceful protest. As more and more rural landscapes become industrialised and begin to be visualised by governments as future energy hubs, governments (as agents of the law) need to recognise that transparent engagement and ongoing accountability are essential keys in defusing land use conflict.

Energy as "crisis" or in crisis?

In both the UK and Australia, governments have framed arguments about shale and coal seam gas (CSG) exploration and production around securing the nation's energy future through the lens of "crisis". In Australia, arguments over energy security have raged for years with one of the main issues of contention being the paradoxical situation in which the nation finds itself. On one hand, it is a world leader in the exportation of gas (about to surpass Qatar) whilst on the other, it is struggling to supply enough gas for domestic purposes at a price that the average consumer can afford (Ripple 2014; Chambers & Creighton 2016). While it has been predicted by the federal Office of the Chief Economist that export volumes of liquefied natural gas (LNG) are set to increase by 28.5% for the period 2017–2018 and beyond, some industry experts have posited that Australia may actually need to import gas for its own uses given that the vast bulk of its supply is exported elsewhere (Office of the Chief Economist in the Department of Industry, Innovation and Science 2016; Cullinane 2017).

In the UK, questions about energy security and the nation's growing dependence on foreign imported energy have featured regularly in the nation's media (see Macalister 2010; *Economist* 2014; Watt 2014; Ambrose 2016; Vaughan 2017). As a way of addressing this concern, former British Prime Minister David Cameron declared that his government was "going all out for shale" (cited in Watt 2014). Likewise, Prime Minister Theresa May stated that her government's priority was to fully develop "the shale industry in Britain" (cited in Vaughan 2017). In support of this vision, and in an effort to negate rising environmental and community concerns, leading UK Government ministers such as Andrea Leadsom from the Department of Energy and Climate Change have publicly stated that "gas is the cleanest fossil fuel and shale gas can provide an effective 'low Carbon Bridge' while we move to renewable energy" (2016).

Statements of this nature have also been touted globally and numerous policies have been produced to fast-track the development of shale and coal seam gas reserves. Essentially, gas is seen as a "bridge fuel", as it is largely recognised as a short-term substitute for aging coal-fired power stations (Kirkland 2010). Nationally, many governments also see gas as a viable "option for cutting power plant emissions and addressing global warming in the short term" (Kirkland 2010) – hence, the wide-spread acceptance and up-take of the "bridge" metaphor. Nowhere, however, is there any attempt to answer the question that preoccupies communities, environmentalists and concerned citizens alike. The question is: how long is the bridge to a renewable future likely to be?

Authors such as Pierce (2012, p.1) have suggested that "the natural gas bridge to carbon-free fuels is likely to be extremely long, at least decades and probably a century", while others such as Levi (2013, p.609) state that:

> In the context of the most ambitious [climate] stabilization objectives (450 ppm CO_2), and absent carbon capture and sequestration, a natural gas bridge is of limited direct emissions-reducing value, since that bridge must be short.

These conflicting views add weight to concerns that resource developments like shale or CSG may not be sufficient in limiting greenhouse gas emissions, and neither will they assist in helping nations meet strict climate change targets. Acknowledging this, the UK House of Commons' Energy and Climate Change Committee (HCECCC) has said that "creating a sustainable energy market will not be the product of one single innovation" and in fact, trying to meet this challenge has created what it has termed the "energy trilemma" (HCECCC 2016, p.6). This involves trying to meet three goals: ensuring energy security, ensuring affordability for consumers and meeting long-term decarbonisation goals (HCECCC 2016, pp.6–7). In a report on the future security of the national electricity market, the Australian Chief Scientist suggested that Australia is also "at a critical turning point" where if its energy is poorly managed, then its long-term future will be "less secure, more unreliable and potentially very costly" (Finkel 2017, p.3). Thus, innovation and diversification of the energy mix are considered to be the key, with the ultimate goal being the orderly transition to a low carbon society (Finkel 2017, p.33).

Despite the acknowledgement by both national governments that a transition away from fossil fuels is needed long-term, it is the different paths offered that have seen a rapid mobilisation of concerned citizens in both countries. In Australia, this has been primarily in response to government attempts to expand exploration, increase the infrastructure footprint and to further develop coal and CSG operations in rural spaces. In the UK, concerns have been specifically associated with the development of shale, biofuels and nuclear reactors, once again slated for rural places. Before discussing the reasons behind mobilisation and the rise in community dissent, however, the next section of this chapter outlines the methodology used to undertake this research.

Methodology

This research emerged from an initial observation back in 2014 that women appeared to be playing a key role in leading resistance movements around shale and CSG development in Western countries such as Britain, the USA and Australia. It should be noted, however, that this is also a growing phenomenon in the Global South (see Arenas 2015; Jenkins 2015; Sultana 2011). A desire to investigate this further, and to unpack the motivations for this in Western countries specifically, led me to undertake qualitative research in the UK (research in the USA is to come at a later date). The experiences of communities in the UK were to be used as a case study and compared with fieldwork previously completed by the author and colleagues in Australia (see Sherval & Graham 2013; Sherval & Hardiman 2014; Farrugia et al. 2018; Sherval et al. 2018). As such, I spent four and a half months in the UK (2016–2017), living in County Durham but visiting actual and potential drill sites in Lancashire and Yorkshire in the north and in Somerset, Sussex, Hampshire and Surrey in the south (see Figure 14.1). In Australia, previous fieldwork had been undertaken in Narrabri and Gloucester in central NSW (2014–2016) (see Figure 14.2). These field site

FIGURE 14.1 Map of UK county site visits. Source: author provided.

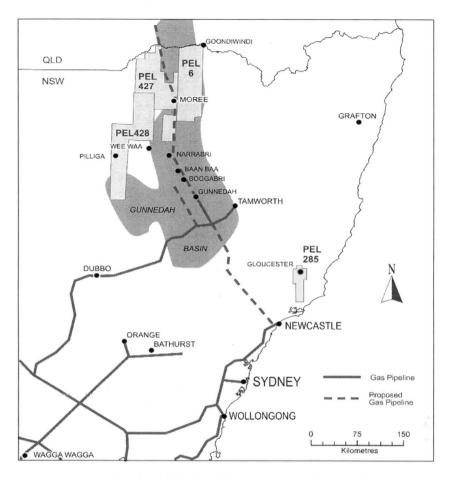

FIGURE 14.2 Map showing NSW site visit areas and petroleum exploration licence areas. Source: author provided.

visits often coincided with protests, marches or camps that were occurring, where I undertook "participant observation" – I became a participant in the context being observed.

Participant observation has long been used in the production of knowledge. Over the years though, the ways it has been approached and employed in research have varied significantly. Confirming this, Rhoads and Wilson (2010, p.27) note that observation as a technique is considered to be a "fluid and dynamic process, a specific practice embedded in a communal sense". In this research, it is embedded in a communal sense of place. Being "in place" or, for Indigenous peoples, being "on" or "with country" acknowledges people's reciprocal relationships and engaged experiences with land and the environment (Bawaka et al. 2015). Thus, observation of places, fracking sites, rallying points where people meet and share their experiences and perceptions, and what Perry (2012) has called their "collective trauma", is an insightful method of engagement. As Denzin and Lincoln (2018, p.10) note, "qualitative research is a situated activity that locates the observer in the world. [It] consists of a set of interpretive, material practices that make the world visible". So, in joining in the holding of banners, marching, singing, listening and participating in discussions and introductions, I was also able to gain a level of acceptance and trust within different activist communities. As Bell et al. (2018, p.137) advise, when we become "engaged witnesses", this enables a richer understanding of the entangled world of both the human and non-human in these spaces.

These observations (which were compiled in a fieldwork diary) were also supplemented by semi-structured interviews with people involved in opposition campaigns and those living in close proximity to actual and potential drilling sites. After gaining ethics approval, contact was initially made via e-mail to publicly available websites such as "Frack off London" and "350.org" seeking participants willing to be interviewed (both are direct-action, grass-roots movements opposed to new coal, oil and gas projects and companies reinvesting in fossil fuels). I was successful in gaining interviewees from these sites who then offered the contact details of others involved in "Frack off" and "Nanna" movements around the UK. This type of "snow-balling" or "referral-sampling" technique relies on tapping into participants' social networks and is a commonly used qualitative method (Browne 2005). From here, I contacted other individuals and groups such as "Frack Free Lancashire", "Frack Free North Yorkshire", "Leith Hill Action Group" and "Frack Free North Somerset" and arranged face-to-face, recorded interviews. After the interviews were transcribed, each participant was given a pseudonym. In total, 14 women were interviewed throughout the UK and 27 in NSW.

As legal geography desires to make the "invisible visible" (Bartel et al. 2013, p.339), these methods, including the use of semi-structured interviews, were deemed the most appropriate methods for engaging with individuals, as they are less formal and intimidating and allow participants themselves to choose locations where they feel most comfortable to speak. Thus, some interviews took place in hotel dining rooms, cafes and homes while others were spent walking and pointing out where fracking sites were to be located. In qualitative research, the "walking interview" is becoming more popular as researchers seek to explore the link between self and place (Jones 2011). Often referred to as "mobile methods" they are used to "reduce the power imbalance and encourage spontaneous conversation" (Kinney 2017, p.1) between participant and researcher. They also assist in building rapport with participants as they allow them greater capacity to shape how an issue is discussed and through what lens their environment and understandings of place are visualised and articulated. Interviews also "capture social life as it happens", as Dowling, Lloyd and Suchet-Pearson (2016, p.5) note. Whether that be in the spaces of action during protests, marches or camps, or following these events, interviews allow for deeper engagement and reflection, as well as an enabling of space where expression of emotion and even distress is possible (Ey 2018).

In the sites where extraction is already occurring or slated to occur, it was possible to see and document not just frustration and concern, but also palpable distress as people questioned the logic used by governments to harness "extreme energy". For many, the risk of such a venture, when the possibility of long-term environmental and social harm was deemed "very high" (Short et al. 2015) caused a complete lack of faith and trust in authority. The law's traditional ignorance of "both its spatial and social contexts" (Bartel et al. 2013, p.340) therefore makes research carried out by a woman, about "women's concerns' over the development of unconventional gas and the exploitation of power by governments and industry, very timely. Given that the field of legal geography desires to highlight law's 'material contexts'" (Bartel et al. 2013, p.341) and seeks where possible to "reform" the law when it is failing its citizens, then case studies such as these help provide evidence of a need for change.

To ensure that the findings of this research were rigorously examined, "thematic analysis" was used to report on the emergent and reoccurring themes found in the data sets. Participants in both the UK and Australia were asked similarly themed questions on: connections to place and community, observed land use change, agents of change (including government and industry), reasons for their activism and their visions for the future. As the questions were open and broad in nature, this allowed participants to elaborate on their own experiences and knowledge of changes in their region, and to articulate their thoughts about current and future land uses. Using NVivo software to aid in sorting and organising the various data sets, key word codes were established and identified in the transcripts of interviews.

Interestingly, despite the geographic distance between the interviewed communities, there were more themes in common than not. The main differences tended to be shaped by whether participants lived in a drought-prone area (where concerns about water resources were highlighted) or whether or not one was involved in agricultural production (where a concern about the continued viability of soils was highlighted). Overall though, the themes that tended to dominate all data sets were connected by concerns about: a lack of transparency by government and industry and the consequences of that, a concern about risk associated with fracking and land use change and overall, a concern about how these new energy pursuits were going to address sustainability and climate change issues into the future. These themes are represented in Table 14.1 and discussed further below, where it is also possible to see the political nature of these issues and recognise in them the "coproduction of subjects and places" as Gibson-Graham (2005, p.131) notes.

Activism and the politics of place

When shale and CSG operations are initially proposed, they are usually framed by governments as providing multiple benefits and being able to "co-exist"

TABLE 14.1 Participant Concerns

Core themes	Common concerns	Evidence
Lack of transparency		
	Loss of trust/ faith in government.	"Governments are the puppets of industry and communities are the ones who are bearing the brunt of their decisions. I have no faith in them anymore" (Suzanna, Teacher, Kirby Misperton, Yorkshire 18 November 2016).
		"Trust is a big word, I am worried by the relationship between government and industry because big bucks talk power … I'm frightened of what could happen, after all, it's our long-term future we're talking about here. No one seems to be representing us" (Denise, Cotton Farmer, Narrabri 17 October 2015).
	Lack of consultation.	"They're essentially paving the way for the fracking industry in our area and that's not democracy, that's not representing your constituents … 92 percent of local people were against Shale coming here and still the local MP, said 'it's coming and you may as well get used to it'" (Joanne, "Yorkshire Nanna", Ryedale, Yorkshire 19 November 2016).
		"The problem here is that the company did not consult with anyone. They wanted to save time and costs. Also, that they got permission to explore in an officially listed 'Area of Outstanding Natural Beauty' is beyond anyone's comprehension. These areas are supposed to be protected" (Jane, Teacher, Leith Hill Action Group, Surrey 26 August 2016).
	No social licence.	"We've all written letters about no social licence, we've had posters done, all sorts of things … No one will stand up and go against the party line. And the worse thing is that they lie about it. They have no conscience and so-called 'consultation' is a joke!" (Louisa, 'Frack-Off Somerset', 9 December 2016).
Risk		
	Lack of scientific understanding of impacts.	"Even though there's a billion reports out there that show how dangerous it is, how can governments just completely ignore it and continue to put local communities under all this stress, it's a disgrace" (Sara, Mother and Farmer, Kirby Misperton, Yorkshire 18 November 2016).
	Threat to ongoing water security.	"The Great Artesian Basin is our lifeblood, and if we lose our groundwater, we simply cannot exist here. Not just the farmers, but communities, towns, vast areas of inland Australia will be uninhabitable. It is our only permanent water supply" (Anne, Farmer, Narrabri NSW, 6 October 2015).

(Continued)

TABLE 14.1 (Continued)

Core themes	Common concerns	Evidence
	Threat to future food security.	"Around us, it's a big agricultural area and we have a lot of tourists who come for the food … there are organic farms, farmers markets … we are now known as one of the food capitals. I can't believe that something like fracking is coming and if it gets a hold, it will industrialise everywhere. I mean who's going to want to come? Nobody's going to want to come to see that and what about our future?" (Joanie, "Yorkshire Nanna", Ryedale, Yorkshire 19 November 2016).
	Threat to rural places, amenity and the environment.	"Would you invest hundreds of thousands of dollars in an area like this if you knew there was going to be a massive hole and gas wells from one end of the valley to other?" (Joyce, Dairy Farmer, Gloucester, NSW 12 July 2014). "We came for a rural change not an industrial change, all our plans are now in ruins and what about the environment? It's a crying shame" (Mary, "Gloucester Nanna", NSW 13 July 2014). "It seems that it's now a crime to care about the destruction of the natural environment. If we don't care, then who will?" (Tanya, "Lancashire Nanna", Preston 12 October 2016).
The Future	Climate change.	"Gas as a bridge or transition fuel, what a load of rubbish! It is catastrophic climate change we are heading towards. You know we've signed the COP 21 agreement in Paris to bring our emissions down? Plus, we've got the Climate Change Act of 2008, it's impossible to meet those targets if we have shale" (Lyn, "Frack-off London", 24 August 2016).
	Long-term sustainability.	"I like to hope that all levels of government will see the writing on the wall and act in a manner that is sustainable that does consider future generations, because to my mind, all this is about those things that are priceless. It's about clean air, clean water, and land to grow clean healthy food. I mean, we only have one Earth" (Caitlin, Farmer, Narrabri 3 October 2015).
	Employment and investment in alternative energy sources.	"The government has kept its foot on the pressure cooker lid of renewables for an age, they can only keep it on there for so long before their foot's going to get blown off! I think renewables have to overtake them … That's where I see the future going because if we keep going on the current trajectory of greenhouse gases … there is no other workable answer" (Sharon, Store-keeper, Narrabri 11 October 2018).

alongside other land uses and land values (Witt, Whitton & Rifkin 2018). Where this vision comes undone, however, is when "host" communities perceive that governments (at all levels)[2] do not have their best interests at heart, and that what they are being asked to accept or "surrender/sacrifice" is not in keeping with the notions of procedural fairness and distributive justice (Witt, Whitton & Rifkin 2018; Whitton et al. 2017). As Cotton (2016, p.8) notes:

> though fracking is determined primarily by the geographic pattern of shale [or CSG] resources, the prioritisation of certain places as extraction sites involves an element of normative political judgement.

He illustrates this further by quoting the former British Energy Secretary Lord Howell, who in 2013 suggested that shale gas was suitable for "desolate" regions of the country, those places he described as "unloved" and "not environmentally sensitive" such as Lancashire (Howell in Cotton 2016, p.8). This is a clear example of regional place-based inequity where those in power and external to an area cannot recognise its value. In Narrabri shire in NSW, many participants spoke of these types of comments as continued evidence of an urban–rural (policy) divide where decisions are made at a distance by city-centric government officials who often have little or no comprehension of their impact locally (Askland et al. 2016). This appears to be a common reality in both the UK and Australia, particularly where those vested with decision-making power are located far away from the results of their decisions. This, however, is only one of many areas of concern shared amongst communities, environmentalists and concerned citizens about the motivations of government to hastily develop unconventional energy sources and choose "appropriate" sites.

As suggested by the comments in Table 14.1, women, in particular, continue to express their concern about threats not only to the places where they live and work, but also to the natural environment itself. Beyond that, they are also concerned about the impacts of climate change and about how to ensure intergenerational equity for their grandchildren and others coming after them. They are concerned not only with defending the places and spaces they consider valuable, but also in responding to the broader issues (such as climate and intergenerational justice) that governments seem reluctant to confront. All participants suggested that such was their commitment to seeing these issues addressed, that often it required them to step out of their comfort zone and "speak up" or "front up" to protest events (see Figure 14.3). For many, standing up was not seen as a choice – it was a requirement of being a good citizen and a good steward in the face of highly crude and divisive contestation around how spaces and places are to be envisaged for the future. Essentially, they saw their participation as one way of tackling what Woods has called (2011, p.259) the "neoliberal rationalities" seeking to "challenge, reassess and reorientate" rural spaces, often without consultation or with respect to policies that once offered environmental protection measures to places as Jane notes in relation to Leith Hill (see Table 14.1).

FIGURE 14.3   Speaking up at an anti-fracking rally in Manchester. Source: author provided.

Most apparent was the wide-spread sense of betrayal felt by many. These feelings were directed against all levels of government, from the local to the national, and it was widely believed that government had abandoned their commitments to meet looming climate change targets and investment in renewables in favour of a metaphor which no one believed in. As Lyn from "Frack-off London" noted in Table 14.1, "Gas as a bridge or transition fuel, what a load of rubbish!". Accompanying these perspectives was also a disquiet about the growing securitisation of drilling sites, the policing of anti-fracking events and the recent prosecution and jailing of protesters (Short & Szolucha 2019). Many felt that this new show of force by government was meant to intimidate and silence criticism, but in reality, it only served to bring into question the government's commitment to protecting the civil rights of all its citizens, rather than just those of the powerful (see Matthews-King 2018).

Ultimately, what this research revealed was that no matter where participants were located, their concern was palpable on many levels. Most, who had never been activists before, saw their participation as a civic duty. Such was the seriousness afforded this role, that all participants spoken to were articulate and well-read on the issues locally, nationally and internationally. They wanted to ensure that, by their actions, there would still be a sustainable future to be had in their respective regions. Nevertheless, the act of "standing up" for many was not easy, and various people spoke of the exhaustion of "keeping up the fight". All, however, remained resolute in their desire to see hydraulic fracturing and the development of shale and CSG bought to a stop. While it is generally accepted amongst participants that meaningful consultation is the goal, there is a widespread desire that other local values be acknowledged, respected and adhered to by both governments and industry alike.

Conclusion

In the areas around the globe where new extractive practices are occurring, place emerges as a site of not only political activism, but also of social transformation (Gibson-Graham 2005, p.131). In legal geography scholarship, case studies of this nature help enlarge our appreciation of how highly politicised materialities such as energy development are intimately connected with both the application and consequences of legal decision-making on the ground in different geographical locations. While critics remain content to dismiss community concerns and the rise of gendered political subjects as mere "NIMBYism" (Sebastien 2017), confrontations over different land uses in reality prove to be much more complex than this. As Willow and Keefer (2015, p.93) suggest

> women's motives for grassroots environmental engagement cannot be reduced to any single or simple expression or refutation of traditionally gendered expectations [or] identities.

Women are in fact called to action "by a dynamic constellation of concerns" and as such, women "understand the catalysts for their actions and the ultimate goals of their ongoing work" (Willow & Keefer 2015, p.93). As Miller (2002, p.217) also notes, when external forces confront and disrupt embodied relations, around the "three cornerstones of belonging – history, people and place", then resistance has to be an expected outcome. Whether governments choose to acknowledge it or not, opposition towards shale and coal seam gas is not waning, and no amount of rhetoric around fossil fuels being "green bridges" to the future will change that. Effective change will only come when governments accept that some communities have their own visions for their future. While this might not be how governments planned to address energy security, new ways of thinking might just be possible if we allow the contradiction and messiness that constitute the world(s) we navigate to take hold (Ey, Sherval & Hodge 2017). In the world of new energy geographies, perhaps the time has finally come for governments to seriously engage with communities and invite them in as partners to help find energy solutions. By empowering communities, a form of "civic engagement" can occur which promotes shared visions, rather than only a single vision, and this surely has to offer a better way towards a decarbonised future than what currently exists.

Notes

1 The author wishes to sincerely thank all of those who participated in her research projects in the UK and Australia. All quotations from participants used throughout the chapter are with permission. Many thanks also to Olivier Rey-Lescure, cartographer at the University of Newcastle, who provided mapping assistance for this chapter. Special thanks to Melina Ey who assisted with the initial peer review of this chapter.
2 Australia has a three-tiered system of government (Federal/National, State and Local). Primary responsibility for the approval of CSG and coal mining activities, and regulation of any environmental impacts associated with the resource sector, rests mainly with state and territory governments. The Australian federal government will only become involved if a CSG development or large coal mine is likely to have a significant impact on a water resource or other matter of national environmental significance. In the UK, while every county and district has a local government that is responsible for a range of environmental matters and community services, by and large, local governments have very few legislative powers and must act within the framework of laws passed by the central Parliament.

References

Ambrose, J. 2016, 'Why the UK is using less energy, but importing more – and why it matters', *The Telegraph*, 18 August, viewed 13 November 2018, <http://www.tele graph.co.uk/business/0/why-the-uk-is-using-less-energy-but-importing-more---and-why-it/>.

Arenas, I. 2015, 'The mobile politics of emotions and social movement in Oaxaca, Mexico', *Antipode*, vol. 47, no. 5, pp. 1121–1140.

Askland, H. H., Askew, M., Hanley, J., Sherval, M., Farrugia, D., Threadgold, S., & Coffey, J. 2016, *Local Attitudes to Changing Land Use – Narrabri Shire*, Full Report for the New South Wales Department of Primary Industries, <https://www.new castle.edu.au/__data/assets/pdf_file/0006/336768/CSRRF_Narrabri-Report_Dece mber2016_240217_Optimised.pdf>.

Bartel, R., Graham, N., Jackson, S., Prior, J. H., Robinson, D. F., Sherval, M., & Williams, S. 2013, 'Legal geography: An Australian perspective', *Geographical Research*, vol. 51, no. 4, pp. 339–353.

Bawaka Country, Wright, S., Suchet-Pearson, S., Lloyd, K., Burarrwanga, L., Ganambarr, R., Ganambarr-Stubbs, M., Ganambarr, B., Maymuru, D., & Sweeney, J. 2015, 'Co-becoming Bawaka: Towards a relational understanding of place/space', *Progress in Human Geography*, vol. 40, no. 4, pp. 455–475.

Bell, S., Instone, L., & Mee, K. 2018, 'Engaged witnessing: Researching with the more-than-human', *Area*, vol. 50, no. 1, pp. 136–144.

Bridge, G. 2009, 'The hole world: Scales and spaces of extraction', *New Geographies*, vol. 2, pp. 43–48.

Browne, K. 2005, 'Snowball sampling: Using social networks to research non-heterosexual women', *International Journal of Social Research Methodology*, vol. 8, no. 1, pp. 47–60.

Chambers, M., & Creighton, A. 2016, 'Benefits of LNG boom will fall to agile', *The Australian*, 28 May, viewed 17 November 2018, <http://www.theaustralian.com.au/business/mining-energy/benefits-of-lng-boom-will-fall-to-agile/news-story/cfd9 4b7ca6073fca6517a4846265d375>.

Cotton, M. D. 2016, 'Fair fracking? Ethics and environmental justice in United Kingdom shale gas policy and planning', *Local Environment: The International Journal of Justice and Sustainability*, vol. 22, no. 2, pp. 185–202.

Cullinane, B. 2017, 'The Australian oil and gas paradox: Will the world's largest LNG exporter become an importer?' *Business News*, 13 February, viewed 17 November 2018, <https://www.businessnews.com.au/article/The-Australian-oil-and-gas-par adox-Will-the-world-s-largest-LNG-exporter-become-an-importer>.

Denzin, N. K., & Lincoln, Y. S. (eds) 2018, *The Sage Handbook of Qualitative Research*, fifth edition, Sage, Los Angeles.

Dowling, R., Lloyd, K., & Suchet-Pearson, S. 2016, 'Qualitative methods I: Enriching the interview', *Progress in Human Geography*, vol. 40, no. 5, pp. 679–686.

Ey, M. 2018, '"Soft, airy fairy stuff"? Re-evaluating "social impacts" in gendered processes of natural resource extraction', *Emotion, Space and Society*, vol. 27, pp. 1–8.

Ey, M., Sherval, M., & Hodge, P. 2017, 'Value, identity and place: Unearthing the emotional geographies of the extractive sector', *Australian Geographer*, vol. 48, no. 2, pp. 153–168.

Farrugia, D., Hanley, J., Sherval, M., Askland, H., Askew, M., Coffey, J., & Threadgold, S. 2018, 'The local politics of rural land use: Place, extraction industries and narratives of contemporary rurality', *Journal of Sociology*, vol. 55, no. 2, pp. 1–17.

Finkel, A. 2017, *Independent Review into the Future Security of the National Electricity Market: Blueprint for the Future*, Department of Energy and Environment, Commonwealth of Australia, Canberra.

Frydenberg, J. 2016, COAG Energy Council Meeting, Canberra, August 19, viewed 06 July 2017 <https://www.environment.gov.au/minister/frydenberg/media-releases/mr20160725.html>.

Ghosen, R. 2009, 'Energy as a spatial project', *New Geographies*, vol. 2, pp. 1–10.

Gibson-Graham, J. K. 2005, 'Building community economies: Women and the politics of place', *Women and the Politics of Place*, Harcourt, W., & Escobar, A. (eds), Kumarian Press, Bloomfield, 130–157.

Graham N. 2011, *Lawscape: Property, Environment and Law*, Routledge, Oxford.

Graham N. 2012, 'Dephysicalisation and entitlement: Legal and cultural discourses of place as property', in *Environmental Discourse in Public and International Law*, Jessup, B., & Rubenstein, K. (eds), Cambridge University Press, Melbourne, pp. 96–119.

Heidegger, M. 2012, cited in Graham N. 2012, "Dephysicalisation and entitlement: Legal and cultural discourses of place as property', in *Environmental Discourse in Public and International Law*, Jessup, B., & Rubenstein, K. (eds), Cambridge University Press, Melbourne, pp. 96–119.

House of Commons, Energy and Climate Change Committee 2016, 'The energy revolution and future challenges for UK energy and climate change policy', *Third Report of Session 2016–2017*, HC 705, London, viewed 19 November 2018, <https://publications.parliament.uk/pa/cm201617/cmselect/cmenergy/705/705.pdf>.

Illich, I. 1983, 'The social construction of energy', in *New Geographies: Landscapes of Energy*, Ghosn, R. (ed), no. 2, pp. 11–19.

Jenkins, K. 2015, 'Unearthing women's anti-mining activism in the Andes: Pachamama and the "mad old woman"', *Antipode*, vol. 47, no. 2, pp. 442–460.

Jones, O. 2011, 'Geography, memory and non-representational geographies', *Geography Compass*, vol. 5, no. 12, pp. 875–885.

Kinney, P. 2017, 'Walking interviews', *Social Research Update*, vol. 67, pp. 1–4.

Kirkland, J. 2010. 'Natural gas could serve as 'bridge' fuel to low-carbon future', *Scientific American*, 25 June 2010, viewed 19 November 2018, <https://www.scientificame rican.com/article/natural-gas-could-serve-as-bridge-fuel-to-low-carbon-future/>.

Kuch, D., & Titus, A. 2014, 'Emerging dimensions of networked energy citizenship: The case of coal seam gas mobilisation in Australia', *Communication, Politics & Culture*, vol. 47, no. 2, pp. 35–59.

Leadsom, A. 2016, 'Government's vision for shale gas in securing home grown energy supplies for the UK', *Speech at the Shale World Conference in London*, May 25 2016, viewed 09 November 2019, <https://www.gov.uk/government/speeches/govern ments-vision-for-shale-gas-in-securing-home-grown-energy-supplies-for-the-uk>.

Levi, M. 2013, 'Climate consequences of natural gas as a bridge fuel', *Climate Change*, vol. 118, no. 3–4, pp. 609–623.

Macalister, T. 2010, 'Government must "take back control" of North Sea oil and gas production', *The Guardian*, 7 March 2010, viewed 05 November 2018, <http://www .guardian.co.uk/business/2010/mar/07/oil-gas-production-north-sea>.

Malin, S. A., Opsal, T., O'Connor Shelley, T., & Mandel Hall, P. 2018, 'The right to resist or a case of injustice? Meta-power in the oil and gas fields', *Social Forces*, vol. 97, no. 4, pp. 1–27.

Mathews-King, A. 2018, 'Fracking protesters "absurdly harsh" jail sentences spark calls for judicial review backed by hundreds of scientists', *The Independent*, 1 October 2018, viewed 2 October 2018 <https://www.independent.co.uk/environment/fracking -protest-cuadrilla-preston-lancashire-drilling-simon-roscoe-blevins-richard-robe rts-rich-a8562276.html>.

Miller, L. 2002, 'Belonging to country – A philosophical anthropology', *Journal of Australian Studies*, vol. 27, no. 76, pp. 215–233.

Neville, K. J., Baka, J., Gamper-Rabindran, S., Bakker, K., Andreasson, S., Vengosh, A., Lin, A., Nem Singh, J., & Weinthal, E. 2017, 'Debating unconventional energy: Social, political, and economic implications', *Annual Review of Environment and Resources*, vol. 42, pp. 241–266.

Obama, B. 2014, 'Full transcript: Obama's 2014 state of the union address' *Washington Post*, January 29 2014, viewed 09 November 2018, <https://www.washingtonpos t.com/politics/full-text-of-obamas-2014-state-of-the-union-address/2014/01/ 28/e0c93358-887f-11e3-a5bd-844629433ba3_story.html?utm_term=.434d6f346 048>.

Office of the Chief Economist in the Department of Industry, Innovation and Science 2016, *Resources and Energy Quarterly – December 2016 Forecast Data*, viewed 10 June 2017, <https://www.industry.gov.au/data-and-publications/resources-and-energy -quarterly-all/resources-and-energy-quarterly-december-2016>.

Perry, S. L. 2012, 'Development, land use, and collective Trauma: The marcellus shale gas boom in rural Pennsylvania', *Culture, Agriculture, Food and Environment*, vol. 34, no. 1, pp. 82–92.

Pierce, R. J. 2012, 'Natural gas: A long bridge to a promising destination', *Utah Environmental Law Review*, vol. 32, no. 2, pp. 245–252.

Reed, H. 2011, 'Mining', in *The Environmental Law Handbook: Planning and Land Use in New South Wales*, fifth edition, Farrier, D., & Stein, P. L. (eds), Thomson Reuters, Pyrmont, Sydney, pp. 717–768.

Rhoads, B. L., & Wilson, D. 2010, 'Observing our world', in *Research Methods in Geography: A Critical Introduction*, Gomes, B. & Jones, J. P. (eds), Wiley-Blackwell, Chichester, pp. 26–40.

Ripple, R. D. 2014, *The Geopolitics of Australian Natural Gas Development*, Research Project for James A. Baker III Institute for Public Policy of Rice University, Houston.

Schlosberg, D., Rickards, L., & Byrne, J. 2018, 'Environmental justice and attachment to place: Australian cases', in *The Routledge Handbook of Environmental Justice*, Holifield, R., Chakraborty, J., & Walker, G. (eds), Routledge, United Kingdom, 591–602.

Sebastien, L., 2017, 'From NIMBY to enlightened resistance: A framework proposal to decrypt land-use disputes based on a landfill opposition case in France', *Local Environment: The International Journal of Justice and Sustainability*, vol. 22, no. 4, pp. 461–477.

Sherval, M., Askland, H. H., Askew, M., Hanley, J., Farrugia, D., Threadgold, S., & Coffey J. 2018, 'Farmers as modern-day stewards and the rise of new rural citizenship in the battle over land use', *Local Environment: The International Journal of Justice and Sustainability*, vol. 23, no. 1, pp. 100–116.

Sherval, M. 2013, 'Arctic Alaska's role in future United States energy independence', *Polar Geography*, vol. 36, no. 4, pp. 305–322.

Sherval, M. 2015, 'Canada's oil sands: The mark of a new "oil age" or a new threat to Arctic security?' *The Extractive Industries and Society*, vol. 2, no. 2, pp. 225–236.

Sherval, M., & Graham, N. 2013, 'Missing the connection: How SRLU Policy fragments landscapes and communities in NSW', *Alternative Law Journal*, vol. 38, no. 3, pp. 176–180.

Sherval, M., & Hardiman, K., 2014, 'Competing perceptions of the rural Idyll: Responses to threats from coal seam gas development in Gloucester, NSW, Australia', *Australian Geographer*, vol. 45, no. 2, pp. 185–203.

Short, D., Elliot, J., Norder, K., Lloyd-Jones, E., & Morley, J. 2015, 'Extreme energy, "fracking" and human rights: A new field for human rights assessments'? *International Journal of Human Rights*, vol. 19, no. 6, pp. 697–736.

Short, D., & Szolucha, A. 2019, 'Fracking lancashire: The planning process, social harm and collective trauma', *Geoforum*, vol. 98, pp. 264–276.

Sultana, F. 2011, 'Suffering *for* Water, suffering *from* water: Emotional geographies of resource access, control and conflict'. *Geoforum*, vol. 42, pp. 163–172.

The Economist 2014, 'The North Sea – Running on Fumes', 1 March 2014, Author, viewed 17 November 2018, <https://www.economist.com/britain/2014/03/01/running-on-fumes>.

Turton, D. J. 2015, 'Unconventional gas in Australia: Towards a legal geography', *Geographical Research*, vol. 53, no. 1, pp. 53–67.

Turton, D. J. 2017, 'Legal determinations, geography and justice in Australia's coal seam gas debate', in *Natural Resources and Environmental Justice: Australian Perspectives*, Lukasiewicz, A., Dovers, S., Robin, L., McKay, J., Schilizzi, S., & Graham, S. (eds), CSIRO Publishing, Clayton South, Victoria, pp. 155–168.

Vaughan, A. 2017, 'Boom in renewables weakens fracking's case in UK says Tory MP', *The Guardian Online*, 27 November 2017, viewed 13 November 2018, <https://www.theguardian.com/environment/2017/nov/26/boom-in-renewables-weakens-frackings-case-uk-tory-mp-north-sea-oil-gas-shale>.

Watt, N. 2014, 'Fracking in the UK: "We're going all out for shale," admits Cameron', *The Guardian*, 13 January 2014, viewed 13 November 2018, <https://www.theguardian.com/environment/2014/jan/13/shale-gas-fracking-cameron-all-out>.

Whitton, J., Brasier, K., Charnley-Parry, I., & Cotton, M. 2017, 'Shale gas governance in the United Kingdom and the United States: Opportunities for public participation and the implications for social justice', *Energy Research & Social Science*, vol. 26, pp. 11–22.

Willow, A. J., & Keefer, S. 2015, 'Gendering extraction: Expectations and identities in women's motives for shale energy opposition', *Journal of Research in Gender Studies*, vol. 5, pp. 93–120.

Witt, K., Whitton, J., & Rifkin, W., 2018, 'Is the gas industry a good neighbour? A comparison of UK and Australia experiences in terms of procedural fairness and distributive justice', *The Extractive Industries and Society*, vol. 5, no. 4, pp. 547–556.

Woods, M. 2011, *Rural*, Routledge, London.

15

EXPLORING THE PRODUCTION OF CLIMATE CHANGE THROUGH THE NOMOSPHERE OF THE FOSSIL FUEL REGIME

Lauren Rickards and Connor Jolley

Introduction

Addressing climate change demands a reconfiguration of often taken-for-granted social relations, especially with respect to the power structures undergirding our energy systems and the associated extraction and accumulation regimes. Social science on the extractive industries and energy sector has burgeoned over the last decade. Much of this research, notably that informed by Marxist analyses, points to the large, entangled structural forces in play, forces that reveal how entwined the mining and energy sectors are with other sectors, including the public sector (e.g. Baer 2016). Other research focuses on particular cases and groups of decision-makers (e.g. Snell 2018), including local groups involved in, for example, contesting a new coal mine or establishing a new renewable energy project by negotiating externally imposed legal requirements and policy settings (e.g. Moffatt and Baker 2013). While both of these areas of research – that on broad-scale structures and that on local cases – are important, they tend to leave intact a major blind spot: the everyday social processes, practices, dynamics and systems, and the individuals and "local" interpersonal interactions, which imagined "high-level" structures actually consist of. In both the public and private sectors, including the law profession, senior decision-makers, associated elites and their organisational milieus are hidden from view (Rickards, Wiseman & Kashima 2014). Just like the public they often seek to inform, social science researchers have enormous trouble accessing, understanding and interrogating the elite decision-making spaces in which the most significant decisions related to energy and climate change are made. Thus, social science researchers also have trouble accessing, understanding and interrogating the logics and presumptions that guide those decisions. These logics and presumptions include not just those arenas in which decisions about "energy" or "climate change" are explicitly made, but the numerous other intersecting areas of society and bureaucracy

– including law, planning and finance – that powerfully shape the high green-house gas emissions trajectory we are collectively on. As some science and technology studies (STS) scholars emphasise – notably those using Geels's (2005, p.369) "multilevel perspective" – researchers and others advocating for positive change need to attend to the "regime" level of society in which "incumbents" perpetuate existing systems to protect their interests. Moreover, such research needs to not only "map" the institutions, networks and alliances involved in particular controversies, or document specific cases in which power is wielded by certain individuals. Rather, what needs to be unpacked is the normalised, invisibilised "background" context in which these controversies and cases are situated, a context in which power operates as a distributed, emergent effect of the "governmental" system at large, as Michel Foucault has influentially argued (Foucault 1998; Foucault 2008).

One social science research response to the limitations of both structural and individual-scale analyses is to try to gain access to "elite decision-makers" in order to put faces to the "faceless" decision-makers internal to formal governmental structures. Some work in legal geography makes inroads in this direction. David Delaney, for example, delves into the workplaces of formal law to try to "enliven legal geography, to 'people' it with characters who are more recognizably like us" (Delaney 2014a, p.4). Delaney draws on critical organisation studies, a field in which some scholars have helped make visible the actual people that organisations consist of, notably organisations with sizable greenhouse gas emissions profiles (e.g. Wright & Nyberg 2015). While important, this methodological angle leaves unexamined many of the social complexities of the fossil fuel regime. In this chapter, we argue that legal geography is especially needed to understand the latter. In particular, there is a need for legal geography methodological approaches that appreciate the distributed, pervasive character of law, whether within inter- and intra-corporate interactions, a local community or the confines of a research project. We point to new methodological challenges and possibilities by drawing on the two concepts of the nomosphere (cf. Delaney 2010) and "politics of aesthetics" (Rancière 2010; Rancière 2013). Both point to the need for more reflexive and non-representational social science that helps advance more progressive, creative and effective research methods.

The nomosphere

Law has a prominent role in the energy sector, perhaps especially that component devoted to the extraction, transportation and combustion of fossil fuels. Being able to negotiate within and around formal laws and regulations is a key condition of production for energy projects, regardless of what stage of its "life cycle" (for example, speculation, expansion or site "rehabilitation") a project is at. A conducive *legal* environment is a condition of production, and the extent to which this is manifest in any one setting strongly shapes whether fossil fuels are transformed into economic resources. As David Delaney (2014a, p.1) puts it, the 'legal' is not simply poured into pre-existing 'spaces' but, rather, is constitutive of spatialities, spatial relationships, spatial performances and experiences, as these, in turn, condition the lived character of the legal.

Beyond formal laws and the associated, often violent enforcement of legal rights, people's "sense" of legality and unspoken rules circumscribe a normative socio-spatial order (Jeffrey 2018). Encapsulating these arguments is the idea of the nomosphere: "the cultural-material environs that are constituted by the reciprocal materialization of the legal and the legal signification of the sociospatial" (Delaney 2004, p.851).

One way the legal co-constitutes space is by shaping ideas about the appropriate *use* of specific spaces. The very idea of "land use" is a social construct that Valverde (2011) suggests arose in the 19th century as part of a broader biopolitical project of ordering space and populations. Above and beyond the specific visions imposed by land use planning policies and regulations *per se*, legal logics help provide narratives that shape how humans interact with different areas of the world in particular ways, converting place and peoples in concert. Like Delaney and others, Cover (1983, pp.4–5) turns to the Greek idea of law as nomos to emphasise its real, world-making role:

> Once understood in the context of the narratives that give it meaning, law becomes not simply a system of rules to be observed, but a world in which we live ... This nomos is as much 'our world' as is the physical universe of mass, energy, and momentum. Indeed, our apprehension of the structure of the normative world is no less fundamental than our appreciation of the structure of the physical world.

Besides being real and important, law is inseparable from the physical world. As Bartel et al. (2013, p.343) argue:

> the physical realm, as manifest in various and highly specific geological, hydrological, atmospheric and climate conditions, determines the possibility and sustainability of laws and economies.

Moreover, the relationship flows the other way, as the role of "land use" illustrates. While formal laws are created in response to the physical potentials perceived in an area at a given point in time, their subsequent effect on land uses and land management mean they are consequently imprinted upon the land, particularly if the result is the extractive land use of open-cut mining. Together, formal and informal laws (or their absence) work with the non-human aspects of an area, such as slopes, fertile soil, water sources or subterranean mineral deposits, to afford certain land uses (as STS scholars would put it). To the extent that those land uses are realised, they, in turn, re-shape perceptions of place, afford some future land uses over others and encourage alterations to the laws applicable to the area to enable those emergent visions of a place. In this way, law and land (use) continually co-evolve, as the evolution of the Latrobe Valley in Victoria, Australia, illustrates over the period its brown coal resources were discovered, made visible, mined, combusted and transformed legally and socially into the region's identity and then curse (Alexandra 2017; Rickards 2017).

The nomosphere carries implications for empirical research in legal geography. We highlight three here. First, law not only varies over space, but helps create these spatial differences as it does so. In formal law, this is exemplified by spatial planning regulations, and the pervasiveness of legally enacted spatial divisions in everyday life. Many spatial borders and boundaries delineate space by physically manifesting legal delineations. As geographers and mobility scholars have documented (Schiller & Salazar 2013; Bauder 2014; Spinney, Aldred & Brown 2015), a consequence of this is that people's identities and rights can shift across space as their legal status differs with location. This underlines the well-known point in human geography that, counter to the common-sense presumption that space is a universal, it is experienced and lived differently by different people, and (as discussed below) can change over time (Lefebvre 2009). Thus, scholars need to attend to how groups' and things' spatio-legal identities and implications alter across space and time as they contribute to shifts in both.

An avenue into such analysis is "follow the thing" studies of the sort pioneered by Ian Cook (2004) to trace the physical journeys, transformations and effects of various things, for example, wheat (Head, Atchison & Gates 2012) and coal (Connor 2016). Climate change demands that what is followed expands beyond graspable things to include powerful but dispersed entities such as smoke and greenhouse gas emissions. Tracing greenhouse gas emissions directly, and comparing ways in which they are accounted for or not, is an exemplary legal geography question, with some molecules allocated to regions or nations on account of where they are produced (e.g. Liu et al. 2019), and some emissions allocated according to where the related goods are consumed (Bergmann 2013; e.g. Meng et al. 2017). The consequent categorisation and ranking of nations, sectors and businesses is an important aspect of the global nomosphere under climate change. Exploring how formal and informal laws (for example, nation's commitments under the United Nations Framework Convention on Climate Change) encourage some accounting methodologies over others, and how these decisions then alter how various groups perceive, engage with and try to shape specific regions, is an important area for legal geography scholarship. In the Latrobe Valley, mentioned above, its status as home to the most polluting power station in the OECD in carbon dioxide terms has contributed strongly to its stigmatisation and subsequent industrial transformation (Birkbeck & Rickards 2015). This is happening at the same time as its ongoing combustion of wet, young coal is producing unregulated mercury emissions that are being ingested by local inhabitants, likely contributing to the heavy health burden they already carry in their bodies (Lipski, Rivers & Whelan 2017). Via the climate system, coal combustion emissions are now feeding back on many lives, physically as well as legally. In the Latrobe Valley, warming is moving the region back towards conditions that may come to resemble those that allowed the tropical rainforest of Gondwana to flourish in this location millions of years ago. Leaves from that time can still be found in the brown coal now dug out from and burnt in the region (Rickards 2018). It is a temporal journey that directs attention to other Gondwanan remnants across the Bass Strait in Tasmania, where the associated newly combustible conditions are exposing ancient trees to fire for the first time (Rickards 2016). This transforms the trees into black soot that not only lodges in human and non-human bodies but travels as far as Antarctica where, distributed across the snow and ice, it absorbs sunlight and adds to climatic warming (Hadley & Kirchstetter 2012, Highwood & Kinnersley 2006). These complex, far-reaching feedbacks pose significant challenges for monitoring protocols and greenhouse gas emission accounting practices that have yet to catch up (Wedderburn-Bisshop, Longmire & Rickards 2015). This points to a disjuncture between the legal and lived landscapes, and the importance of recognising that the nomosphere is as much about the absence of formal laws as the presence of appropriate ones.

Second, the nomospheric character of life means that the legal must be sought above and beyond the formal letter of the law, including, for example, in popular discourses about certain areas and their presumed land uses. With the latter shaped in part by the past, present and envisaged physical affordances of an area, this points to the need for interdisciplinary analysis blending social sciences. As part of this, economic analysis is important to the extent that it is a powerful determinant of the envisaged affordances of an area (including its potential for infrastructural enhancement) and its consequent nomospheric qualities. Economic geography is especially valuable for the way it underlines that the relevant characteristics of a place are not simply its spatially delineated, absolute characteristics, but its relative qualities. That is, space and law are not simply co-constituted on a place-by-place basis but in constant inter-relation with other places. This applies whether the relationship is: one of co-dependence as in global supply networks (e.g. Liu & Müller 2013); one of tele-connected weather patterns of the sort increasingly evident under climate change (e.g. van Noordwijk et al. 2014); or one of competition, as in the constant corporate comparison of different sites for their transnational operations. Once established in a place, transnational corporations, including those in the fossil fuel industry and related high emissions industrial supply chains, frequently use the threat of withdrawing their operations from a region to press governments for a more favourable legal environment or at least exemptions. Threats of capital flight have been made by various energy-intensive businesses in response to the purported unreliability and expense of electricity in Australia (Weller 2018) and exemplified by Engie's closure of the Hazelwood coal mine and power station in the Latrobe Valley. These tactics illustrate how negotiations over legal-economic agreements can construct relative nomospheric settings, whether the comparison is between extant regions, or the present and possible futures.

Methodologically, this points to the need to combine legal and economic analysis. Case study-based scholarship could usefully track financial agreements, arrangements and transfers and how they are argued, altered and justified relative to possible corporate conditions elsewhere. There is also a need for multi-sited and/or comparative research that tries to capture the networked and relational character of what Delaney (2010, p.59) calls "nomic settings". Analysis could encompass different points along supply chains (for example, bauxite, alumina and aluminium) or alternative sites for energy sources (for example, different coal mines, coal regions or alternative, renewable sources elsewhere). Looking at the "various per capita benefits from land use", for example, Luke Bergmann and Mollie Holmberg (2016, p.932) argue that "uneven development under neoliberal globalisation results in strong global net redistributions" of such benefits, "especially from Global South to Global North". They conclude that:

> from the perspective of capital investment, the median square meter of global land use contributes to futures of human populations outside, not inside, of the country of that land.

More work is needed that examines to whom the benefits of the nomospheric shaping of land flow, where these beneficiaries are located relative to the site of extraction/impact/labour and what the justice issues are.

Third, there is a need to attend to the neglected temporal dimension of the nomosphere. Neither space nor the law are static, despite common cartographic understandings of space as a mere container and the frequent conflation of law(s) with timeless facts of nature. Rather, space and the legal are both processual, and their co-constitution is dynamic. This points to the need for historical and longitudinal analyses in legal geography. An insight into this is provided by the Latrobe Valley, discussed above. The area has progressively become known as Victoria's brown coal region, as the coal production and combustion in the area has come to dominate physically, socially and politically. Now normalised, this centralisation of coal in the constitution of the region has been to the detriment of alternative primary land uses and identities, such as ones centred on its water or forest resources (Rickards 2017; Alexandra 2017). How regions are defined and differentiated over time illustrates the intersection of legal-administrative, physical and sociocultural readings of space. There is great scope for legal geography to engage more directly with regional studies on these questions. Regional studies (of a post-structural ilk) have a sophisticated understanding of the ambiguous, political notion of "the region" and how particular regional identities are contested and change over time. Conceptualising this as evolution in the nomosphere of "a region" promises to be productive for both regional studies (which has a limited view of law) and legal geography. A methodological approach in some regional studies that holds promise for legal geography is demographic analysis and related statistics on resources, services, infrastructure, etc. Not only could analysis could be undertaken by scholars, but it could be used as a target for research in order to better understand its evolving internal logic and governmental use. By unpacking a region into population and other categories and tracking their change over time, demographic and other statistical analyses help scholars "see" regions "like a state". This borrows from James Scott's famously phrased bureaucratic and governmental perspective of the British Empire in high modernity (Scott 1998), with the historical specificity of this modernist lens underlining the need to see the nomosphere itself as dynamic. In other words, such analyses encourage a deeper understanding of a regional nomosphere by providing insight into the evolution of both regional characteristics, such as population distribution and employment trends, and of the lenses used to view, frame and govern them. These lenses, in turn, feed back on their characteristics. Awareness of not just the sequential but relational positioning of regions in these entwined processes is key (see, for example, the Australian Government's (2017) *Transitioning Regional Economies* report), providing an important indication of the sort of nomospheric differences that are likely to be accentuated over time.

The politics of aesthetics

A crucial aspect of nomic settings is the way that particular elements of the world are, relative to others, made sensible. In aesthetics, sensible means both detectable and common sense (normal, expected, appropriate). Although the term aesthetics is often erroneously thought to be limited to questions of beauty and taste, in its original Greek form it referred to sensual perception and experiential knowledge. In social theory aesthetics similarly refers to how the world is partitioned into foreground and background, or socially visible and invisible. As a primary theorist of "the politics of aesthetics", Jacques Rancière (2013, p.11) defines aesthetics as:

> the systems of *a priori* forms determining what presents itself to sensory experience. It is a delineation of spaces and times, of the visible and the invisible, of speech and noise, that simultaneously determines the place and the stakes of politics as a form of experience.

The point is that "sensory experience is not given, but is instead contingent, an effect of power relations that give each person and thing their proper space and time" (Grove & Adey 2015, p.79). What we perceive and experience is highly socialised and politically shaped through institutions such as education, the media, entertainment (e.g. Grayson 2017) and policies (e.g. resilience agendas, see Grove & Adey 2015). We are implicitly trained to feel and know the world in some ways and not others. This "training" can be considered an informal type of law of the sort Delaney's (2010) idea of the nomosphere emphasises. Indeed, Delaney's introduction of the nomosphere, and the emergence of the field of legal geography more generally, can be understood as an effort to make visible and sensible an (informal) element of law that is generally overlooked/invisible. In *The Ideology of the Aesthetic*,

Terry Eagleton refers to aesthetics as "lawfulness without law", a political unconscious that helps "secure the consensual hegemony which neither the coercive state nor a fragmented civil society can achieve" (Eagleton 1988, pp.330, 332).

Rancière (2010, pp.36–7) views two distinct forms of politics. The first, a social order referred to as "police" is the internal consensus according to which forms of action and speech are legitimised through processes of exclusion and inclusion within established social and cultural aesthetic regimes. It is in this way that what is considered intelligible, knowable and thinkable, within the space or language of a community, is "partitioned" so that some things are sensed, while others are not. The second form of politics, which Rancière considers the "real" form and refers to as "dissensus", is collective action that disrupts hierarchies within established regimes by making visible those spaces which have previously been obscured to the senses and, in the process, reconfigures the partitions that define and produce communities. As Rancière (2010, p.37) writes, "[p]olitics, before all else, is an intervention in the visible and the sayable". While the legal sphere in and of itself is not the focus of Rancière's analysis, Viktorija Kalonaityte (2018) observes that the ways in which the law renders social phenomena legible or illegible means that all of those striving to shape change in society cannot avoid engaging with law. We focus now on three ideas in Rancière's work – the police order, political acts and aesthetic acts – and the way in which each opens up valuable new methodological possibilities for legal geography.

Police order

As mentioned above, what Rancière (2010, pp.36–7) called "police" – the implicit framework that legitimises some forms of speech and action and not others – is a powerful aspect of the social realm. In particular, it forcefully shapes how we think about the world, including classing certain things and beings as "resources" and positioning resource extraction as incontrovertible within society. From the perspective of typical Environmental Impact Assessments, such as that conducted for the proposed Adani mega-coal mine in Queensland, in Australia (Jolley & Rickards 2019), the impacts of resource extraction and combustion are framed by the police order as predominantly local, and even then, only a small proportion of impacts are in view. Yet it is increasingly clear that when all local impacts are considered, multiplied across space, and combined with direct effects on global systems such as the climate, there is a far larger, and predominantly obscured situation in play – one in which the chemical composition of the Earth's atmosphere has no prior analogue. Mining moves more sediment than the combined river systems of the world, and virtually all freshwater is channelled into human uses (Intergovernmental Panel on Climate Change 2018; Zalasiewicz et al. 2017).

Awareness of the police order calls for methodologies that make visible the social, cultural and political processes that generate both resources for capitalism and claimants for Western law (Kalonaityte 2018). This includes the way scale is truncated and lines of causation are erased to reduce the political visibility of large, long-distant relations (for example, the role of distant finance and political power in a local controversy) and negative effects (think climate change). This is evident in the Adani case, where mine proponents cast it as a local-scale, narrowly economistic issue. In a recent paper undertaken by the authors (Jolley & Rickards 2019), an analysis of interview data and media reports on the Adani mine controversy focused on how scale was deployed in legal representations of the issue. It found the controversy is contested, in large part by attempts to expand or contract the scale at which it is represented. The mine proponents were adept at promoting the Adani project at local and regional scales, which emphasised economic benefits by "stacking" (Sica 2015, pp.445–6) representations of these scales together in ways that situated their intended audiences into familiar landscapes (Jolley & Rickards 2019). In doing so, alternative scalar representations, for instance those which emerged from urban centres in contrast to the "real" concerns of regional Australia, were actively delegitimised in ways that sought to exclude such actions from "presenting themselves to the sense" (Rancière 2013). The highest profile of these examples was the attempt to repeal sections of the *Australian Environmental Protection and Biodiversity Conservation Act 1999* (Cth) (EPBC Act) that provides legislated third-party standing rights to appeal environmental approvals granted under the Act. Court proceedings brought under the Act were branded as illegitimate acts of "lawfare" (Brandis 2015) that subverted the rights of elected representatives to exercise executive power. Although ultimately unsuccessful, the attempted amendments marked a hostile escalation in rhetoric and tactics against the environmental movement and illustrate how scale is employed to construct and deconstruct legitimacy in spatial controversies (for further examples see Jessup 2013; Andrews & McCarthy 2014; Turton 2015; Williams 2016; Della Bosca & Gillespie 2019). From a methodological perspective, this underlines the need for analytical frames that attend to the scales being implicitly invoked by different parties and for different ends, including the privileging of some scales via their institutionalisation in formal legal processes such as those that designate legal standing.

Political acts

Some political interventions may on the surface seem to subvert certain police orders but actually take on their own exclusionary forms. Kalonaityte (2018) observes in relation to Rancière that the political subject emerges in dissensus to the police order as a manifestation of those who have no part in the order. That is, a new political subject emerges when it "exceeds those conceptual and legal categories that separate and exclude" (Kalonaityte 2018, p.522). When political claims are made by or on behalf of such emergent political subjects, or for constituencies otherwise excluded by the police order, they can be viewed as political acts that challenge the assumptions and logics of partition and often take many forms. Examples offered by Kalonaityte (2018) include insurrectionary movements (as the School Climate Strikes illustrate) artistic interventions and, relevant to our discussion, litigation. Kathryn Yusoff extends the reading of Rancière to incorporate the biopolitics of aesthetics, writing that:

> Aesthetics is, in Foucault's terms, fundamentally biopolitical. Alongside this connection between aesthetics and politics as a space in which ecologies are made, there are various kinds of loss and violence that are an attendant part of anthropogenic-induced climate change, which generate a social urgency to these questions of representation and violence, aesthetics and existence. (Yusoff 2010, p.78)

Attending to the biopolitical aesthetics of climate change means facing the biopolitics of what we make visible – and thus of what we help "make live and let die", as Foucault (2003, p.241) famously put it. Much of how the problem of global climate change is made visible relies on how the "global" is produced as a scale of representation by the institutional chains through which it is mediated, resulting in an often technoscientific rendering that privileges certain ontological orientations at the exclusion of others (Blok 2010). It is partly out of concern about the exclusions of a global lens that scholars critique climate and earth system science as a form of "Earth System governmentality" (Lövbrand, Stripple & Wiman 2009, pp.7–8). While the exclusion of the global scale from analyses such as Environmental Impact Assessments is problematic, so too are the exclusions of the global scale to the extent that they lead to generalisations about humans and the world which obscure significant differences.

Methodologically, this means it is vital to critically analyse the categorisations being used in legal controversies to better understand how political subjects are being constituted and to what ends. As an example, s. 487 of the EPBC Act (1999) extends the definition of a "person aggrieved" in the *Administrative Decisions (Judicial Review) Act 1977* (Cth), allowing individuals and organisations to seek judicial review of administrative decisions made in respect to matters of national environmental significance. As it states:

(2) An individual is taken to be a person aggrieved by the decision, failure or conduct if:
  (a) the individual is an Australian citizen or ordinarily resident in Australia or an external Territory; and
  (b) at any time in the 2 years immediately before the decision, failure or conduct, the individual has engaged in a series of activities in Australia or an external Territory for protection or conservation of, or research into, the environment.
(3) An organisation or association (whether incorporated or not) is taken to be a person aggrieved by the decision, failure or conduct if:
  (a) the organisation or association is incorporated, or was otherwise established, in Australia or an external Territory; and

(b)   at any time in the 2 years immediately before the decision, failure or conduct, the organisation or association has engaged in a series of activities in Australia or an external Territory for protection or conservation of, or research into, the environment; and

(c)   at the time of the decision, failure or conduct, the objects or purposes of the organisation or association included protection or conservation of, or research into, the environment. (EPBC Act, s. 487)

While a wide-standing clause and very useful for those who fall within its remit, the overall effect of the test is to formalise in the eyes of the law a political subject that is determined by their proximity to "the environment". In doing so, it includes some actors, but also necessarily closes off spaces of legitimacy to those who do not recognisably meet its criteria. Moreover, the federal government more recently sought to repeal this section of Act and to revert to older, more conservative interpretations of an aggrieved person as someone directly impacted (economically) by a decision, where directly impacted refers to spatial proximity. While this effort to repeal s. 487 was withdrawn when it became clear it would not be passed by the Senate, the government has pursued the idea that a legitimate environmental subject is only one engaged in "real" environmental conservation work, by trying, for example, to redefine, delegitimate and undermine environmental charities through alterations to the federal tax code. Under proposed changes, environmental charities would only qualify for tax-free status (which is vital for their financial viability) if 50% of their income was used for "practical" conservation and remediation work (Staples 2017). The intended effect is to starve environmental advocacy, such as that performed by groups opposed to the Adani mine, of the tax-deductible donations many rely on. Actors whose interests in the Adani project are excluded from the formal spaces of law and the territorially bounded spaces of electoral representation are thus increasingly forced toward using dissensus, or as what Kalonaityte (2018, p.523) frames as an "insurrectionist movement", in an attempt to make their interest "visible and sayable" (Rancière 2013, p.37). In the Adani case, this is evident in the emergence of a heterogeneous coal opposition movement and its evolution over time towards more militant direct-action approaches as formal legal avenues for delaying or stopping the project have been closed off one-by-one. All of these struggles over legal standing and legitimacy underline the need for researchers to think hard about what counts as the collective, and what counts as legitimate, in all work on and influenced from afar by the fossil fuel regime.

Aesthetic acts

The transformational change needed to tackle the incumbent fossil fuel regime means that research needs to not just expose and expand existing categories but produce new ways of thinking – indeed, new ways of sensing. To this end, the third key area of aesthetics we consider are aesthetic acts. Rancière thinks about aesthetic acts "as configurations of experience that create new modes of sense perception and induce novel forms of political subjectivity" (Rancière 2013, p.3). Carroll Clarkson expands on this definition by framing "aesthetic acts" as relational, in the sense that they have the capacity to alter how a subject perceives themselves and their standing *in relation to* others by altering "the way a community delineates itself in terms of what it perceives to be significant, or even noticeable at all" (Clarkson 2014, p.3). Resonating with Sara Ahmed's (2006) discussion of orientation, this is about sensing as both attuning oneself to new aspects of the world and altering one's sense of how they are positioned in the world.

An illustration of this is the way groups are becoming sensitised to heat. One of the effects of climate change in Australia and many parts of the world is increasingly intense periods of environmental heat, such as the most recent summer period (2018–19) which was characterised by "an unusual extended period of heatwaves over much of Australia" (Australian Government 2019, p.5). More than just a weather condition, heat is experienced by animals, including humans, at an intimate bodily level, in part because it is also produced internally (Oppermann et al. 2018). Arguably, one of the implications of climate change is increased societal awareness about our bodily vulnerability to extreme heat, our bodily production of heat and the way heat affects our engagement with the world. Since changes to such engagement include falls in labour productivity under hot conditions, the heat–body relation has emerged as a capitalist concern, reinforcing the capitalist logic that underpinned the invention of artificial temperature-controlled air conditioners as a governmental means of creating ideal working conditions for office workers (Rickards & Oppermann 2018). Outdoor workers – such as construction, utility and emergency service workers involved in the existing fossil fuel system – are partially defined by their exclusion from such temperature-controlled workspaces and from the environmental protection offered by indoor work sites more generally. To the extent outdoor work involves manual labour, outdoor workers are both especially exposed to extreme environmental heat and produce high levels of bodily heat through their work. In recognition of this dual vulnerability, labour unions have started becoming more vocal about protecting workers from dangerously hot working conditions (e.g. Jarrett 2018). While the political subjectivity labour unions invoke is not novel per se, based as it is on the classic Cartesian divide between manual and mental workers, what is novel is the (re)emergent specific categorisation of indoor and outdoor workers as a key political division. Methodologically, it is significant to note that there are parallels here with the hierarchical division of labour that can characterise academic research. Students and assistants are often relegated to fieldwork while more senior academics stick with analysis and writing in the comfort of their offices. As a result, the two groups *sense* the research environment in fundamentally different embodied ways, with likely but unknown consequences for what the research concludes. What, for example, would it mean if research was analysed with a greater sensitivity to the environmental conditions that characterised data collection, including the physical discomfort and fatigue of the research assistants involved? What may they have missed in the data collection because they were too tired, and what may have the research project thus missed because others were not there with them? These are some of the practical methodological questions that a changing climate forces academic research – including legal geography – to confront.

In response to the difficult working conditions that many casual academic staff have to endure, including but far from limited to their physical environment, some are becoming more vocal in their protestations (e.g. National Tertiary Education Union 2019). In Kalonaityte's (2018, p.525) reading of Rancière, generating "noise and affect" in support of an excluded constituency, as unions often do, is not sufficient for political acts. She argues instead for the need for logos, or legitimate speech, in the form of "judicially legal claims" (p.522) to complement forms of protest and produce real societal change (such as improvements in casual academics' contractual relations and status). While this can be the case, it is important not to divorce protest from legitimate speech. One of the striking things about academic resistance to the casualisation and marketisation of higher education is how it is frequently expressed *via* conventional "legitimate speech" outlets such as journal articles (e.g. Bone, Jack & Mayson 2018; O'Dwyer, Pinto & McDonough 2018), books (e.g. Waters 2017) and conferences (e.g. The Australian Sociological Association 2019), testing the boundaries of academic institutions' expectations around legal conduct and contractual obligations.

At the same time, the need for any legitimate speech can be overstated, and the power and analytical importance of affect should not be dismissed. Luke Bennett and Antonia Layard (2015) suggest that a turn away from traditional discursive textual analyses to non-representational methods focused on notions such as affect, can usefully expand the frontiers of legal geography. One way to approach questions of affect is through the idea of atmosphere – an increasingly popular metaphor in non-representational theory for the fluid, dynamic, ubiquitous yet difficult-to-sense nature of how affect is transmitted in communicative ways untethered from "the narrowly discursive or symbolic constitution of text and talk" (Bissell 2010, p.271). Philippopoulos-Mihalopoulos (2013, p.36) argues that law is like atmosphere because it is ubiquitous in every spatial dimension of life yet is rendered invisible like air. Law, he suggests is "there but not there, imperceptible yet all-determining" and as such "the law determines an atmosphere by allowing certain sensory options to come forth while suppressing others" (Philippopoulos-Mihalopoulos 2013, p.36). Within this realm (which Australian legal geographer Nicole Graham (2003) evocatively terms the "lawscape") sensory inputs normatively guide and manipulate subjects while obscuring the relationship between "sense, meaning and feeling", and the way these relations are used to govern subjects (the police order) (Philippopoulos-Mihalopoulos 2013, p.39). In this reading, "(a)tmospheres, just as the law … remind us of their presence while at the same time allowing us to be lulled into safe oblivion" (Philippopoulos-Mihalopoulos 2013, p.43). At the same time, the air pollution being exacerbated directly and indirectly by the fossil fuel regime and its effects on wildfires, and made worse by hot conditions, means that atmospheres are increasingly not presumed to be safe, pointing to one of the ways many groups are being sensitised to their material environment.

Attending to non-representational aspects of the lawscape does not mean abandoning methods traditionally associated with legal geography. Anderson and Ash (2015, p.35) argue that non-representational theory (NRT) involves not methods *per se* as much as "research styles" that account for how "background" social phenomena have the capacity to both affect and be affected, at the same time as breaking with the epistemological reductionism that demands backgrounded context to either explain something or be explained away. As they put it:

> Non-representational methods do not refer to a separate set of methods neatly distinguished from methods that are now supposedly deficient. Instead, we take non-representational methods to name a set of ways of approaching a phenomenon, of relating or not to the weight or touch of something, which intensify the problems that the object of inquiry poses for social analysis. We could say, then, that a non-representational method involves an intensification of problems and requires staying with those problems for a while. (Anderson & Ash 2015, p.48)

That said, making sense of the non-representational does require an embrace of novelty and experimentation in order to capture and analyse what can be fleeting and ephemeral moments. This might involve experimenting with empirical methods through interdisciplinary collaborations (Michels 2015), through the reproduction of sensory experiences as part of iterative methodological processes (Adey et al. 2013) or in creative analytical approaches to teasing out non-linear causalities (Anderson & Ash 2015). For legal geography, a project which is characterised by interdisciplinarity (Delaney 2014b) and a desire to push the research horizon beyond the limitations imposed by reductionist textual methods through which the law is traditionally understood, such experimental approaches have great potential.

A leading thinker on experimental methods such as speculative fabulation, Donna Haraway, similarly calls on researchers to "stay with the trouble" (Haraway 2010; Haraway 2016). By this she means we should not turn away from the awkward, painful questions posed by the Anthropocene – a moment of reckoning for Western imperialism, its standard Cartesian ontology, and our political and practical fossil-fuelled worlds. Instead we need to heed feminist scholars in attending far more carefully to our material context, including the multitudinous non-living and living things with which we share the Earth and our bodies (e.g. Alaimo 2018). Such a direction opens up valuable questions about felt and political subjectivity, including the sovereignty of an individual human who is never simply human but is always more-than-human in a more bacterial sense among others (Lorimer 2012), or the political legitimacy of humans whose bodies and homes are riddled with pollutants like mercury that are not (yet) visible to the epidemiological or regulatory eye of the law (e.g. Pitkanen 2017).

Basic to exploring the material-affective co-production of atmospheres is to trace the relationships between physical atmospheres (such as those polluted by coal combustion) and felt atmospheres (such as those in marginalised, disadvantaged polluted areas), as the case of the Latrobe Valley illustrates (Duffy et al. 2017). Anderson and Ash (2015, p.36) suggest that one of the main questions that scholarship on affective atmospheres needs to address is how we might "become sensitive to the causal powers of phenomena that exert a force, but may be vague and diffuse, ephemeral and indeterminate". The entwinement of air pollution, embodied harms, economic and political disadvantage, social stigma and, increasingly, wildfires in fossil fuel landscapes exemplifies the need for such scholarship. Once again, the need to work across the boundaries of physical sciences (such as atmospheric science, fire science and public health) and social sciences and humanities (such as human geography and anthropology) is evident. In their work on Latrobe Valley community members' experience of the 2014 Hazelwood coal mine fire, Duffy and Whyte (2017, p.431) draw on chemical analyses, epidemiological studies and qualitative research to underline the role of affective and material flows, and historical as well as present-day events, in understanding people's experiences. Helping extend work on the emotional geographies of the extractive sector in Australia (e.g. Ey, Sherval & Hodge 2016) thematically and methodologically, they use "poetic transcription" to powerfully represent the data from interviewees affected by the fire and capture some of their collective emotional and affective experiences in a creative, distilled manner. This includes how the coal mine fire revived a long-standing "sense of abandonment" (Duffy & Whyte 2017, p.435) that was triggered in the Valley in the 1990s when the nomosphere of Victoria was reconfigured along more neo-liberal lines, and the related "poorly managed process of privatising the electricity industry left a significant proportion of the local population without work and without the prospect of finding work locally" (Weller 2017, p.382). Local anger at past treatment reflects also the pain of closing and removing the entire town of Yallourn in the 1980s to make way for the expansion of one of the coal mines (see Wadley & Ballock 1980), and at the prevalence and official neglect of asbestos-related deaths among ex-power station workers (Hunter & Lamontagen 2008). These layers of simmering resentment are generally backgrounded in contemporary research in the Valley, which tends to focus on more near-term and large-scale issues such as climate change. It is skilfully surfaced in Duffy and Whyte's transcriptions, such as the following stanza (2017, p.432):

> We've been decimated
> and
> it's just like nobody
> really gives a fuck about us
> seven thousand jobs

The pain expressed about these past harms provides vital context for then under-standing the research interviewees' experiences of the fire, which burned for more than 45 days thanks to a bungled government and corporate response that revealed, with the help of a subsequent parliamentary inquiry, previously unseen gaping holes in the governance of fire prevention in private coal mines (Doig 2015). As another stanza expresses it (Doig, p.434):

> Your eyes are burning
> You're coughing
> you can't breathe
> we were just treated
> like we were whinging,
> complaining about nothing
> they were evacuating
> government departments
> but we were all left here
> We just
> keep being lied to
> over and over and over
> That's where the trauma came in
> an
> absolute
> feeling
> of
> abandonment
> It didn't matter
> if we lived or died
> we were expendable,
> seems to be
> that's still the case.

Poetic transcription itself can be thought of as an aesthetic act on behalf of the researcher for the way in which it brings to the fore the affective dimension of a situation and helps represent participants in a new more powerful, political way. In helping generate a sense of empathy for the participants, the process helps develop the public profile of the participants as a group in a way that offers new understanding and empathy.

We come then to the role of scholar activism in legal geography and its role in helping generate Rancière's "real politics" (dissensus) to help expose and change what is visible and sayable. In the Latrobe Valley, community groups have fought to have injustices in the Valley recognised. One emerged in the 1980s to bring to visibility asbestos pollution and related deaths (Hunter & Lamontagne 2008). Another, called "Voices of the Valley", emerged in 2014 in response to the coal mine fire (Yell & Duffy 2018). While initially a source of "noise and affect", such a group helped to give inhabitants of the Valley a new, albeit contested and not wholly shared, political subjectivity. Recognising the power of epidemiology in the existing legal common sense, they then drew on the services of an epidemiologist in Queensland to have deaths in the area analysed. The resultant report, which indicated that approximately 12 people had died indirectly as a result of the fire's air pollution, was then used, with the help of lawyers in Environmental Justice Australia, to push for a second parliamentary inquiry into the fire, directly challenging the first inquiry's conclusion that the fire had been unfortunate but not fatal (Environmental Justice Australia 2014). By making visible the deceased subjects, and the material links between the fire and the community's health, Voices of the Valley and those who assisted them intervened to newly politicise the region and its pasts and futures. This deliberate politicisation has contributed to the region becoming a strong ongoing concern for the Victorian government and wider community, who are now more aware than ever that their power use is entwined with the lives of distant others in the Latrobe Valley (Duffy et al. 2017).

Conclusions

In this chapter we have argued that a legal geography that attends to the relational, distributed, aesthetic character of the law – that is, of the nomosphere – is needed to understand and address the complexities of the fossil fuel regime and its connections to climate change. Methodologically, the overarching implication is that research is a political tool, one that can help expose and alter what is visible or sensible, and thus, what problems are recognised and what arguments and responses are legitimated. To expose and alter the distribution of the sensible in progressive ways – ways that help get to the heart of the climate change challenge – reveals methods that help researchers stay with the material-discursive-aesthetic character of the world as needed. Some such approaches have been proposed here, and more can be found within legal geography scholarship (e.g. Della Bosca & Gillespie 2019), albeit often using different terms. Nevertheless, far more experimentation is needed, perhaps especially in the messy world of fossil fuel extraction. This messy world is so readily backgrounded or reduced to a set of abstract arguments or small set of competing actors, instead of being recognised as the huge, contested machine shaping our lives and environments, our spaces and laws, in far-reaching ways, that it is. As the cases of the closed Hazelwood coal mine in the Latrobe Valley and the proposed new Adani coal mine in Queensland illustrate, sensitivity to the subjects, objects, lives, scales, affects, emotions and qualities visible in formal and informal decisions is vital. This requires not just new methods but utilising existing methods for new ends and adopting overall a refined analytical lens. More than anything this requires reflexivity among researchers as to what political and aesthetic acts, orders and regimes their work is contributing to and how.

References

Adey, P., Brayer, L., Masson, D., Murphy, P., Simpson, P., & Tixier, N. 2013, '"Pour votre tranquillité": Ambiance, atmosphere, and surveillance', *Geoforum*, vol. 49, pp. 299–309.

Administrative Decisions (Judicial Review) Act 1977 (Cth).

Ahmed, S. 2006, *Queer Phenomenology: Orientations, Objects, and Others*, Duke University Press, Durham.

Alaimo, S. 2018, 'Material feminism in the anthropocene', in *A Feminist Companion to the Posthumanities*, Åsberg, C., Braidotti, R. (eds), Springer, Switzerland, pp. 45–54.

Alexandra, J. 2017, 'Water and coal – Transforming and redefining 'natural' resources in Australia's Latrobe Region', *The Australasian Journal of Regional Studies*, vol. 23, no. 3, pp. 358–381.

Anderson, A., & Ash, J. 2015, 'Atmospheric methods', in *Non-Representational Methodologies: Re-Envisioning Research*, Vannini, P. (ed), Routledge, New York, pp. 34–51.

Andrews, E., & McCarthy, J. 2014, 'Scale, shale, and the state: Political ecologies and legal geographies of shale gas development in Pennsylvania', *Journal of Environmental Studies and Sciences*, vol. 4, no. 1, pp. 7–16.

Australian Environmental Protection and Biodiversity Conservation Act 1999 (Cth) (EPBC Act).

Australian Government 2017, *Transitioning Regional Economies*, 15 December, Productivity Commission, Canberra.

Australian Government 2019, *Special Climate Statement 68 – Widespread Heatwaves during December 2018 and January 2019*, 14 March, Bureau of Meteorology, Canberra.

TheAustralian Sociological Association 2019, 'TASA precarious work scholarship fund', viewed 15 February 2019 <https://tasa.org.au/tasa-awards-prizes/precarious-work-conference-scholarships/>.

Baer, H. A. 2016, 'The nexus of the coal industry and the state in Australia: Historical dimensions and contemporary challenges', *Energy Policy*, vol. 99, no. 2, pp. 194–202.

Bartel, R., Graham, N., Jackson, S., Prior, J.H., Robinson, D.F., Sherval, M., & Williams, S. 2013, 'Legal geography: An Australian perspective', *Geographical Research*, vol. 51, no. 4, pp. 339–353.

Bauder, H. 2014, 'Domicile citizenship, human mobility and territoriality', *Progress in Human Geography*, vol. 38, no. 1, pp. 91–106.

Bennett, L., & Layard, A., 2015, 'Legal geography: Becoming spatial detectives', *Geography Compass*, vol. 9, no. 7, pp. 406–422.

Bergmann, L., & Holmberg, M. 2016, 'Land in motion', *Annals of the American Association of Geographers*, vol. 106, no. 4, pp. 932–956.

Bergmann, L. 2013, 'Bound by chains of carbon: Ecological-economic geographies of globalization', *Annals of the Association of American Geographers*, vol. 103, no. 6, pp. 1348–1370.

Birkbeck, M., & Rickards, L. 2015, *A Tale of Two Power Stations: Comparing Australian and European Approaches to Regulation of the Impacts of Coal on Health*, Environmental Justice Australia, Melbourne.

Bissell, D. 2010, 'Passenger mobilities: Affective atmospheres and the sociality of public transport', *Environment and Planning D: Society and Space*, vol. 28, no. 2, pp. 270–289.

Blok, A. 2010, 'Topologies of climate change: Actor-network theory, relational-scalar analytics, and carbon-market overflows', *Environment and Planning D: Society and Space*, vol. 28, no. 5, pp. 896–912.

Bone, K., Jack, G., & Mayson, S. 2018, 'Negotiating the greedy institution: A typology of the lived experiences of young, precarious academic workers', *Labour & Industry: A Journal of the Social and Economic Relations of Work*, vol. 28, no. 4, pp. 225–243.

Brandis G. 2015 (Attorney-General of Australia) 2015, *Government Acts to Protect Jobs from Vigilante Litigants*, media release, 18 August, Canberra, viewed 26 February 2017, <https://parlinfo.aph.gov.au/parlInfo/download/media/pressrel/4020386/upload_binary/4020386.pdf;fileType=application%2Fpdf#search="media/pressrel/4020386">.

Clarkson, C. 2014, *Drawing the Line: Toward an Aesthetics of Transitional Justice*, Fordham University Press, New York.

Connor, L. H. 2016, 'Energy futures, state planning policies and coal mine contests in rural New South Wales', *Energy Policy*, vol. 99, pp. 233–241.

Cook, I. 2004, 'Follow the thing: Papaya', *Antipode*, vol. 36, no. 4, pp. 642–664.

Cover, R. M. 1983, 'The supreme court, 1982 term – Foreword: Nomos and narrative', *Harvard Law Review*, vol. 97, no. 4, pp. 4–68.

Delaney, D. 2004, 'Tracing displacements: Or evictions in the nomosphere', *Environment and Planning D: Society and Space*, vol. 22, no. 6, pp. 847–860.

Delaney, D. 2010, *The Spatial, the Legal and the Pragmatics of World-Making: Nomospheric Investigations*, Routledge, New York, pp. 1–33.

Delaney, D. 2014a, 'At work in the nomosphere: The spatio-legal production of emotions at work', in *The Expanding Spaces of Law: A Timely Legal Geography*, Braverman, I. Blomley, N., Delaney, D., & Kedar, A. (eds), Stanford University Press, California, pp. 239–262.

Delaney, D. 2014b, 'Legal geography I: Constitutivities, complexities, and contingencies', *Progress in Human Geography*, vol. 39, no. 1, pp. 96–102.

Bosca, H. D., & Gillespie, J. 2019, 'The construction of "local" interest in New South Wales environmental planning processes', *Australian Geographer*, vol. 50, no. 1, pp. 49–68.

Doig, T. 2015, *The Coal Face*, Penguin Books, Sydney, Australia.

Duffy, M., & Whyte, S. 2017, 'The Latrobe valley: The politics of loss and hope in a region of transition', *Australasian Journal of Regional Studies*, vol. 23, no. 3, pp. 421–446.

Duffy, M., Wood, P., Whyte, S., Yell, S., & Carroll, M. 2017, 'Why isn't there a plan? Community vulnerability and resilience in then Latrobe valley's open cut coal mine town', in *Responses to Disasters and Climate Change*: *Understanding Vulnerability and Fostering Resilience*, first edition, Companion, M. & Chaiken, M. S. (eds), Routledge in association with GSE Research, pp. 207–217.

Eagleton, T. 1988, 'The ideology of the aesthetic', *Poetics Today*, vol. 9, no. 2, pp. 327–338.

Environmental Justice Australia, 'Hazelwood mine fire submission to the coroner', viewed 10 June 2019 <https://www.envirojustice.org.au/sites/default/files/files/envirojustice_Hazelwood_mine_fire_submission_to_Coroner.pdf>.

Ey, M., Sherval, M., & Hodge, P. 2016, 'Value, identity and place: Unearthing the emotional geographies of the extractive sector', *Australian Geographer*, vol. 48, no. 2, pp. 153–168.

Foucault, M. 1998, 'Governmentality', *The Foucauldian Effect: Studies in Governmentality*, Burchell, G., Gordon, C. R., & Miller, P. (eds), University of Chicago Press, Chicago, pp. 102–103.

Foucault, M. 2003, *Society Must Be Defended: Lectures at the Collège de France, 1975–1976*, Picador, New York.

Foucault, M. 2008, *The Birth of Biopolitics: Lectures at the Collège de France, 1978–1979*, Palgrave Macmillan, New York.

Geels, F. 2005, 'Co-evolution of technology and society: The transition in water supply and personal hygiene in the Netherlands (1850–1930) – A case study in multi-level perspective', *Technology in Society*, vol. 27, no. 3, pp. 363–397.

Graham, N. 2003, 'Lawscape: Paradigm and place in Australian property law', PhD thesis, University of Sydney.

Grayson, K. 2017, 'Capturing the multiplicities of resilience through popular geopolitics: Aesthetics and homo resilio in Breaking Bad', *Political Geography*, vol. 57, pp. 24–33.

Grove, K., & Adey, P. 2015, 'Security and the politics of resilience: An aesthetic response', *Politics*, vol. 35, no. 1, pp. 78–84.

Hadley, O., & Kirchstetter, T. 2012, 'Black-carbon reduction of snow albedo', *Nature Climate Change*, vol. 2, pp. 437–440.

Haraway, D. 2010, '*When Species Meet*: staying with the trouble', *Environment and Planning D: Society and Space*, vol. 28, pp. 53–55.

Haraway, D. 2016, *Staying with the Trouble: Making Kin in the Chthulucene*, Duke University Press, Durham.

Head, L. M., Atchison, J. M., & Gates, A. 2012, *Ingrained: A Human Bio-geography of Wheat*, Ashgate, Burlington.

Highwood, E., & Kinnersley, R. P. 2006, 'When smoke gets in our eyes: The multiple impacts of atmospheric black carbon on climate, air quality and health', *Environment International*, vol. 32, no. 4, pp. 560–566.

Hunter, C., & Lamontagne, A. D. 2008, 'Investigating 'Community' through a History of Responses to Asbestos-Related Disease in an Australian Industrial Region', *Social History of Medicine*, vol. 21, no. 2, pp. 361–379.

The Intergovernmental Panel on Climate Change (IPCC) 2018, 'Summary for policymakers' in *Global Warming of 1.5°C. An IPCC Special Report on the Impacts of Global Warming of 1.5°C above Pre-industrial Levels and Related Global Greenhouse Gas Emission Pathways, in the Context of Strengthening the Global Response to the Threat of Climate Change, Sustainable Development, and Efforts to Eradicate Poverty*, Masson-Delmotte, V., Zhai, P. Pörtner, H.-O., Roberts, D., Skea, J., Shukla, P. R., Pirani, A., Moufouma-Okia, W., Péan, C., Pidcock, R., Connors, S., Matthews, J. B. R., Chen, Y., Zhou, X., Gomis, M. I., Lonnoy, E., Maycock, T., Tignor, M., & Waterfield, T. (eds), World Meteorological Organization, Geneva, pp. 1–31.

Jarrett, V. 2018, 'Union boss fires up on worker safety in CQ heat', *The Morning Bulletin*, 14 February 2018, viewed 7 March 2019 <https://www.themorningbulletin.com.au/news/union-boss-fires-up-on-worker-safety-in-cq-heat/3335353/>.

Jeffrey, A. 2018, 'Legal geography 1: Court materiality', *Progress in Human Geography*, vol. 43, no. 3, pp. 565–573.

Jessup, B. 2013, 'Environmental justice as spatial and scalar justice: A regional waste facility or a local rubbish dump out of place?' *McGill International Journal of Sustainable Development Law and Policy*, vol. 9, no. 2, pp. 69–107.

Jolley, C. & Rickards, L. 2019, 'Contesting coal and climate change using scale: emergent topologies in the Adani mine controversy', *Geographical Research*, doi: https://doi.org/10.1111/1745-5871.12376.

Kalonaityte, V. 2018, 'When rivers go to court: The Anthropocene in organization studies through the lens of Jacques Rancière', *Organization*, vol. 25, no. 4, pp. 517–532.

Lefebvre, H. 2009, *State, Space, World: Selected Essays*, University of Minnesota Press, Minneapolis.

Lipski, B., Rivers, N., & Whelan, J. 2017, *Toxic and Terminal: How the Regulation of Coal-Fired Power Stations Fails Australian Communities*, Environmental Justice Australia, Carlton.

Liu, G., & Müller, D. B. 2013, 'Mapping the global journey of anthropogenic aluminium: A trade-linked multilevel material flow analysis', *Environmental Science & Technology*, vol. 47, no. 20, pp. 11873–11881.

Liu, X.-Y., He, K.-B., Zhang, Q., Lu, Z.-F., Wang, S.-W., Zhang, Y.-X., & Streets, D. 2019, 'Analysis of the origins of black carbon and carbon monoxide transported to Beijing, Tianjin, and Hebei in China', *Science of The Total Environment*, vol. 653, pp. 1364–1376.

Lorimer, J. 2012, 'Multinatural geographies for the anthropocene', *Progress in Human Geography*, vol. 36, no. 5, pp. 593–612.

Lövbrand, E., Stripple, J., & Wiman, B. 2009, 'Earth System governmentality: Reflections on science in the Anthropocene', *Global Environmental Change*, vol. 19, no. 1, pp. 7–13.

Meng, J., Mi, Z., Yang, H., Shan, Y., Guan, D., & Liu, J. 2017, 'The consumption-based black carbon emissions of China's megacities', *Journal of Cleaner Production*, vol. 161, pp. 1275–1282.

Michels, C. 2015, 'Researching affective atmospheres', *Geographica Helvetica*, vol. 70, no. 4, pp. 255–263.

Moffatt, J., & Baker, P. 2013, 'Farmers, mining and mental health: The impact on a farming community when a mine is proposed', *Rural Society*, vol. 23, no. 1, pp. 60–74.

National Tertiary Education Union 2019, 'NTEU protests casualisation at Universities Australia Conference', 15 February 2019 <https://www.nteu.org.au/tas/article/NTEU-protests-casualisation-at-Universities-Australia-Conference-21238>.

O'Dwyer, S., Pinto, S., & McDonough, S. 2018, 'Self-care for academics: A poetic invitation to reflect and resist', *Reflective Practice*, vol. 19, pp. 243–249.

Oppermann, E., Strengers Y., Maller, C., Rickards, L., & Brearley, M. 2018, 'Beyond threshold approaches to extreme heat: Repositioning adaptation as everyday practice', *Weather, Climate, and Society*, vol. 10, no. 4, pp. 885–898.

Philippopoulos-Mihalopoulos, A. 2013, 'Atmospheres of law: Senses, affects, lawscapes', *Emotion, Space and Society*, vol. 7, no. 1, pp. 35–44.

Philippopoulos-Mihalopoulos, A. 2015, *Spatial Justice: Body, Lawscape, Atmosphere*, first edition, Routledge, Oxon.

Pitkanen, L. 2017, 'The state comes home: Radiation and in-situ dispossession in Canada', *Political Geography*, vol. 61, pp. 99–109.

Rancière, J. 2010, *Dissensus: On Politics and Aesthetics*, Bloomsbury, New York.

Rancière, J. 2013, *The Politics of Aesthetics: The Distribution of the Sensible*, Continuum International Publishing Group, New York.

Rickards, L. 2016, 'Goodbye Gondwana? Questioning disaster triage and fire resilience in Australia', *Australian Geographer*, vol. 47, no. 2, pp. 127–137.

Rickards, L. 2017, 'Regional futures', *Australasian Journal of Regional Studies*, vol. 23, no. 3, pp. 295–304.

Rickards, L. 2018, 'Aluminium dreams', *Australian Book Review*, no. 405, pp. 29–30.

Rickards, L., & Oppermann, E. 2018, 'Battling the tropics to settle a nation: Negotiating multiple energies, frontiers and feedback loops in Australia', *Energy Research & Social Science*, vol. 41, pp. 97–108.

Rickards, L., Wiseman, J., & Kashima, Y. 2014, 'Barriers to effective climate change mitigation: The case of senior government and business decision makers', *Wiley Interdisciplinary Reviews: Climate Change*, vol. 5, no. 6, pp. 753–773.

Schiller, N. G., & Salazar N. B. 2013, 'Regimes of mobility across the globe', *Journal of Ethnic and Migration Studies*, vol. 39, no. 2, pp. 183–200.

Scott, J. C. 1998, *Seeing Like a State: How Certain Schemes to Improve the Human Condition Have Failed*, Yale University Press, New Haven.

Sica, C. E. 2015, 'Stacked scale frames: Building hegemony for fracking across scales', *Area*, vol. 47, no. 4, pp. 443–450.

Snell, D. 2018, '"Just transition"? Conceptual challenges meet stark reality in a "transitioning" coal region in Australia', *Globalizations*, vol. 15, no. 4, pp. 550–564.

Spinney, J., Aldred, R., & Brown, K. 2015, 'Geographies of citizenship and everyday mobility', *Geoforum*, vol. 64, pp. 325–332.

Staples, J. 2017, 'Environmental NGOs, Public Advocacy and Government, *Pearls and Irritations*', 26 July 2017, viewed 7 March 2019 <https://johnmenadue.com/joan-staples-environmental-ngos-public-advocacy-and-government/>.

Turton, D. J. 2015, 'Unconventional gas in Australia: Towards a legal geography', *Geographical Research*, vol. 53, no. 1, pp. 53–67.

Valverde, M. 2011, 'Seeing like a city: The dialectic of modern and premodern ways of seeing in urban governance', *Law & Society Review*, vol. 45, no. 2, pp. 277–312.

van Noordwijk, M., Namirembe, S., Catacutan, D., Williamson, D., & Gebrekirstos, A. 2014, 'Pricing rainbow, green, blue and grey water: Tree cover and geopolitics of climatic teleconnections', *Current Opinion in Environmental Sustainability*, vol. 6, pp. 41–47.

Wadley, D., & Ballock, M. 1980, 'Satisfaction and positive resettlement: Evidence from yallourn, latrobe valley, Australia', *Journal of the American Planning Association*, vol. 46, no. 1, pp. 64–75.

Waters, J. 2017, *The Toxic University: Zombie Leadership, Academic Rock Stars and Neoliberal Ideology*, Taylor & Francis, London.

Wedderburn-Bisshop, G., Longmire, A., & Rickards, L. 2015, 'Neglected transformational responses: Implications of excluding short lived emissions and near-term projections in greenhouse gas accounting', *The International Journal of Climate Change: Impacts & Responses*, vol. 7, no. 3, pp. 11–27.

Weller, S. A. 2018, 'Globalisation, marketisation and the transformation of Australia's electricity sector Australia's electricity sector', *Australian Geographer*, vol. 49, no 3, pp. 439–453.

Weller, S. A. 2017, 'The geographical political economy of regional transformation in Latrobe Valley', *Australasian Journal of Regional Studies*, vol. 23, no. 3, pp. 382–399.

Williams, S. 2016, 'Space, scale and jurisdiction in health service provision for drug users: The legal geography of a supervised injecting facility', *Space and Polity*, vol. 20, no. 1, pp. 95–108.

Wright, C., & Nyberg, D. 2015, *Climate Change, Capitalism and Corporations: Processes of Creative Self-Destruction*, Cambridge University Press, Cambridge.

Yell, S., & M. Duffy 2018, 'Community empowerment and trust: Social media use during the Hazelwood mine fire', *The Australian Journal of Emergency Management*, vol. 33, no. 2, pp. 66–70.

Yusoff, K. 2010, 'Biopolitical economies and the political aesthetics of climate change', *Theory, Culture & Society*, vol. 27, no. 2–3, pp. 73–99.

Zalasiewicz, J., Waters, C. N., Summerhayes, C. P., Wolfe, A. P., Barnosky, A. D., Cearreta, A., Crutzen, P., Ellis, E., Fairchild, I. J., Gałuszka, A., Haff, P., Hajdas, I., Head, M. J., Ivar Do Sul, J. A., Jeandel, C., Leinfelder, R., McNeill, J. R., Neal, C., Odada, E., Oreskes, N., Steffen, W., Syvitski, J., Vidas, D., Wagreich, M., & Williams, M. 2017, 'The working group on the anthropocene: Summary of evidence and interim recommendations', *Anthropocene*, vol. 19, pp. 55–60.

PART 5

In memoriam

16

SPACE, SCALE AND JURISDICTION IN HEALTH SERVICE PROVISION FOR DRUG USERS

The legal geography of a supervised injecting facility

Stewart Williams[1]

Introduction

In most countries the possession and use of drugs such as heroin, methamphetamines and cocaine are illegal. However, there are places inside the nation-state where this prohibition is lifted. For people who inject drugs (PWID), the provision of supervised injection facilities (SIFs) allows consumption of what otherwise remain prohibited substances.[2] Since the first of these legally sanctioned facilities opened in Switzerland in 1986 their number has continued to grow.[3] While spurred by public health arguments, the delivery of such services to PWID has been hampered because it has also been influenced in significant ways in terms of law, crime and policing.

The Medically Supervised Injecting Centre (MSIC) in Sydney, Australia, is the only official SIF in the Southern Hemisphere. It has been contentious despite operating within the law since 2001 and gaining more permanence as it moved beyond trial status in 2010. The story behind the trials and tribulations of this service holds insights into the challenges and opportunities for delivering health services to PWID. Our analysis focuses on the protracted debate over the establishment of the MSIC as we look at how the key stakeholders and their arguments have linked different places from which the law is variously spoken and enacted.

In the first section ("SIFs, public health and the law") of this paper, we situate our case study with respect to the relationships among SIFs, public health and law. We then describe in the second section ("The legal geography approach") our analytical framework as that of legal geography, outlining the focus of inquiry and methods used. The third section ("Case study: the contested history of the MSIC's establishment") provides an overview of the MSIC's contested history involving diverse and variously placed stakeholders. In the fourth section ("Discussion: the role of jurisdictional space and scale"), we discuss how the different spaces and scales of the MSIC's legal framing influenced the debate about its proposed and ongoing delivery. In sum, the paper makes important contributions to both legal geography and public health literatures. First, it illustrates the complexities of jurisdiction as evinced in the actual practice of law; second, it then reveals the value of these insights with potential application in advancing the delivery of harm reduction, notably health services for PWID.

SIFs, public health and the law

In the 1980s and 1990s SIFs began operating in Switzerland, Germany and the Netherlands in response to increased levels of injecting drug use and HIV infection. Recognition of the role of injecting drug use in blood-borne virus (BBV) transmission then figured in combating the spread of hepatitis C among PWID. Efforts to establish services such as needle and syringe programmes (NSPs) as well as SIFs intensified at this time as the quality, affordability and availability of heroin and hence its use and adverse health impacts were at a peak in Europe, North America and Australia. A substantial body of research links SIFs to reductions in the rates of fatal overdose and BBV transmission and improvements in the health and wellbeing of PWID including referral into treatment programmes (Fry, Cvetkovski & Cameron, 2006; Hedrich, 2004; Hedrich, Kerr & Dubois-Arber, 2010; IWG, 2006; Kimber, Dolan, Van Beek, Hedrich & Zurhold, 2003).

Given their success, SIFs are becoming more numerous mostly in Europe. The exceptions are the MSIC, which opened in Sydney in 2001, and InSite, which opened in Vancouver in 2003. Proposals for establishing these SIFs have been based on public health arguments, especially those of harm reduction which aims "to reduce the adverse health, social and economic consequences of the use of legal and illegal psychoactive drugs without necessarily reducing drug consumption" (IHRA, 2010, n.p.). Those opposing such services for PWID, including in Australia, have typically taken the illegality of drugs as a founding premise.

Decisions about health service provision tend to be devolved from the federal level to state and territory governments in Australia. Since the early 1990s, three such governments have attempted on several occasions to establish SIFs. While proposals for the MSIC in Sydney (in New South Wales or NSW) eventually garnered enough support to proceed, those advanced for Canberra (in the Australian Capital Territory or ACT) and Melbourne (in Victoria) have been consistently quashed including most recently in 2003 and 2011, respectively (Fitzgerald, 2013; Gunaratnam, 2005; Mendes, 2002). One early international comparison noted that Australian proposals were "the subject of considerable controversy and debate, and have been met with some resistance" (Elliot Malkin & Gold, 2002, p.20). A persistent intractability has rendered the MSIC's existence in NSW fraught, and encouraged policy reversals by the governments of Victoria and the ACT (Bessant, 2008; Fitzgerald, 2013; Schatz & Nougier, 2012).

Health service provision for PWID is highly contentious and politicised because of the moral ambiguity that emerges with illicit drugs and the clear-cut stance held by the police on their possession and use. Opposition from communities and governments to proposed services such as NSPs as well as SIFs has often succeeded in North America, Australia and even Europe through alignment with the dominant legal position (Bernstein & Bennett, 2013; Bessant, 2008; Davidson & Howe, 2013; Fitzgerald, 2013; Houborg & Frank, 2014; Tempalski, Friedman, Keem, Cooper & Friedman, 2007; Zampini, 2014). While arguments about establishing SIFs are often finally determined on legal rather than public health grounds, the outcomes vary among and within nation-states because the law's interpretation and application is spatially contingent. A legal geography approach is therefore useful for examining how particular decisions get made about such services in light of their jurisdictional framing.

The legal geography approach

Legal geography is a substantive area of research conducted for over two decades from diverse disciplinary perspectives (see, e.g. Blomley, 1994, 2011; Blomley, Delaney & Ford, 2001; Cooper, 1998; Delaney, 1998, 2003; Holder & Harrison, 2003). It has increasingly evinced how law and space warrant closer attention because of a mutual constitution that is powerful and reaches everywhere. In an introductory overview of legal geography's corpus, some of its key contributors note how "nearly every aspect of law is either located, takes place, is in motion, or has some spatial frame of reference ... Likewise, every bit of social space, lived places and landscapes are inscribed with legal significance" (Braverman, Blomley, Delaney & Kedar, 2014, p.1). However, Braverman et al. (2014) do not simply celebrate legal geography. In a constructive critique, they identify weaknesses in the bulk of scholarship undertaken to date, including its focus on where but not how law happens. Still, advances in recent work have explored legal geography's variously material and discursive, performative and relational assemblages (e.g. Blomley, 2013, 2014; Delaney, 2010; Graham, 2011; Riles, 2011; Valverde, 2009, 2011). The continuous making and remaking of different legal realities subsequently invites a scholarly reorientation to include, for example, a focus on variations over time as well as space, and an embrace of more diverse empirical materials and sophisticated theorisations (Bartel et al., 2013; Braverman, Blomley, Delaney & Kedar, 2014; Graham, 2011).

Such developments are exemplified here with this investigation into the MSIC in Australia. SIFs have been examined in terms of their geographic location and legal determination, but not explicitly using a legal geography framework.[4] Here we look at the MSIC's provision as the historically and geographically complex and contingent result of a contest among diverse stakeholders with assorted views on drugs and drug treatment. Notably, these stakeholders, the debates had, decision-making processes used and outcomes reached were always already situated in relation to laws variously holding sway over different places. So, in this analysis, we focus on the role of jurisdiction understood in terms of administering legal governance territorially and thus as a problem of space and scale.

Early work in legal geography was at pains to elucidate law's power to shape legal subjects and practices dependent on their location. It therefore focused on identifying the enunciation of law and its impacts within those bounded spaces of jurisdiction where a court is empowered to hear and determine legal disputes. This interest in the law's territorialisation was useful for explaining what is allowed to happen where, for example, in terms of property rights and judgments (see, for example, Clark, 1982, 1985; Ford, 1999; Frug, 1996; Neuman, 1987). Yet such work is problematic, despite reflecting the law's geographical imaginary, as its conceptual foundation relies on an overly simplistic nesting of spaces that are assumed to be tightly bound units, mutually exclusive and hierarchically ordered. In reality, the practice of law is quite different. Legal scholarship and jurisprudence have therefore entertained endless debate in Australia, for example, about the High Court's capacity to impose on state and territory judicial function via judicial review, advisory opinion and protection of the rule of law (Fearis, 2012; Goldsworthy, 2014; Irving, 2004), and elsewhere, for example, regarding the power of local legislatures and officials to challenge a country's exercising its national laws and meeting international obligations (Butt, 2010) or the scope for extra-territorial jurisdiction in regional conventions (Miller, 2010). More accurate and nuanced interpretations of how space and scale function in the actual practice as well as geographical imaginary of law have likewise been informing the legal geography literature. One early but important observation, for example, states:

> sociolegal life is constituted by different legal spaces operating simultaneously on different scales and from different interpretive standpoints. So much is this so that in phenomenological terms and as a result of interaction and intersection among legal spaces one cannot properly speak of law and legality but rather of interlaw and interlegality … Interlegality is a highly dynamic process because the different legal spaces are nonsynchronic and thus result in uneven and unstable mixings of legal codes. (de Sousa Santos, 1987, p.288)

That socio-legal life is constituted through heterogeneous assemblages reinforces the need to see space less as reified in the fixed, container-like objects of territory and more as networks momentarily connecting phenomena in open, dynamic relations of force and flow, proximity and distance. It also challenges the traditional scalar architecture of hierarchically nested, areal units which situates the local and its supposedly lesser matters inside provinces and regions subsumed by the "bigger" concerns and powers of national and transnational spheres. Indeed, "legal powers and legal knowledges appear to us as always already distinguished by scale … [as it] organizes legal governance, initially, by sorting and separating" (Valverde, 2009, p.141). Scale is not an ontological reality but an epistemological device manifesting in legal practice as jurisdiction, and thus best understood through empirical studies:

Rather than treating it as a thing in the world, our task should become that of tracing the ways in which scale solidifies and is made "real", and under what conditions, and making sense of the work such solidifications do. (Blomley, 2013, p.8)

In this paper, we look at how different geographical imaginaries and legal practices have shaped the MSIC's establishment in Australia. Methodologically, we follow Braverman (2014) and Watkins and Burton (2013) in using a multi-disciplinary approach to examine the processes and outcomes obtained in this particular case. The materials available from archival research include public health policies and proposals, political statements, media reports, legislation, case law, drug service evaluations, organisation websites and research papers. In addition to doctrinal research in law, the most suitable methods include the close textual reading and coding of narrative analysis. We deploy them in our case study, as have done Zadjow (2006) and Fitzgerald (2013), to examine the main arguments framing the debate over the establishment of the MSIC. Like these researchers, we recognise the dominance of legal narratives in the SIF debate, but we add new insights here by focusing on the role of jurisdictional spaces and scales.

Case study: the contested history of the MSIC's establishment

The MSIC was established after a SIF was proposed for Sydney's Kings Cross. This red-light district was at the centre of Australia's 1990s heroin epidemic and infamous for its escalating levels of public injecting and drug overdose. Yet issues of policing, crime and law dominated the ensuing, highly politicised debate. Understanding what subsequently played out therefore requires attention to the country's juridico-political institutions and governance structure.

The Commonwealth of Australia is a federation of six states and two territories which are in turn made up of local government areas or councils. The *Customs Act 1901* (Commonwealth) regulates the importation of drugs, enforced by the Australian Federal Police, but each of the states and territories has a judiciary and legislature hence its own police and laws governing the manufacture, possession, distribution and use of drugs both legal and illegal. They are likewise responsible for the delivery of health services albeit with funding mostly provided by the federal government. The MSIC's establishment was driven by stakeholders including the state government of NSW, but with connections to other jurisdictions reflecting the three levels of governance in Australia (see Figure 16.1). Levels of legal jurisdiction follow those of political governance in Australia with a similarly hierarchical system of courts and tribunals applying a single body of common law. Highest up are those superior courts comprising the High Court and Federal Court, and of which the former can hear appeals from all other courts in Australia and determines constitutional matters, whereas the latter's original jurisdiction is to hear criminal and civil cases concerned with

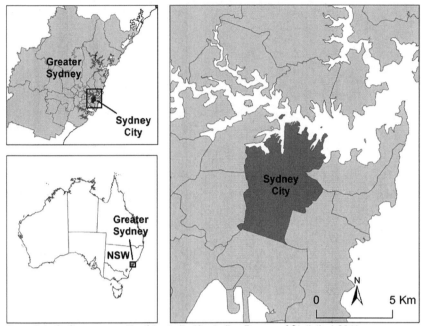

Base data © Commonwealth of Australia (Australian Bureau of Statistics) 2011.
© Robert J. Anders 2014

FIGURE 16.1   The Australian subnational jurisdictions of NSW and Sydney City.

Commonwealth law. And then there is the Supreme Court, the highest court in each state and territory, which hears the most serious or complex of criminal offences and civil disputes with lesser cases devolved to the lowest inferior court of record, the Magistrates Court, although in some states there is also an intermediate County or District Court.

The first proposal to trial a SIF in Sydney resulted from the NSW state government's 1997 Royal Commission into the NSW Police Service. In his report, Justice James Wood condoned the closure of illegal shooting galleries in the Kings Cross area of inner Sydney because of their links to organised crime and police corruption. He also found that "health and public safety benefits [of establishing a SIF] outweigh the policy considerations against condoning otherwise unlawful behaviour" (Wood, 1997, p.222). In his recommendation to trial a SIF, Wood delegated the licencing and supervision of such a facility to the NSW Department of Health pending an amendment to the *Drug Misuse and Trafficking Act 1985* (NSW).

The proposal was dismissed though in 1998 as the members of a Joint Select Committee into Safe Injecting Rooms, established in 1997 by NSW Premier Bob Carr, voted six to four against it. Their final report (Parliament of New South Wales, 1998) recognised that a SIF would result in fewer charges of self-administration (of drugs), taking up less police and court time, but concerns arose around the problem of complicity as discretion would have to be exercised in policing the area around any such facility. In effect, the report "redefined the central rationale for SIFs and the deeper systemic links between the drug market and police corruption had all but disappeared" (Fitzgerald, 2013, p.83). The law thus cleansed of any malfeasance in this context subsequently became available for framing the SIF debate in Australia in a very particular manner. It was readily used under the conservative rule of Prime Minister John Howard (1996–2007) as a time when prevention rather than treatment characterised national drug policy (Bessant, 2008).

Meanwhile, there was growing support for a SIF in King's Cross. A community group was formed in late 1998 by recovering drug users and parents of drug users (some whose children had died from drug overdoses) led by the Reverend Ray Richmond of the Uniting Church's Wayside Chapel in Kings Cross. Although spurred by the local manifestation of a worsening public health crisis, the group comprised some well-connected individuals. They included NSW Legislative Assembly member Clover Moore (later Lord Mayor of Sydney), former NSW parliamentarian Ann Symonds (chair of the 1998 parliamentary inquiry and cofounder of the Australian Parliamentary Group for Drug Law Reform) and two doctors of whom one was internationally respected harm reduction advocate Alex Wodak (director of the Alcohol and Drug Service, St Vincent's Hospital, Sydney).

The group was aware that a drug summit planned by the NSW government did not include SIFs on any agenda given the recent parliamentary inquiry outcome. Members therefore agreed to commence the temporary operation of an illegal SIF (called the Tolerance or T-Room) in the Wayside Chapel. It was opened in May 1999 as a public event coinciding with the drug summit. The room operated for a few days until closed by police with several arrests made, but all charges were subsequently dropped. Most importantly, the group's act of civil disobedience afforded a media presence advancing their cause (Wodak, Symonds & Richmond, 2003).

There was then an unexpected turn at the NSW Drug Summit with a trial SIF supported (NSW Government, 1999). The recommendation moved by Clover Moore was seconded by Ingrid van Beek (Director of Kings Cross's Kirketon Road Centre which delivers health services for PWID, including an NSP). The Sisters of Charity, a religious organisation, were invited by the NSW government to run the MSIC as Sydney's first official SIF trial, but in October they were instructed to withdraw by Cardinal Ratzinger (Prefect of the Congregation for the Doctrine of the Faith in Rome). The directive seemed an unprecedented intervention into state affairs (Totaro, 1999), but its concerns were spiritual not political. The Vatican then decreed, after some deliberation, that no Catholic organisation should participate in such a trial as it involved cooperation with "grave evil" that was understandably illegal.

The NSW government instead approached the Uniting Church of Australia, which applied in June 2000 to operate the MSIC for an 18-month trial period.

The NSW government granted the license in October 2000 having amended the *Drug Misuse and Trafficking Act 1985* (NSW) via Schedule 1 of the *Drug Summit Legislative Response Act 1999* (NSW). Amid ongoing debate, the Uniting Church defended its position theologically on moral grounds (as had the Vatican). However, in "upholding the ultimate sanctity of human life" it was emphatic about "acting completely within the law" (Herbert & Talbot, 2000, n.p.).

The MSIC opened in May 2001 under trial conditions (on short-term basis subject to rigorous evaluations). The legislation was subsequently extended on three separate occasions until this trial status was lifted in November 2010 with enactment of the *Drug Misuse and Trafficking Amendment (Medically Supervised Injecting Centre) Bill 2010* (NSW). The MSIC operates now on a continuing basis without the uncertainty of having to reapply every four years for the legislative change needed to extend its duration as a trial.

Discussion: the role of jurisdictional space and scale

Policy decisions and legislation enabling the provision of SIFs have been enacted with little consistency in Australia.[5] Much remains the remit of states and territories, but their exercise of jurisdictional power over one bounded geographical area is not total or mutually exclusive. With the MSIC, the NSW government was pitted against the Australian federal government but entwined stakeholders at all levels of governance. Arguments from "above" and "below" (in the traditional hierarchy of scale) had varying effect because their power and influence were not always limited to or determined by any one space or scale of jurisdiction.

Scalar interventions from above

State and territory governments wanting to initiate new health services for PWID in Australia have often faced federal government resistance on legal grounds, including the opposition to SIFs then led by Prime Minister John Howard. Continuing to promote his "Tough on Drugs" strategy adopted in 1997, he stated:

> The Federal Government also believes that the introduction of injecting rooms or a heroin trial or both would be damaging to the Australian community insofar as such a step would signal that illicit drug use is acceptable. (Howard, 2000, n.p.)

Under Howard, national drug policy favoured law enforcement over health programmes, and the funding level for prevention was and has since remained several times greater than for treatment (Bammer, Hall, Hamilton & Ali, 2002; Gunaratnam, 2005; Moore, 2005; Ritter, McLeod & Shanahan, 2013). The MSIC's establishment in such conservative times is remarkable. It eventuated because the contestation was reduced to legal arguments had across multiple jurisdictions, none of which necessarily had any greater reach or authority. Supranational jurisdictions and powers did not over-rule others when the federal government linked its case to Australia's international obligations. Being a signatory to the three main international drug control treaties administered by the UN has complicated the SIF debate in Australia.[6] The UN's International Narcotics Control Board (INCB) is a quasi-judicial entity requiring these treaties to be observed, and its annual reports regularly criticise efforts to establish SIFs (Schatz & Nougier, 2012). In 2000, when the MSIC's trial operation was meant to commence, an INCB spokesperson reportedly stated:

> Any national, state, or local authority that permits the establishment and operation of such drug injection rooms also facilitates illicit drug trafficking. (Yamey, 2000, p.667)

Also, Australia each year produces almost half the world's supply of licit opiates, which is regulated by the INCB (Williams, 2010, 2013). Therefore, when Prime Minister Howard berated NSW Premier Carr in the media for supporting the MSIC, he mentioned the INCB and possible UN sanctions on the Australian opiates industry (Nolan, 2003).

The UN encourages members to take a stance against SIFs, but contrary decisions at national and subnational jurisdictional levels do not necessarily contravene these international treaties (Elliott, Malkin & Gold, 2002; Gunaratnam, 2005; Malkin, Elliott & McRae, 2003). As the UN's own legal advice to the INCB (2002, p.5) states:

> It might be claimed that [establishing SIFs] is incompatible with the obligations to prevent the abuse of drugs, derived from article 38 of the 1961 Convention and article 20 of the 1971 Convention. It should not be forgotten, however, that the same provisions create an obligation to treat, rehabilitate and reintegrate drug addicts, whose implementation depends largely on the interpretation by the Parties of the terms in question.

Indeed other international conventions can be taken as demanding such initiatives. Harm reduction advocates have long drawn on human rights law and jurisprudence to balance the drug conventions (Barrett, 2010; Bewley-Taylor, 2005; Bewley-Taylor & Jelsma, 2012; Elliott, Csete, Wood & Kerr, 2005; Malkin, 2001). Bodies such as the UN Human Rights Commission and the World Health Organisation have enshrined in law those principles entitling individuals to the highest levels of health and wellbeing, but variations arise in how it gets interpreted and applied. International law, whether it concerns drug control or human rights, permits some autonomy to the signatories as written into the overarching Vienna Convention on the Law of Treaties 1969. While every international treaty is binding under the Vienna Convention, this latter's articles provide the signatories with several escape clauses. For example, a nation-state "may not invoke the provisions of its internal law as justification for its failure to perform a treaty" (Article 27) but the treaty is to be interpreted and applied simply "in good faith" (Article 26). Likewise, a nation-state must apply a treaty to "its entire territory" unless, that is, "a different intention appears from the treaty or is otherwise established" (Article 29). So, international law need not always prevail with its power imagined as imposing comprehensively on the nation-state. On the other hand, legal arguments put forward nationally as well as internationally can still be challenged and overturned at subnational jurisdictional levels.

Scalar interventions from below

The MSIC's establishment was a state government initiative shaped by local factors. The spread of SIFs outwards from Europe to Australia and Canada via global networks and mobile policy circuits has always been contingent on the particularities of place (McCann, 2008; McCann & Temenos, 2015). Similarly, top-down approaches to drug regulation typified by international treaties and national prohibition can be enhanced by involving non-state or third-party actors (Ritter, 2010).

After the state drug summit, the NSW government decided to work with non-government organisations wanting SIFs established, "providing there is support for this at the community and local government level" (NSW Government, 1999, p.46). The MSIC's establishment was therefore carefully managed here. For example, Part 5, 36Q (1) and (2) of the amended *Drug Misuse and Trafficking Act 1985* (NSW) enabled the development to proceed outside the usual statutory planning processes that require approval from local authorities administering the *Environmental Planning and Assessment Act 1979* (NSW). A Community Consultation Committee was formed though, representing local residents, drug users and their families, the Kings Cross Chamber of Commerce and Tourism, local health and social welfare services, the police and local and state governments.

The committee, charged with identifying the best location for a SIF, examined 39 sites over six months (MSIC, 2014). In 2000, a public presentation of two possible sites incited locals to form the Potts Point Community Action Group. The cry for a less residential location was answered when local businesses agreed to a site in the main thoroughfare of Kings Cross. Its location also made sense in being close to the Kirketon Road Centre's NSP outlet, as well as having the highest prevalence of heroin deaths in Australia (MSIC, 2014; Wodak, Symonds & Richmond, 2003). While the MSIC's current location was thus decided, some local businesses were displeased and took action albeit unsuccessfully to the NSW Supreme Court. In *Kings Cross Chamber of Commerce and Tourism Inc v The Uniting Church of Australia Property Trust (NSW) & Ors* [2001] NSWSC 245, representations for the plaintiff referred to state and federal law and even constitutional concerns based on the *Commonwealth of Australia Constitution Act 1900* (Imp). In practice, the case was confined to a far more parochial matter.

In summing up the case (*Kings Cross Chamber of Commerce and Tourism Inc v The Uniting Church of Australia Property Trust (NSW) & Ors* [2001] NSWSC 245, p.2), Justice Sully stated:

> The sole function and duty of the Court is to examine and construe the terms of the license as issued; and the procedures by means of which the application for the license was assessed and granted; and then to come to a reasoned answer to the question whether the license has been properly issued according to law. The substance of this particular case was subordinate to other legal decisions made at state level, but it illustrates the importance of attending to the technicalities of law practiced on the ground where matters of local jurisdiction can be critical.

With Clover Moore elected Sydney Lord Mayor in 2004, the local government of Sydney City Council has long supported SIFs and now promotes itself as the MSIC's "home" (SCC, 2014a). Council acknowledges in its Drug and Alcohol Strategy objective of "Advocating to other levels of government ... for the continued operation of the Medically Supervised Injecting Centre" (SCC, 2007, p.26) that multiple spaces and scales of jurisdiction pertain here, but elsewhere highlights its own special jurisdictional position:

> Local governments are uniquely positioned to address drug harm with key partners because the impacts of drug use are felt at a community level the most. Councils can also respond to specific problems more swiftly than other levels of government. (SCC, 2014a, n.p.)

Sydney City Council's regulatory interests concern order and safety. These issues and their importance for the success of the MSIC are reflected in its first evaluation (MSIC Evaluation Committee, 2003). The MSIC's impacts on law and order (public injecting, drug-related loitering and property crime, and community attitudes) as well as public health (opioid overdoses; BBV incidence, prevalence and transmission; client health and service use) were found there to be positive, and increasingly so in subsequent evaluations.

The second of four evaluations administered by the National Centre in HIV Epidemiology and Clinical Research focused solely on community attitudes. It revealed from telephone surveys of Kings Cross conducted in 2000, 2002 and 2005 that the proportion of respondents agreeing with the MSIC's establishment had generally increased over this time to 73% of residents and 68% of business operators (NCHECR, 2006). Concerns persisted around crime and safety, negative image for the area and discarded syringes (NCHECR, 2006).

Fears that the MSIC's establishment would lead to more crime happening in its vicinity were not realised (Fitzgerald, Burgess & Snowball, 2010). The final, most comprehensive evaluation of the MSIC thus focused on "public amenity" rather than crime per se, and the report detailed "substantial" decreases in (observed and self-reported) public injecting and "steady" declines in (observed and collected) amounts of discarded injecting equipment (KPMG, 2010).

Discarded syringes, like public injecting, pose broad health risks for society, but are best tackled locally. They are a priority for Sydney City Council, which exercises its "regulatory and enforcement actions ... to improve the safety and amenity of residents and visitors" (SCC, 2014b, n.p.) and manages 65 community syringe disposal bins and a 24-hour needle clean-up hotline funded under the NSW government's Community Sharps Management Program (SCC, 2014a). Its support for the MSIC is understandable as discarded syringes pose one of council's "biggest problems" (SCC, n.d., n.p.).

Conclusion

In this case study, we have used a legal geography approach and the method of narrative analysis to examine how jurisdiction influenced the protracted contest over the MSIC's establishment in Sydney, Australia. We note the importance of attending to the socio-spatial imaginaries and empirical practices of law as our findings challenge the traditionally accepted reification of different jurisdictional spaces and scales into the hierarchically ordered nesting of discreet areal units.

In some countries, establishing SIFs is seen to have depended on a sympathetic government implementing change to national legislation (Houborg & Frank, 2014; Zampini, 2014). In the case of the MSIC, however, the Australian federal government's stance against SIFs even when calling on the supposedly higher authority of international treaties was resisted and beaten by the state government of NSW which supported such a trial. Furthermore, the lowest level of jurisdiction in the form of the Sydney City Council has been critical to the ongoing delivery of this health service. While the values of harm reduction and the lofty ideals of human rights to good health have informed the establishment of SIFs around the world, in this instance it is the municipal governance of such matters as public injecting and discarded syringes that has driven continuing support for the MSIC.

Our research demonstrates legal geography's utility as an approach eminently suitable for understanding how law is made manifest in space. It especially demands our rethinking the traditional architecture of space and scale with respect to the exercise of juridico-political power which, as our case study has shown, can be exerted sideways and upwards (and not simply or only downwards) from one jurisdiction to another. The territorialisation of legal power has long been envisaged as jurisdiction, but its hierarchical ordering of space and scale is in practice far more complex and even contradictory than has normally been conceded in the geographical imaginary of law.

There are also implications around health service provision for PWID. Local stakeholders play an important role in the fate of proposals for such services, and the North American experience has led other researchers to suggest that the proponents of NSPs and SIFs form coalitions with higher order (national and international) actors to progress their objectives (McCann, 2008; Tempalski, Friedman, Keem, Cooper & Friedman, 2007). Efforts to establish and then optimise the delivery of such health initiatives on the grounds of harm reduction are often seen to be countered by the enactment of law or undermined by the juridico-political structures of government (Fischer, Turnbull, Poland & Haydon, 2004; Houborg & Frank, 2014; Zampini, 2014).

However, as we have shown in this Australian case, and Bernstein and Bennett (2013) intimate in the Canadian context, there are significant opportunities to progress the provision of health services for PWID through collaboration among local stakeholders, including not least those responsible for the regulation and enforcement of public order, amenity and safety. This is so, precisely if also perhaps surprisingly, because they have the capacity to intervene in legal processes at other jurisdictional levels as any one or more of the whole suite of laws along with its many actors, instruments and practices, variously described as being international, federal, state or local, can at any time influence what happens in a particular place or territory irrespective of its spatial extent and scalar position.

Notes

1 This paper has benefitted variously from conversations initially had with Associate Professor Jason Prior, feedback offered by participants at the Inaugural Australian Legal Geography Symposium held at the University of Technology, Sydney (12–13 February 2015), reviews received from two referees for Space and Polity and the support of the Institute of Australian Geographers.
2 SIFs are also called safe or safer (as well as supervised) injecting spaces, places, sites or centres. They are the main type of facility known as drug consumption rooms (DCRs) and distinct from alternatives such as safe or supervised inhalation rooms which cater for people who use drugs by means other than injecting.
3 Recent counts include that provided by Hedrich et al. (2010, p.307) who state: "By the beginning of 2009 there were 92 operational DCRs in 61 cities, including in 16 cities in Germany, 30 cities in the Netherlands and 8 cities in Switzerland".
4 One exception is Prior and Crofts' (2015) analysis of the MSIC understood as a space of sanctuary.
5 The variation among states and territories regarding the provision of SIFs has also been apparent with NSPs, methadone maintenance treatment programmes and funding for drug user organisations in Australia.
6 These treaties comprise the Single Convention on Narcotic Drugs of 1961 as amended by the 1972 Protocol, the Convention on Psychotropic Substances of 1971 and the Convention against Illicit Traffic in Narcotic Drugs and Psychotropic Substances of 1988.

References

Bammer, G., Hall, W., Hamilton, M., & Ali, R. (2002). Harm minimisation in a prohibition context – Australia. *Annals of the Academy of Political and Social Sciences*, 582, 80–98.
Barrett, D. (2010). Security, development and human rights: Normative, legal and policy challenges for the international drug control system. *International Journal of Drug Policy*, 21, 140–144.

Bartel, R., Graham, N., Jackson, S., Prior, J. H., Robinson, D. F., Sherval, M., & Williams, S. (2013). Legal geography: An Australian perspective. *Geographical Research*, 51, 339–353.

Bernstein, S. E., & Bennett, D. (2013). Zoned out: 'NIMBYism', addiction services and municipal governance in British Columbia. *International Journal of Drug Policy*, 24, e61–e65.

Bessant, J. (2008). From 'harm minimization' to 'zero tolerance' drugs policy in Australia: How the Howard government changed its mind. *Policy Studies*, 29, 197–214.

Bewley-Taylor, D. (2005). Emerging policy contradictions between the United Nations drug control system and the core values of the United Nations. *International Journal of Drug Policy*, 16, 423–431.

Bewley-Taylor, D., & Jelsma, M. (2012). *The UN Drug Control Conventions: The Limits of Latitude*. Series on legislative reform of drug policies No. 18. Amsterdam: Trans National Institute.

Blomley, N. (1994). *Law, Space and the Geographies of Power*. New York, NY: Guilford Press.

Blomley, N. (2011). *Rights of Passage: Sidewalks and the Regulation of Public Flow*. Oxford: Routledge.

Blomley, N. (2013). Performing property, making the world. *Canadian Journal of Law and Jurisprudence*, 26, 23–48.

Blomley, N. (2014). What sort of legal space is a city? In A. M. Brighenti (Ed.), *Interstices: The Aesthetics and Politics of Urban In-betweens* (pp. 1–20). Farnham: Ashgate.

Blomley, N., Delaney, D., & Ford, R. (2001). *The Legal Geographies Reader: Law, Power, and Space*. Oxford: Blackwell.

Braverman, I. (2014). Who's afraid of methodology? Advocating a methodological turn in legal geography. In I. Braverman, N. Blomley, D. Delaney, & A. Kedar (Eds.), *The Expanding Spaces of Law: A Timely Legal Geography* (pp. 120–141). Stanford: Stanford University Press.

Braverman, I., Blomley, N., Delaney, D., & Kedar, A. (2014). Expanding the spaces of law. In I. Braverman, N. Blomley, D. Delaney, & A. Kedar (Eds.), *The Expanding Spaces of Law: A Timely Legal Geography* (pp. 1–29). Stanford: Stanford University Press.

Butt, A. (2010). Regional autonomy and legal disorder: The proliferation of local laws in Indonesia. *Sydney Law Review*, 32, 177–191.

Clark, G. (1982). Rights, property, and community. *Economic Geography*, 58, 120–138.

Clark, G. (1985). *Judges and the Cities: Interpreting Local Autonomy*. Chicago, IL: University of Chicago Press.

Cooper, D. (1998). *Governing Out of Order: Space, Law and the Politics of Belonging*. London: Rivers Oram Press.

Davidson, P. J., & Howe, M. (2013). Beyond NIMBYism: Understanding community antipathy toward needle distribution services. *International Journal of Drug Policy*, 25, 624–632.

Delaney, D. (1998). *Race, Place and the Law: 1836–1948*. Austin: University of Texas Press.

Delaney, D. (2003). *Law and Nature*. New York, NY: Cambridge University Press.

Delaney, D. (2010). *The Spatial, the Legal and the Pragmatics of World-making: Nomospheric Investigations*. Abingdon: Routledge.

Elliott, R., Csete, J., Wood, E., & Kerr, T. (2005). Harm reduction, HIV/AIDS, and the human rights challenge to global drug control policy. *Health and Human Rights*, 8, 104–138.

Elliott, R., Malkin, I., & Gold, J. (2002). *Establishing Safe Injection Facilities in Canada: Legal and Ethical Issues*. Toronto: Canadian HIV/AIDS Legal Network.

Fearis, E. (2012). Kirk's new mission: Upholding the rule of law at the state level. *Western Australian Jurist*, 3, 61–101.

Fischer, B., Turnbull, S., Poland, B., & Haydon, E. (2004). Drug use, risk and urban order: Examining supervised injection sites (SISs) as 'governmentality'. *International Journal of Drug Policy*, 15, 357–365.

Fitzgerald, J. L. (2013). Supervised injecting facilities: A case study of contrasting narratives in a contested health policy arena. *Critical Public Health*, 23, 77–94.

Fitzgerald, J. L., Burgess, M., & Snowball, L. (2010). *Trends in Property and Illicit Drug Crime around the Medically Supervised Injecting Centre: An Update (Crime and Justice Statistics: Bureau Brief 51)*. Sydney: NSW Bureau of Crime Statistics and Research.

Ford, R. (1999). Law's territory (a history of jurisdiction). *Michigan Law Review*, 97, 843–930.

Frug, J. (1996). The geography of community. *Stanford Law Review*, 48, 1047–1108.

Fry, C., Cvetkovski, S., & Cameron, J. (2006). The place of supervised injecting facilities within harm reduction: Evidence, ethics and policy. *Addiction*, 101, 465–467.

Goldsworthy, J. (2014). Kable, Kirk and judicial statesmanship. *Monash Law Review*, 40, 75–114.

Graham, N. (2011). *Lawscape*. New York, NY: Routledge.

Gunaratnam, P. (2005). *Drug Policy in Australia: The Supervised Injecting Facilities Debate (Asia Pacific School of Economics and Government Discussion Papers)*. Canberra: Australian National University.

Hedrich, D. (2004). *European Report on Drug Consumption Rooms*. Lisbon: European Monitoring Centre for Drugs and Drug Addiction.

Hedrich, D., Kerr, T., & Dubois-Arber, F. (2010). Drug consumption facilities in Europe and beyond. In T. Rhodes & D. Hedrich (Eds.), *Harm Reduction: Evidence, Impacts, and Challenges* (pp. 306–331). Lisbon: European Monitoring Centre for Drugs and Drug Addiction.

Herbert, H., & Talbot, W. (2000). *Theological Perspectives on the Medically Supervised Injecting Centre to be Operated by the Uniting Church Board for Social Responsibility*. Sydney: Uniting Church of Australia.

Holder, J., & Harrison, C. (2003). *Law and Geography*. Oxford: Oxford University Press.

Houborg, E., & Frank, V. A. (2014). Drug consumption rooms and the role of politics and governance in policy processes. *International Journal of Drug Policy*, 25, 972–977.

Howard, J. W. (2000, December 13). Illicit drugs policy. Media release. Retrieved from http://pmtranscripts. dpmc.gov.au/browse.php?did=11562.

IHRA. (2010). What is harm reduction? International Harm Reduction Association. Retrieved from http://www.ihra.net/files/2010/08/10/Briefing_What_is_HR_English.pdf.

INCB. (2002, September 30). Flexibility of treaty provisions as regards harm reduction approaches, prepared by the Legal Affairs Section of the United Nations Drug Control Programme, E/INCB/2002/W.13/SS.5.

Irving, H. (2004). Advisory opinions, the rule of law, and the separation of powers. *Macquarie Law Journal*, 6, 105–134.

IWG. (2006). *The Report of the Independent Working Group on Drug Consumption Rooms*. York: Joseph Rowntree Foundation.

Kimber, J., Dolan, K., Van Beek, I., Hedrich, D., & Zurhold, H. (2003). Drug consumption facilities: An update since 2000. *Drug and Alcohol Review/Harm Reduction Digest*, 22, 227–233.

KPMG. (2010). *Further Evaluation of the Medically Supervised Injecting Centre during Its Extended Trial Period (2007–2011) Final Report*. Sydney: Author.

Malkin, I. (2001). Establishing supervised injecting facilities: A responsible way to help minimise harm. *Melbourne University Law Review*, 25, 680–756.

Malkin, I., Elliott, R., & McRae, R. (2003). Supervised injection facilities and international law. *Journal of Drug Issues*, 33, 539–578.

McCann, E. (2008). Expertise, truth, and urban policy mobilities: Global circuits of knowledge in the development of Vancouver, Canada's 'four pillar' drug strategy. *Environment and Planning A*, 40, 885–904.

McCann, E., & Temenos, C. (2015). Mobilizing drug consumption rooms: Inter-place connections and the politics of harm reduction drug policy. *Health & Place*, 31, 216–223.

Mendes, P. (2002). Drug wars down under: The ill-fated struggle for safe injecting facilities in Victoria, Australia. *International Journal of Social Welfare*, 11, 140–149.

Miller, S. (2010). Revisiting extraterritorial jurisdiction: A territorial justification for extraterritorial jurisdiction under the European Convention. *European Journal of International Law*, 20, 1223–1246.

Moore, T. J. (2005). *Monograph No. 01: What is Australia's 'Drug Budget'? The Policy Mix of Illicit Drug Related Government Spending in Australia (DPMP Monograph Series)*. Melbourne: Turning Point Alcohol and Drug Centre.

MSIC. (2014). Background and evaluation. Medically Supervised Injecting Centre. Retrieved from http://www.sydneymsic.com/background-and-evaluation

MSIC Evaluation Committee. (2003). *Final Report of the Evaluation of the Sydney Medically Supervised Injecting Centre*. Sydney: Author.

NCHECR. (2006). *Sydney Medically Supervised Injecting Centre Interim Evaluation* Report no. 2: *Evaluation of Community Attitudes towards the Sydney MSIC*. Sydney: Author.

Neuman, G. (1987). Territorial discrimination, equal protection, and self-determination. *University of Pennsylvania Law Review*, 135, 261–382.

Nolan, T. (2003, December 1). *Bob Carr Attacks John Howard over Heroin Injecting Room Politics*. ABC Radio.

NSW Government. (1999). *NSW Drug Summit 1999: Government Plan of Action*. Sydney: Author.

Parliament of New South Wales. (1998). *Report on the Establishment or Trial of Safe Injecting Rooms. Joint Select Committee into Safe Injecting Rooms*. Sydney: Author.

Prior, J. H., & Crofts, P. (2015). Shooting up illicit drugs with God and the State: The legal–spatial constitution of Sydney's Medically Supervised Injecting Centre as a sanctuary. *Geographical Research Advance Online Publication*. doi:10.1111/1745-5871.12171.

Riles, A. (2011). *Collateral Knowledge: Legal Reasoning in the Global Financial Markets*. Chicago, IL: University of Chicago Press.

Ritter, A. (2010). Illicit drugs policy through the lens of regulation. *International Journal of Drug Policy*, 21, 265–270.

Ritter, A., McLeod, R., & Shanahan, M. (2013). *Monograph no. 24: Government Drug Policy Expenditure in Australia – 2009/10 (DPMP Monograph Series)*. Melbourne: Turning Point Alcohol and Drug Centre.

SCC. (2007). *Drug and Alcohol Strategy*. Sydney: Author.

SCC. (2014a). *Drug Safety*. Sydney: Author. Retrieved from http://www.cityofsydney. nsw.gov.au/ community/health-and-safety/alcohol-and-drugs/drug-safety.

SCC. (2014b). *Compliance Policy*. Sydney: Author. Retrieved from http://www. cityofsydney.nsw.gov.au/ council/our-responsibilities/policies.

SCC. (n.d.). *Clean Streets*. Sydney: Author. Retrieved from http://www.cityofsydney.n sw.gov.au/live/wasteand-recycling/clean-streets.

Schatz, E., & Nougier, M. (2012). *Drug Consumption Rooms: Evidence and Practice*. London: International Drug Policy Consortium.

de Sousa Santos, B. (1987). Law: A map of misreading: Toward a postmodern conception of law. *Journal of Law and Society*, 14, 279–299.

Tempalski, B., Friedman, R., Keem, M., Cooper, H. J., & Friedman, S. R. (2007). NIMBY localism and national inequitable exclusion alliances: The case of syringe exchange programs in the United States. *Geoforum*, 38, 1250–1263.

Totaro, P. (1999, October 29). *Pope Vetoes Nuns' Injecting Room Role*. Sydney Morning Herald.

Valverde, M. (2009). Jurisdiction and scale: Legal 'Technicalities' as resources for theory. *Social and Legal Studies*, 18, 139–157.

Valverde, M. (2011). Seeing like a city: The dialectic of modern and premodern ways of seeing in urban governance. *Law and Society Review*, 45, 247–312.

Watkins, D., & Burton, M. (2013). *Research Methods in Law*. London: Routledge.

Williams, S. (2010). On islands, insularity and opium poppies: Australia's secret pharmacy. *Environment and Planning D: Society and Space*, 28, 290–310.

Williams, S. (2013). Licit narcotics production in Australia: Geographies nomospheric and topological. Geographical Research, 51, 364–374.

Wodak, A., Symonds, A., & Richmond, R. (2003). The role of civil disobedience in drug policy reform: How an illegal safer injection room led to a sanctioned Medically Supervised Injecting Centre. *Journal of Drug Issues*, 33, 609–623.

Wood, J. R. T. (1997). *Royal Commission into the NSW Police Service Final Report. Volume II: Reform*. Sydney: Government of New South Wales.

Yamey, G. (2000). UN condemns Australian plans for 'safe injecting rooms'. *British Medical Journal*, 320, 667.

Zadjow, G. (2006). The narrative of evaluations: Medically supervised injecting centers. *Contemporary Drug Problems*, 33, 399–426.

Zampini, G. F. (2014). Governance versus government: Drug consumption rooms in Australia and the UK. *International Journal of Drug Policy*, 25, 978–984.

PART 6

Conclusion

17

CONCLUSION

Legal geography futures

Tayanah O'Donnell, Daniel F. Robinson
and Josephine Gillespie

Futures

The uniqueness of our "brand" of Australian and Asian-Pacific legal geography scholarship is our acute and deliberate contribution to what we have categorised as "environmental" legal geography scholarship. The framing of this book, into thematic sections, is reflective of this intent. But, more so, this book offers a deliberate engagement with legal geography method, an endeavour hitherto not yet undertaken.

How has this book achieved these two complementary goals? First, we have deliberately sought out contributions that reflect not only the utility of a case study foundation for legal geography analysis, but contributions where authors have thoughtfully engaged with one or more discrete methods, and illustrated the utility of those methods to their legal geography scholarship. Second, we have invited scholars from a range of disciplines, locations and career stages who are engaged in this theory/method intersection within legal geography. In doing so, this collection affirms that legal geography has drawn its own lines on the scholarly map, and will continue to do so. As is tradition, legal geographers continue to compel us to consider that in the "world of lived social relations and experience, aspects of the social that are analytically identified as either legal or spatial are conjoined and co-constituted" (Braverman et al. 2014, p.1). This collection evidences this.

To enable a book such as this one, we have the remarkable work of early legal geographers, who have provided a strong foundation for our theoretical and empirical endeavours to contribute. As outlined in the introduction, these works include Nicholas Blomley's (1994) seminal work *Law, Space and the Geographies of Power*, along with Professor Gordon Clark, a distinguished Australian geographer, who was an early legal geographer with his 1985 book, *Judges and the Cities*. Contributions by Bennett and Layard (2015); Braverman et al. (2014); Delaney (extensive, but in particular: 2010; 2015a; 2015b; 2017), Blomley (also extensive, but see 1994; 2004a;

2004b; 2005a; 2005b; 2008a; 2008b; 2010; 2011; 2013; 2014; 2016a; 2016b, along with Blomley et al. 2001); and Massey (1992) have all played a critical part in the development of the legal geography approach. Innovative research that has more recently emerged, foreshadowed in detail in Chapter 1, has built the imbrication not only of law, space and place, but also of an emerging interdisciplinary scholarship that engages reflexively to respond to the greatest challenges of our times. Several chapters in this collection attest to this.

Critical to the exercise of doing legal geography, our contributors focus on the ways in which different methods in research practice, and different methodological concern, sharpen any legal geographical analysis. Our contributors unravel human-environment problems, placing the people–place–law nexus front and centre, by embracing diverse tools. From historical, archival analysis to insightful contemporary engaged co-produced research design, this collection takes us on a journey through the diverse tools available to legal geography researchers. We are not restricted to the socio-legal perspective alone but extend our methodological reach to embed time and space in work that critiques how our lived places and experiences are co-produced.

Diversity in scholarship, especially in a field such as a legal geography, must be seen as a strength as to the identity of our field. The Australian and Asian-Pacific perspectives contained in this book apply different methods best able to interrogate the "world-making" (following Delaney 2010) in those places underpinned through legal-socio-spatial processes. We repeat that we do not claim that this book is an exhaustive list of *all* possible methods in legal geography. This book is, however, a significant and comprehensive attempt. Cumulatively, this collection reveals the critical relevance of an explicit legal geography approach, grounded in a common concern with relationality, while recognising and advocating that the interpretation of legal rules is an essential component of such scholarship. We are not alone in this endeavour (see Cuomo & Brickell 2019). What does this mean for legal geography scholarship in 2020 and beyond?

To date in legal geography globally, we have seen a commitment to the pursuit of spatially informed social and environmental justice (see Delaney 2015b). This is reiterated in the work of the authors in this book, where we see a range of explorations of the way law may be used or abused in the name of social or environmental causes, and the response of different actors to legal categorisations or processes, such as climate protestors (see Sherval's contribution) or lawyers (see Turton's chapter), known "experts" (McFarland's chapter) or even plants (Bartel's chapter).

Methodologically, we are seeing an emphasis on reflexive and critical research that uses a case study lens to grapple with legal layering (see Gillespie's work), legal transplants (see Spencer's chapter), positionality (O'Donnell's chapter) and related legal complexities. The authors in this volume have put forward their approaches and ideas for "doing legal geography" reflexively and ethically, towards a more informed understanding of the mutual co-constitution of law and spaces/nature. We consider that these provide a suite of tools to be used in our thinking about law, space, scales, nature and justice. These include a range of approaches, most qualitative (e.g. interviews, surveys, ethnography, legal discourse analysis, case law analysis, comparative and doctrinal law analysis), but some also quantitative (e.g. patent land-scaping/mapping, such as in Robinson et al.'s chapter), to help the future of our field navigate the spatial-legal pathways ahead. The contribution by Calyx Jessup and Sihombing is an exciting addition, blending multiple methods and broaching new frontiers for legal geography research endeavours.

Within the legal geography field, there has been active engagement with the serious challenges arising from settler colonialism across a range of legal categories or fields, and in situations of legal pluralism (Robinson & Graham 2018). Some of our papers here continue this engagement with Indigenous peoples and local communities, their rights struggles and their environmental custodianship. We see efforts to recognise customary law and land/resource rights of Indigenous peoples and local communities set against the spatio-temporal colonial challenges that occur through systems like Native Title in Australia (Godden's chapter), set against the *Treaty of Waitangi* in Aotearoa New Zealand (Bargh and van Wagner's chapter), in land rights cases under the paternal state in Indonesia (Calyx, Jessup and Sihombing's contribution) and in postcolonial settings with the influence of international laws (see Robinson et al.'s chapter). We also see cross-cultural issues and effects for local communities arise across different readings and understandings arising from laws and legal actions (see Schenk's and Gillespie's contributions).

The immediate and urgent challenges of the Anthropocene are also embedded throughout in many chapters, including those focused on climate change action (Sherval's chapter), water use (Graham's chapter), climate change adaptation (O'Donnell's chapter), energy conflicts and energy futures (see the chapter by Rickards and Jolley) and rethinking biodiversity in light of extinction crises (per the Bartel and Robinson et al. chapters). Like much work across geography in general, there is a demand for urgent critical action on global environmental challenges. The legal geography work here undertakes important critical analysis of the politics of legal framings, legal interpretations and, importantly, areas of inaction/regressive actions relating to energy and climate change.

Across all chapters we see a continued commitment to the critical interrogation of the law and the politics of law-making as well as "world-making". The perception of law as separate and closed is a recurring thematic in which legal geography has sought a re-thinking and an "opening". Parallel to, or embedded within, this opening is an attempt to open the legal categorisation of space and place to create new possibilities, and more environmentally and socially just futures. The contributions here each put forward their own cases towards a more inclusive and thoughtful future for our region, and indeed our planet. We leave you then on this note of hope for the future.

References

Bennett, L., & Layard, A. 2015, 'Legal geography: Becoming spatial detectives', *Geography Compass*, vol. 9, no. 7, pp. 406–422.

Blomley, N. 1994, *Law, Space, and the Geographies of Power*, The Guilford Press, New York.

Blomley, N. 2004a, *Unsettling the City*, Routledge, United Kingdom.

Blomley, N. 2004b, 'The boundaries of property: Lessons from Beatrix Potter', *Canadian Geographer*, vol. 48, no. 2, pp. 91–100.

Blomley, N. 2005a, 'Flowers in the Bathtub: Boundary crossings at the public–private divide', *Geoforum*, vol. 36, no. 3, pp. 281–296.

Blomley, N. 2005b, 'Remember property?', *Progress in Human Geography*, vol. 29, no. 2, pp. 125–127.

Blomley, N. 2008a, 'The spaces of critical geography', *Progress in Human Geography*, vol. 32, no. 2, pp. 285–393.

Blomley, N. 2008b, 'Simplification is complicated: Property, nature and the rivers of law', *Environment and Planning A*, vol. 40, pp. 1825–1842.

Blomley, N. 2010, 'Cuts, flows and the geographies of property', Law, Culture and the Humanities, vol. 7, no. 2, pp. 203–216.

Blomley, N. 2011, *Rights of Passage: Sidewalks and the Regulation of Public Flow*, Routledge, United Kingdom.

Blomley, N. 2013, 'Performing property, making the world', *Canadian Journal of Law and Jurisprudence*, vol. 26, no. 1, pp. 23–48.

Blomley, N. 2014, 'Disentangling law: The practice of bracketing', *Annual Review of Law and Social Science*, vol. 10, no. 1, pp. 133–148.

Blomley, N. 2016a, 'The boundaries of property: Complexity, relationality, and spatiality', *Law and Society*, vol. 50, no. 1, pp. 224–255.

Blomley, N. 2016b, 'The territory of property', *Progress in Human Geography*, vol. 40, no. 5, pp. 593–609.

Braverman, I., Blomley, N., Delaney, D., & Kedar, A. (eds), 2014, *The Expanding Spaces of Law: A Timely Legal Geography*, Stanford University Press, Redwood City, California, pp. 120–141.

Clark, G. 1985, *Judges and the Cities. Interpreting Local Autonomy*, University of Chicago Press, Chicago.

Cuomo, D., & Brickell, K. 2019, 'Feminist legal geographies', *Environment and Planning A*, vol. 51, no. 5, pp. 1043–1049.

Delaney, D. 2010, *The Spatial, the Legal and the Pragmatics of World-Making: Nomospheric Investigations*, Routledge, Oxon.

Delaney, D. 2015a, 'Legal geography I: Constitutivities, complexities, and contingencies', *Progress in Human Geography*, vol. 39, pp. 96–102.

Delaney, D. 2015b, 'Legal geography II: Discerning injustice', *Progress in Human Geography*, vol. 40, pp. 267–274, <DOI:10.1177/0309132515571725>.

Delaney, D. 2017, 'Legal geography III: New worlds, new convergences', *Progress in Human Geography*, vol. 41, no. 5, pp. 667–675.

Massey, D. 1992, 'Politics and space/time', *New Left Review*, no. 196, pp. 65–84.

Robinson, D. F., & Graham, N. 2018, 'Legal pluralisms, justice and spatial conflicts: New directions in legal geography', *Geographical Journal*, vol. 184, no. 1, pp. 3–7.

INDEX